Perspectives on Product Stewardship

Perspectives on Product Stewardship

Scott Cassel

PRODUCT STEWARDSHIP INSTITUTE
SINCE 2000

With editorial contributions by Amy Cabaniss

Bernan Press

Lanham • Boulder • New York • London

Bernan Press

An imprint of The Rowman & Littlefield Publishing Group, Inc.
4501 Forbes Boulevard, Suite 200, Lanham, Maryland 20706
www.rowman.com

86-90 Paul Street, London EC2A 4NE

British Library Cataloguing in Publication Information available

Library of Congress Cataloging-in-Publication Data available

ISBN 9781641433174 (paperback) | ISBN 9781641433181 (ebook)

™ The paper used in this publication meets the minimum requirements of American National Standard for Information Sciences—Permanence of Paper for Printed Library Materials, ANSI/NISO Z39.48-1992.

This book is dedicated to my wife, Susan Podziba, who encouraged me to leave a secure state job to start a nonprofit organization based on an idea. Our long walk in the woods around the Mount Washington Hotel in Bretton Woods, New Hampshire, led to the creation of the Product Stewardship Institute some 22 years ago.

Inspiration: *Harold and the Purple Crayon*
By Crockett Johnson

Harold created his own reality. With his purple crayon, he drew a walkway for his exploration into the world. He drew a moon by which to see at night, a sailboat when caught at sea, and a balloon with a basket when falling through the air. Harold used his crayon to create solutions to problems as they arose. He kept drawing until he got home and drifted off to a peaceful sleep in his bed, one that he, of course, drew for himself.

Contents

Preface

The industry guys in their suits and ties were on a roll, flipping page after page of their printed slide presentation as they spoke. They came to our state government office to show us how much mercury their companies had removed from the manufacture of fluorescent lights and household batteries. As the director of waste policy and planning for the Massachusetts Executive Office of Energy and Environmental Affairs (EEA), I was tasked with developing and implementing policies to reduce the amount of household and commercial waste disposed of in our state landfills and waste-to-energy plants, including products containing toxic materials.

In the 1990s, mercury pollution was a prime concern for our agency. Research detailed the impacts on the Massachusetts environment of mercury emitted primarily from power plants, transportation vehicles, and the disposal of consumer products such as fluorescent lights and batteries. For this reason, we were interested in hearing from representatives from the lighting and battery industries to discuss strategies to collect and recycle their mercury-containing products, which at that time were mainly fluorescent tubes, small button-cell batteries used for hearing aids and watches, and single-use batteries that powered radios, tape players, toys, and flashlights.

Since Governor William Weld, under whose administration I worked, was keen on public-private collaboration, our approach was to develop a partnership with the battery and fluorescent lamp industries. At this meeting, we hoped to gain agreement from these two industries to voluntarily collect and recycle their products, which would reduce the amount of mercury emitted into the environment. To make our case, we sought to convey that there was a problem worth fixing. Prior to the meeting, I conferred with our agency scientists, who had conducted extensive research and masterfully reduced voluminous amounts of data into simple charts, graphs, and diagrams that told the story: mercury in Massachusetts, and throughout the Northeast, was already a serious health risk, and any amount of additional mercury emitted from consumer products into the environment only intensified that risk.

At the meeting, we laid out the research detailing the significance of consumer products releasing mercury into the air while being processed in waste-to-energy plants, a primary waste management practice in Massachusetts. Mercury emitted into the air can fall back into lakes, rivers, and oceans, where it is transformed by natural processes into a more toxic form called methylmercury, which is then absorbed by small organisms that are eaten by fish. Since big fish often eat little fish, the bigger the fish you eat, the more mercury you're likely to ingest. If someone eats too much of certain fish, they can experience serious health impacts, including brain damage.[1] Since mercury was already ubiquitous in Massachusetts water bodies, state health advisories warned against eating too much of specific types of fish (e.g., swordfish, shark, pickerel, bass).[2]

After outlining the problem to our industry colleagues, we offered to work collaboratively on a range of voluntary initiatives that would showcase their leadership to combat this major environmental and public health problem. The industry executives, along with the lobbyist who introduced them, quickly returned the volley to my side of the court. They told us that they had removed nearly all mercury from button cell batteries, and that mercury was no longer added to single-use batteries. In fact, they said that the amount of mercury left in batteries was no longer an environmental problem. In addition, they had developed a "green tip" fluorescent tube that contained only one-third the amount of mercury as before. It was so low, they said, that our agency should allow these products to be buried in landfills and burned in waste-to-energy plants. After all, they would still meet the federal US Environmental Protection Agency standards.

I was taken aback by this response, given all the compelling data we had laid out. Although eliminating added mercury from batteries might mean that *new* batteries no longer posed a mercury exposure risk, that was not the case for the *older* products that would still need to be disposed of, along with the lower-mercury fluorescent lamps. But these executives claimed to have already solved the problem themselves and there was no need for them to collect their products after consumer use. They wanted us to agree that they had lowered the mercury risk from their products to levels that did not pose health or environmental concerns when disposed of. We then tried a new approach—thanking the industry for the changes that they had made—and intentionally omitting that they had acted only in response to multiple states enacting laws that banned mercury in those products. We again explained our

---

1. "Health Effects of Mercury," Massachusetts Department of Environmental Protection, accessed February 19, 2023, https://www.mass.gov/guides/massdep-mercury-information#-health-effects-of-mercury.

2. "Mercury in Fish Poster," Maine Department of Health & Human Services, accessed February 19, 2023, https://www.maine.gov/dhhs/mecdc/environmental-health/eohp/fish/hgposter.htm.

agency position that it was important to *further* reduce mercury emissions into the environment. If their products were at all contributing to the health risk, didn't they have a responsibility to make sure their products were collected and safely disposed of after consumers were done with them?

My guests did not blink. Right on cue they repeated their position and doubled down. They explained how they had spent millions of dollars to make product design and manufacturing changes to reduce mercury from their products, and their scientists had worked for years to come up with safe solutions. *And don't forget that Osram Sylvania and Philips employ thousands of people in Massachusetts, and they will be forced to lay off employees if the company must collect their products.*

We were only 30 minutes into the meeting, but my sense of optimism for a voluntary agreement dissipated like a snuffed-out candle, soft wisps of smoke dancing aimlessly into the air. The facts about mercury pollution were not in dispute. The industry executives simply did not want to do any more than what they believed their companies had to do legally and politically. We shook hands and they left the meeting. I was left to ponder what it would take to get these and other consumer product companies to take responsibility for reducing the impacts of the products they sold in the marketplace.

Then it hit me. We were asking Big Industry to spend money on a problem they had already decided had been solved by reducing the amount of mercury in their products to levels that they believed shielded them from additional regulation. But another thing occurred to me as well. We, officials from the state of Massachusetts—one small state in the Northeast—were asking these two industries to change their policy approach for the entire United States. If they agreed to set up a system to collect and recycle fluorescent lights and batteries in Massachusetts, other states would demand the same. They could not say mercury was a problem in one state and not in others. Massachusetts would set a precedent.

After that meeting, I came to an important realization. The outcome we hoped to achieve was not going to happen based on data and a request for voluntary efforts. Our agency knew what needed to be done to protect communities from product impacts. We had funded and implemented dozens of statewide programs to remove toxic products from the municipal waste stream. We were making a difference, but our technical experts were telling us that the health and environmental risks were still too great. We needed to divert more toxic products from disposal to truly protect public health, but we did not have the money or policy leverage to reach that goal.

I had asked the lamp and battery manufacturers to help because government efforts alone were not effective enough. The industry's response that day is similar to what I still often encounter from corporate executives nearly

30 years later—they will not take full responsibility for their product and packaging impacts unless legally forced to do so. As I now know, most US companies fight regulation with all the financial resources, political tools, and media savvy available to them. Yes, times have changed somewhat, and today many more companies seek to align sustainability principles with corporate objectives. Back then, though, emerging from that meeting, I knew we needed a much stronger approach than a plea for voluntary help.

Our drive to use material resources sustainably, then recover them for remanufacturing back into new products in the "circular economy," is a survival instinct. When mercury or plastics fill our rivers and oceans and enter our bodies, we know we need to change what we are doing. When the oil, gas, and coal that we extract threatens our lives, we know we need to realign our priorities. When governments struggle to protect their citizens from the harm emanating from their own consumerism, something needs to change.

Extended Producer Responsibility (EPR) laws play a vital role in bringing the environment and economy into balance. They tap into our ancient social contract for how we should treat the earth and one another. People who make products have a primary responsibility to reduce impacts through the entire product life cycle.

Our federal waste management policies require companies to reduce impacts within their facilities to protect worker safety. They require companies to adhere to air and water emission limits. Companies abide by federal laws that protect consumers from unsafe products in their use. But there are no federal laws that require companies to reduce the impacts of the products they create once consumers are done with them. This burden has fallen to municipal and state officials, and it has become untenable.

Product stewardship plays a significant role in bringing our consumption of natural resources and emission of greenhouse gases back into balance with the earth's ability to absorb these impacts. It seeks to do this, in part, by requiring producers to internalize the cost of *preventing* pollution, rather than continuing to rely on taxpayers to assume the economic, social, and environmental costs of *cleaning up* pollution. Product stewardship provides a tangible thread of accountability that highlights the important role each of us has in achieving our shared goals of a healthy environment and a just and thriving economy, including producers, retailers, collectors, recyclers, governments, environmental advocates, and the public.

This book explains how the massive increase in consumer product waste, its toxicity, and the complexity of materials have created an unbearable financial and management burden for municipal officials in the United States. It traces the growth of recycling from volunteers to government officials, along with the funding challenges faced by municipal programs in the midst of recycling market fluctuations, fragmented waste management systems, and an increase in non-recyclable packaging. These burdens have grown so great that there is now widespread recognition that a paradigm shift toward EPR is needed. It's hard to fathom that it has been about 50 years since the country's first curbside recycling programs were hailed as a major innovation. We have now outgrown that era.

The following pages document my personal journey in the recycling field, from volunteer to government official, which provided me with a clear view about why change to the US recycling system was desperately needed. This story covers the circumstances that led to the creation and growth of the Product Stewardship Institute (PSI), an organization I founded, which has built capacity for product stewardship in the United States around the concept that producers of products and packaging—not governments, taxpayers, or ratepayers—are most responsible for the health, economic, and environmental impacts of consumer products across their life cycle, from mining and manufacture to reuse, recycling, composting, and disposal.

The book also provides my perspective on the first two decades of the product stewardship movement in the United States, where we are now on this journey, and expected trends for the future. It outlines the basic principles of product stewardship and EPR, along with the roles played by each stakeholder group. And it traces the origin of the concept of responsibility, how it was incorporated into the environmental and EPR movements, and the reaction to it by corporations and others. Readers will also learn about the major EPR laws developed in Europe and Canada that have provided the blueprint for EPR laws in the United States. In addition, I share the process that I developed, and PSI uses, to create effective EPR laws through collaborative stakeholder engagement, as well as the importance of program evaluation to measure progress and establish best practices.

The conditions for collaboration are often not ideal. Transition to new policies is usually met with resistance. It takes time for relationships to evolve, and institutions must adapt to a new reality. I hope that readers will realize that there is a methodology and creativity in creating good public policy, and that policy is critical to the success of innovative technologies. To illustrate the techniques that PSI uses, I reference many of the nearly 25 product categories on which we work. The last section of the book includes three case studies—on paint, batteries, and packaging—that explain the environmental and economic problems they cause after consumer use, the actions taken by PSI and others to address these challenges, and the transition in stakeholder positions over time. Each stakeholder process designed by PSI is based on a similar approach but is also unique based on the circumstances and stakeholders involved.

The product stewardship movement in the United States has been fueled by the dedication, hard work, and experience of tens of thousands of individuals, each with their own skills, knowledge, and perspective. Every action taken has led to a cultural shift in our country as to how we take responsibility for managing our natural resources. Collectively, these people have contributed to the passage of numerous laws and the initiation of countless voluntary programs that have returned millions of tons of valuable materials back into the emerging circular economy; reduced greenhouse gases; created thousands of jobs; and saved hundreds of millions of dollars. But this is only the beginning. Eventually, all products, no matter how large or small, must be designed, manufactured, and managed in ways that tread lightly on the earth.

My colleague and editor, Amy Cabaniss, offered me the opportunity to write this book and share my personal journey to finding ways to engage with all those interested in reducing the impacts from consumer products and packaging. I have been supported and influenced by countless people, most notably PSI's staff, board of directors, and colleagues in the United States and around the world. I am fortunate that so many of my colleagues have become my friends, and that I found people to work with who hate waste as much as I do. We are a special breed. Just think, we could have spent our careers scoping out parcels of beautiful landscapes for conservation. Instead, we visit landfills, waste-to-energy plants, and recycling and composting facilities, and plot strategies to reduce waste. We are all part of the continuous cycle of ideas. May this book provide you with both insight and inspiration to contribute your own ideas to help heal the earth.

PART I

Chapter 1

Building Capacity for Product Stewardship in the United States

SEEK AND YE SHALL FIND

We have all had experiences we might call prescient. When seeking an answer, if we keep it in mind but don't concentrate too hard on it, our answer might seem to magically appear. And that is exactly what happened to me one day in 1998 when I least expected it. I was listening to Ron Driedger of the British Columbia Ministry of Environment and Climate Change Strategy explain how his agency managed a range of consumer products through a policy called extended producer responsibility, or EPR. Ron was invited to give the keynote speech at the conference of a national association of professionals who safely manage hazardous household products. As president of this association, the North American Hazardous Materials Management Association (NAHMMA), I had a front-row seat.

Driedger began in a slow, steady, matter-of-fact manner. His words and their meaning grew stronger and stronger, sentence by sentence. Instead of relying on a myriad of voluntary industry efforts, the ministry passed laws requiring brand owners or manufacturers (producers, they called them) to set up and finance systems to collect and safely manage paint, pharmaceuticals, and other consumer products.

Under the British Columbia system, producers proposed performance goals to ministry officials who had the authority to approve or strengthen them. For example, brands would be required to collect and recycle a percentage, by weight, of products they sold into the market. The government also didn't tell companies *how* to reach those goals but left it to the industry to propose its own pathway. The agency had authority to approve the implementation plan or require further initiatives. Under British Columbia's EPR laws, the rates of product collection and recycling or safe disposal far exceeded what we were able to achieve in Massachusetts. And, they didn't use public money or have dozens of government employees implement the program. The industry funded and ran these programs themselves, with government oversight and technical assistance.

In British Columbia, Ron continued, the producers set up a nonprofit organization called a "producer responsibility organization" (PRO) to calculate and allocate program costs among its members. The PRO used those funds to contract with service providers to collect and process the materials, roll out a comprehensive province-wide communication and outreach program to inform residents, and collectively report program results to the ministry. The ministry, in turn, played an oversight role: setting goals for materials collected and recycled; approving the PRO's plan for collection, processing, and education; and holding accountable the companies that did not contribute their share to fund the program.

With each of Ron's words, it was becoming crystal clear to me that British Columbia had found a way to pay for programs that we could not afford in Massachusetts. They got better results by requiring the producers to do what they do well—run efficient and effective operations to serve consumers. The ministry maintained the authority to do what government does best, which is to set program parameters, oversee program implementation to ensure goals are met, and enforce a level playing field for all producers.

This was the answer I had been seeking, and it came in three letters: EPR. It seemed simple, yet powerful. Listening to Ron speak was like experiencing the proverbial lightning bolt or finding a key to a door in an escape room of clever design. While the term *extended producer responsibility* does not exactly glide off the tongue, the concept played like a sweet musical note in my brain. It was clear at that point what I had to do next.

I went back to Massachusetts and told my boss, Gina McCarthy, then undersecretary at the Massachusetts Executive Office of Energy and Environmental Affairs, that I wanted to import this EPR concept from our Canadian colleagues to the United States. I wanted the Commonwealth of Massachusetts to lead a national movement that would provide a strong voice for state and local governments that lacked funding and political power to protect their people and environment from the impacts of increasingly complex consumer products.

Gina, the former administrator of the US Environmental Protection Agency (US EPA) in the Obama administration and White House National Climate Advisor for the Biden Administration, promptly asked me to develop a business plan to prove the worth of my concept. Since I was finishing my fourth statewide solid waste master plan, I had every incentive not to write a fifth. So, I set out to shake loose from the well-worn path of government-funded and -operated programs to forge a new direction that would require companies making products and packaging to also be responsible for managing their collection and reuse, recycling, or safe disposal.

THE BIRTH OF AN ORGANIZATION

I had never written a business plan, but I had two essential ingredients—a big problem that I knew firsthand and a brilliant, already established solution with real-world results. The only thing missing was knowing whether my colleagues throughout the country would support this outside-the-box EPR idea and join in starting an organization dedicated to its implementation.

To test the concept within the United States, I contacted nearly every state solid waste director (my counterparts) and asked them to consider a new policy approach that transferred financial and management responsibility from governments and taxpayers to producers. Since Ron Driedger also referred to EPR as "product stewardship," and this seemed easier to understand, I began to use this term. To my ears, the term implied that we all have some responsibility for the safe management of products, even if producers have a much greater role. Consequently, I thought the term *product stewardship* would not be as off-putting to industry as extended producer responsibility, which seemed to point a finger squarely, and solely, at the producer. I knew we needed a dialogue, and I wanted to use words that encouraged discussion.

The deep personal relationships I built across the country through my NAHMMA presidency gave me access to many colleagues to whom I could pitch this new idea of EPR. I started by asking them to name their top waste management problems and describe what made them problematic. Consumer electronic waste topped the list. It was ubiquitous, toxic, and growing exponentially. Leftover paint comprised half of municipal household hazardous waste budgets, even as most waste paint is still stored indefinitely or disposed of in the trash or down the drain. Scrap carpet is bulky, difficult to manage, and tough to recycle. Used tires are found everywhere—in streams, vacant lots, and roadside—and state-legislated fee systems promote burning tires for fuel rather than more sustainable reuse and recycling options. Finally, the low recovery rate of mercury-containing products like thermostats, fluorescent lights, and batteries was inadequate given the adverse health and environmental impacts of mercury. While each state solid waste director had a different set of priorities, several product categories—electronics, paint, carpet, tires, and mercury—floated to the top of everyone's list. By prioritizing products based on the impacts they posed, I came to understand the significant external costs the public was forced to bear at the behest of producers.

With new knowledge about national solid waste program priorities and with the support of my colleagues across the country, I began to develop a business plan for the creation of a new national US organization that would connect trash problems to product stewardship solutions. Essentially, this effort would seek to have producers assume greater responsibility for financing and managing their products after consumers were done with them. As had happened in British Columbia, we wanted to save municipal taxpayer money for more appropriate uses while protecting the environment. I banged out draft after draft of the business plan, learning as I wrote, until it was completed. I named this new organization the Product Stewardship Institute (PSI) to convey an objective policy think tank whose advocacy would be based on research, data, and facts. I also wanted the central tenet of product stewardship to be front and center.

To me, the term *product stewardship* implies that many stakeholders have a responsibility to ensure that products are properly managed. State government agencies set the parameters of laws and ensure that programs work as intended through oversight and enforcement. Local governments often provide collection sites for a range of used products and educate their residents about the problems with waste and what they can do about it. Retailers educate their customers about the products they buy and can also offer them convenient opportunities to bring products back to their store for recovery, thereby providing a community service and building customer loyalty that can increase sales. Collectors pick up scrap materials and transport them to processors and recyclers, which turn that material into feedstock for new products and packaging. And consumers have a responsibility to use the programs that are available to them.

Producers, however, have the most prominent responsibility—to finance and manage the programs needed to keep waste from being disposed of in landfills or incinerators. After all, they are the ones that create and put these products on the market in the first place, profiting from their sale, and they have the power to redesign their products to be more reusable, recyclable, and less toxic. Product stewardship shifts the costs of recycling (but not yet disposal) from taxpayers and municipal governments to producers. This core concept is key to understanding the significant economic value of product stewardship to communities.

During my calls to state waste policy officials, after understanding their priority wastes, I asked them a key question: would you travel to Boston to discuss your major solid waste challenges with others? Fortunately, nearly all my colleagues enthusiastically committed to attend this emerging national "trash therapy session." There, they would learn about a new policy approach that had the potential to ease their pain.

Prior to the inaugural conference, my colleagues and I debated whether to limit attendance to state and local government officials or to open it up to industry and other stakeholders. As a trained mediator, my natural inclination is to include all key stakeholders needed to resolve an issue. But the thought of having battery and fluorescent lamp manufacturers at our conference, when I knew they did not want to take further responsibility for their products, was unnerving. We were building a new organization and we needed to work together with people who shared the same vision. We didn't want to spend time convincing reluctant industries that we needed an association or that product stewardship was a valid policy. Besides, producers and other companies had their own associations to protect their interests, but government officials did not have an organization that supported their product stewardship efforts.

With the decision made to invite only government officials, I turned my attention to designing a conference that would highlight problems with the US waste management system and propel product stewardship forward. My colleagues and I wanted the Product Stewardship Institute to be the national voice for state and local governments on product stewardship issues. PSI would coordinate governments to speak with one voice and develop model state and federal policies to be replicated across the country. By aligning governments, we would create the political power to encourage or compel producers to take responsibility for reducing the impacts of their products and packaging. Although Massachusetts alone could not sway producers to take responsibility for reducing those impacts, the combined strength of many states could have this result.

Coincidentally, a national product stewardship association would also make it more efficient for industry to negotiate with governments. Instead of spending time and resources negotiating individually with officials in 50 different states, companies could work directly with one entity like PSI that provided a single forum for the discussion of all product issues with all key stakeholders. Instead of the risk of a piecemeal policy landscape, companies could operate under harmonized policies across states, creating greater certainty in the marketplace. Even though state-specific issues would still need to be negotiated, this would take place within a wider national context in which many issues are common to all states.

On December 5 and 6, 2000, at a nondescript Holiday Inn in downtown Boston, Massachusetts, Environmental Affairs Secretary Bob Durand announced the creation of the Product Stewardship Institute (PSI) at the very first US Product Stewardship Forum. With this announcement, I ended my career as a state government official and began another as the head of a new national nonprofit organization.

In attendance at the inaugural event were over 100 state and local government officials from more than 20 states. I decided to kick off the conference with the trailblazers from agencies and organizations that had already begun to lay the stepping-stones for product stewardship in the country. Prime among them were Chris Taylor from the Oregon Department of Environmental Quality, Sherry Enzler and Garth Hickle from the Minnesota Pollution Control Agency, Clare Lindsay, a US Environmental Protection Agency official, Catherine Wilt from the University of Tennessee, and David Stitzhal from the Northwest Product Stewardship Council. Oregon and Minnesota had each laid out their own early principles of product stewardship, from which we began to develop a national unified set of principles that would soon become the guiding compass for the movement.

These speakers were followed by sessions highlighting each of the top five product priorities, with electronics as the plenary topic and the other four as break-out sessions from which participants reported back to the full conference group. As I facilitated the electronics session that first day, I could feel the urgency for a national solution take hold. Through my state agency, we had funded the first electronics recycling programs in Massachusetts, and that experience was now combined with the vast and overwhelming knowledge from state and local experts across the country. The room vibrated with a palpable intensity as hand after hand shot into the air to speak, building on previous comments from other participants and setting a new direction for the country. For the break-out sessions on paint, carpet, tires, and mercury products, I selected facilitators who were already involved in product stewardship and had experience convening stakeholders.

Breaks between conference sessions gave me a chance to meet the faces behind the voices of people I had invited by phone. Not surprisingly, some of those from more politically liberal states had already heard about product stewardship and were spirited in their interest to form a national coalition to drive it forward. But many others were hearing about this concept for the first time. One new person stood out to me. He was a long-haired, bearded guy who dressed informally and spoke simply and intelligently. Jim Hull was the solid waste director for the Missouri Department of Natural Resources and he came to Boston to hear more about this new waste policy direction. By the end of the first day, Hull was intrigued and enthusiastic. The movement depended on integrating the viewpoints and expertise of people like Hull from more politically conservative states like Missouri, with those from regions like the West Coast and Northeast that held more progressive views. I thought to myself, if Jim Hull and others from the heartland of our country would come to Boston at the invitation of someone they did not know and be impressed by what they heard, this idea of product stewardship had promise.

The successful first day ended, giving way to a challenge on day two. Several attendees, moved by the powerful first day's plenary on electronics, had taken steps to initiate a national dialogue on scrap electronics funded by the US EPA and facilitated by others. While PSI would attend that dialogue—and work hard to ensure that state and local governments were at the table and their voices heard—I learned that day that carving out a place for PSI would be an ongoing challenge.

Even so, the creation of PSI had tremendous support from state and local officials who wanted to be part of a new national organization focused solely on product stewardship. These trailblazers understood that their agency's interest would be bolstered by the strength of a cohesive organization. They sensed there was an urgent need to create a paradigm shift in how waste was managed, and they viewed PSI as a means for enhancing and scaling up their work. Thus, PSI was born.

Over the next 22 years, as PSI grew to support the interests of our state and local government members across the country, as well as our partners from companies and environmental groups, challenges old and new have risen and receded like the tides, shaping the contours of the product stewardship movement. Movements are messy, shaped by the actions and interests of a vast network of entities and individuals. Although the challenges have been profuse, the hard-earned accomplishments have been exhilarating and most of the relationships have been both professionally and personally rewarding, with many even turning into cherished friendships.

BUILDING CAPACITY FOR PRODUCT STEWARDSHIP

PSI's inaugural conference in 2000 proved that US state and local governments were ready to change how waste was managed. I was fortunate that my former employer supported my yearning to manifest my passion for product stewardship in the creation of a new organization. Months before, the president of the statewide university system, William Bulger, and Secretary of Environmental Affairs Bob Durand had developed a comprehensive agreement to jointly address critical state environmental issues. The business plan I had written for a new product stewardship organization fortuitously fit right into their goals. PSI was given a home at the University of Massachusetts in Lowell and seed funding from the state environmental agency. I was in the right place at the right time.

To capitalize on this momentum, we quickly assembled a founding steering council comprised of state and local government officials to guide the organization. The US product stewardship movement was initially fueled by the vision and energy of these officials. Residing within the university allowed us to start up immediately with the infrastructure already in place. We had the autonomy to develop the organization that we wanted with immediate fiscal accountability and without university interference. Not only did we save precious time developing bylaws, obtaining insurance, and handling other administrative tasks, but we could immediately start implementing what we had initiated at the national forum. Within months we had an office, logo, website, and above all, an identity.

Next came the hard part. There was no road map for starting a national organization to work on product stewardship. We did not have a to-do list coming out of the conference and there were many directions we wanted to run in. I was full of energy and ready to get going. But starting a new organization is like being dropped into a dense forest with no trail signs and *then* having to develop the compass to get yourself out. Even if you know what you want to do, you need to know where you want to go and how to get there.

We started by gaining consensus on basic principles that would guide, and help us articulate, the product stewardship goals we wanted to achieve and the essential elements of our approach. Fortunately, Oregon and Minnesota had already developed their own basic principles of product stewardship. With those efforts as a starting place, I worked with the Steering Council to hammer out a unified and more comprehensive *Principles of Product Stewardship* in 2001. These principles, which were later refined in 2011, became the compass for PSI's work over the next two decades. They still provide the essence of what effective US product stewardship systems should contain.

With these principles in place, we began the arduous task of educating stakeholders about the concept of product stewardship, how it related to them, and what they could do in support. This educational process was the first step in building the capacity for product stewardship in the United States. In the first few years after the inaugural national conference, PSI was flooded with requests to speak at state recycling conferences, government association meetings, and other venues. Government officials understood the problems they faced by footing the waste management bill for companies. Product stewardship offered them a new lens through which they could view their predicament. No longer did they feel obligated to carry a burden they could not support. Product stewardship provided government officials with a tangible system for implementing change that could help them better achieve their goals by placing greater responsibility on producers. Many industry associations extended speaking invitations as well. They wanted to keep an eye on this new movement and its potential impact on their members. They were conducting reconnaissance.

During PSI's initial years, I zigzagged around the country like Johnny Appleseed, dropping product stewardship seedlings in California, Oregon, Washington, Missouri, Oklahoma, Nebraska, Iowa, Minnesota, Illinois,

Ohio, Michigan, Indiana, Wisconsin, Florida, North and South Carolina, and the entire Northeast. I later took a membership swing through Mississippi, Louisiana, Texas, New Mexico, and Arizona. When I look back, there were only a handful of states to which I did not travel. The pace was exhausting.

During PSI's first four years at the University of Massachusetts in Lowell, we were active on products identified at the inaugural US product stewardship forum as priorities—electronics, paint, thermostats, and fluorescent lamps—but had opportunities to work on other products, such as pressurized gas cylinders (used in camping stoves and barbecue grills) and radioactive devices (used in smoke detectors, exit signs, and industrial applications). By convening state and local government officials with producers, retailers, and other key stakeholders, PSI was able to identify common problems and develop agreements on joint solutions.

Our early work on these product issues began to form a cohesive network of state and local government officials across the country who used the same product stewardship language and took a similar approach to solving waste management problems. They also became accustomed to PSI's facilitated dialogue process (discussed in chapter 6), which provides key stakeholders with opportunities to voice their interests and help solve problems within a well-structured series of meetings. These officials slowly became the product stewardship brain trust around which PSI consolidated knowledge that was continuously renewed and refined through each subsequent event, conference call, and meeting. This initial group of government servants propelled the burgeoning movement forward.

Building capacity for product stewardship meant expanding our founding membership exponentially. To spread the word, PSI staff presented at conferences of the major associations for state and local government leaders, including the Environmental Council of the States (ECOS), National Conference of State Legislatures, National League of Cities, US Conference of Mayors, National Association of Counties, and National Caucus of Environmental Legislators. To gain PSI agency memberships, I sought formal letters from environmental agency secretaries, commissioners, and directors who supported our product stewardship principles and wanted to work with us. Requiring a formal signature meant that lower-level agency staff needed to gain support throughout the agency, creating its own capacity-building mechanism. We provided tools for staff to educate their colleagues throughout the agency and to present to management.

We got a big boost from two people in particular. One was Ron Hammerschmidt, former chair of the ECOS waste committee and, at the time, head of the environment division at the Kansas Department of Health and Environment (KDHE). Hammerschmidt was a colleague of Massachusetts Secretary of Environmental Affairs Bob Durand and took a keen interest in PSI and product stewardship. Hammerschmidt agreed to write a letter on our behalf to other state officials across the country in support of PSI's principles of product stewardship. More importantly, he encouraged them to sign a formal letter in support of membership. I tried an offbeat marketing technique—including a Frisbee made of recycled plastic with the PSI logo and website address—to accompany Hammerschmidt's letter.[1] This letter and Frisbee, which everyone remembered receiving, helped generate numerous early memberships.

A second membership boost was provided by Renee Cipriano, then director of the Illinois Environmental Protection Agency (IEPA) who later took over as chair of the ECOS waste committee. At the annual meeting, Cipriano gave me time on a packed one-hour agenda to make the pitch for product stewardship. I had traveled hundreds of miles to speak for five minutes, and it was worth it. Immediately following my rapid-fire pitch, Cipriano announced to the room of state environmental leaders that she was signing up her Illinois agency as a full member and they should do so as well. This endorsement, particularly from a Republican leader, further boosted the ranks of PSI state members. We still have the original letters from 47 states that eventually signed on as PSI members, including KDHE and IEPA.

After four years at the university, the PSI Steering Council and I decided to create an independent nonprofit organization, which provided autonomy from academic constraints that hampered an entrepreneurial organization seeking rapid change. We developed bylaws, secured insurance, set up operational systems, and established a new office. PSI moved from a campus building in Lowell, one hour north of Boston, into downtown Boston—a more convenient location for meeting stakeholders and forging coalitions. Boston is a hub for the Northeast, one of the country's most progressive regions. EPR laws continue to be developed and passed in states stretching from Maine and Vermont to Connecticut, Rhode Island, New York, New Jersey, Maryland, and into Washington, DC.

Boston is also filled with numerous colleges and universities, many with strong environmental policy programs, like the Massachusetts Institute of Technology, Tufts University, Boston University, and Northeastern

1. I want to pay tribute to Chipper "Bro" Bell, world freestyle Frisbee champion, who designed these recycled Frisbees for Patagonia as their longtime brand ambassador. Chipper made the Frisbees with recycled content to replace the numerous ones he broke practicing and performing his routines. Living in California, Bell was aware of the dangers of oil spills. The 1969 oil spill off the coast of Santa Barbara remains the largest oil spill to have occurred in the waters off California, and now ranks third after the 2010 Deepwater Horizon and 1989 Exxon Valdez spills. As plastic is oil-based, Bell took personal responsibility by partnering with Wham-O to manufacture a line of new Frisbees with a high percentage of recycled content plastic. (While the original recycled Frisbees contained recycled plastic flake *from broken Frisbees*, new ones are no longer made from Frisbees but still use recycled plastic.) Bell headlined one of PSI's early national conferences with a freestyle Frisbee demonstration.

University. PSI has consistently attracted top-quality environmental professionals interested in living in a vibrant city with public transportation routes throughout the surrounding area. Sustaining and growing the product stewardship movement requires a continual supply of passionate, highly trained staff, and Boston provides a wonderful launching pad for the next generation. Many of our young staff have gone on to obtain advanced degrees or moved into product stewardship positions with consulting firms, government agencies, companies, and nonprofits. Although the need to rehire and retrain is arduous, as one of our industry partners observed with a smile, "PSI is one of the best EPR job training centers in the United States." We agree.

Before we go further into the work of the product stewardship movement, I'd like to take a step back and convey my own inspiration for getting hooked on waste reduction. I want to mention my own path not because it is anything special but because it is not. I was one of many in the late 1970s and early 1980s who were searching for a way to integrate an awareness of what nature meant to us with practical ways to make changes beyond our personal lives. Reducing our own waste requires a change of habit. Putting in place systems for others to reduce their waste requires changing the deep-rooted systems of which we are all a part. My journey and contribution to waste reduction was no different than the tens of thousands of people who took their own journeys. All of us have roots that make us who we are. Our unique histories are a place from which each of us starts. I suppose that, for me, that moment was when Mark Pasquale, a hometown friend, called me Johnny Natural after I objected to his tossing trash from our car into the brush off the New Jersey Turnpike. I was 19 years old.

Chapter 2

Don't Let Good Things Go to Waste

AN ETHIC OF RESPONSIBILITY

I grew up hanging clothes on a line between two oak trees. I carried the clothespins in a plastic milk jug cut to make a hand-held bin. I squeezed toothpaste from the bottom of the tube, turned off lights, wore clothes handed down from my older brothers, and ate what was on my plate. It was expected in my house that food, clothes, electricity, and other resources were not to be wasted.

My parents amplified this message with stories of growing up during World War II, when the country rallied together to save balls of string, metal, and other materials to be remanufactured for the war effort. The ethic of conservation was all around me when I was young, and I became aware of waste from an early age. My father, an Eagle Scout, instilled in me the value of service by leading community projects that enriched the lives of others. It wasn't too great a leap for me to drive the recycling truck in college and serve as a board member of our neighborhood food co-op and mainstay at its recycling stations. I eventually made a career of managing garbage—rather, resources.

My upbringing gave me the understanding that waste is not a given, and that I can take action to reduce what I throw away. I saw waste reduction as a personal responsibility. My stuff-saving instincts were amped up during a college gap year traveling coast to coast in a Pontiac T-37 with my buddy Jackson, a bearded collie and sheepdog mix. Jack Kerouac catapulted me into the varied expanse of the United States—to the falls of Niagara, the furious wind of the Great Plains, the abyss of the Grand Canyon, and the endless crests of the Great Sand Dunes National Monument. Hugging Route 1, with the sweet scent of California in the air, Jack and I traveled past Big Sur to

Santa Barbara. During six months of living there in a house near the beach, an environmental consciousness that had lain dormant in my New Jersey soul seeped in. It became apparent that our nation's natural heritage was something I could spend a lifetime protecting.

I returned to college in Philadelphia to finish my undergraduate degree in environmental studies and geology, and I landed my first paid job with the Pennsylvania Environmental Council, a statewide environmental advocacy group. I was an information specialist and educated the public, businesses, and governments about how to reduce solid and hazardous waste, including household hazardous waste (HHW). At that time, in 1983, recycling was done by volunteers, well before the beginning of most city and town programs. For years, my house porch had become the neighborhood newspaper drop-off site. My housemate, Art, and I would load the weekly stash of newspapers into Art's repurposed Checker cab, drive them to Ecology Food Co-op in West Philadelphia, and unload them into a huge trailer. When it was full, the trailer was hauled to a recycling plant and an empty one was left in its place. My wife, Susan, likes to remind me that, back then, I told her that one day our recyclables would be picked up at the curb. Most people paid this forecast no mind. To me and other natural-born recyclers, that change seemed inevitable.

RECYCLING: FROM VOLUNTEERS TO GOVERNMENT

Although we loved the comradery of recycling, it was time consuming and exhausting. It taught us a strong lesson—recycling could be started, but not sustained, by volunteers. Recognizing the need for paid professionals, our food co-op's recycling committee lobbied the City of Philadelphia to develop a new recycling director position. I attended meetings with Mayor Wilson Goode and his staff to convince the city to establish this position. Under the leadership of Jennifer Nash, then executive director of the Clean Air Council, we put pressure on the city to start a recycling program rather than support the construction of a new trash incinerator. We believed that, if residents could recycle enough waste, the city would not need a disposal facility that would commit the city to feed it with garbage for decades. We also knew that individual residents could not be expected to take on the responsibility of recycling without a city-managed program to support their efforts. We needed government authority and resources to coordinate and fund a recycling program. Although it took another six years, Philadelphia became one of Pennsylvania's first municipalities to provide curbside recycling service, starting in 1989.[1]

---

1. Catalina Jaramillo, "Waste Not: Philadelphia's Route to Better Recycling," PBS WHYY (March 2, 2017), https://whyy.org/articles/waste-not-philadelphia-s-route-to-better-recycling.

The evolution of recycling in Philadelphia—from volunteers to a government program—was emblematic of the recycling revolution that was taking place all across the country. This societal shift represented mainstream acceptance of recycling. It was an affirmation of the efforts of volunteer recyclers who had labored long hours over many years for love of the environment. It was proof that the deep values we held and promoted were now accepted by more people and a wider variety of individuals, including those in governing positions. We had reached a watershed moment when government finally embraced our environmental values and agreed to use its power and resources to start and maintain recycling programs. Government was taking on a burden and responsibility that we, as volunteers, had previously been carrying alone.

Government-run programs were a prime reason for the significant increase in recycling nationwide, and many governments hired former recycling volunteers to lead those efforts. From 1980 to 2000, the amount of municipal solid waste (MSW) that was recycled increased from 10 percent to 28 percent.[2] Governments purchased and distributed recycling bins, educated residents about how to recycle, and collected and transported recyclables to markets or hired companies to perform those services for them. As agency officials assumed their new recycling responsibilities, they took pride in their programs and the new social contract they'd made with their citizens. Governments, as representatives of the people, performed the new recycling services their residents demanded. They also created a significant number of jobs, since many more jobs are created by recycling than by trash disposal.

By providing coordinated recycling services to each individual household, local governments made recycling more convenient for their residents. Convenience was key to reducing the burden on individuals; it made recycling possible. In many areas, curbside recycling pickup became as convenient as curbside garbage pickup, and drop-off recycling programs were similarly convenient for those who hauled their garbage to the town landfill or transfer station. Even today, though, many communities do not offer recycling or trash service to their residents, leaving each household to subscribe with a private collector. Governments that did assume waste management responsibilities now had the professional staff and outreach materials to educate citizens about how to recycle and why it mattered.

Once recycling was convenient, it allowed citizens to take personal responsibility for reducing waste and protecting the environment. It was a public expression of an individual's care for the environment and one that became a visible norm in many neighborhoods. If your recycling bin was not lined up

2. US EPA, National Overview: Facts and Figures on Materials Wastes and Recycling, https://www.epa.gov/facts-and-figures-about-materials-waste-and-recycling/national-overview-facts-and-figures-materials#:~:text=The%20total%20generation%20of%20municipal,25%20million%20tons%20were%20composted.

at the curb or you were not seen lugging recyclables to the drop-off center, friends and neighbors noticed your inaction. Recycling also became a tangible act that introduced the next generation to environmentalism. It became a cherished environmental anthem because people understood it was better for materials to be cycled back into products than to be buried or burned as waste. Recycling is still one of the top actions people mention when asked how they show their concern for the environment,[3] and many want more recycling.[4]

Over time, government agencies and private companies made significant investments in recycling, which evolved into a big business, employing tens of thousands of people in the public and private sectors and becoming a major economic driver.[5] It was also a big budget item for municipalities. With this financial investment came an expectation that citizens would use the new recycling programs, and residents were now being held accountable for maximizing financial investments. It became government's job to encourage, cajole, or even compel the public to recycle.

A LOCAL RECYCLING COMMITTEE

In 1986, I moved to Brookline, Massachusetts, and soon joined the volunteer solid waste advisory committee (SWAC), working with others to increase recycling and decrease waste. One way to motivate citizens to recycle is to offer financial incentives and one of the best incentive programs then was a new concept called "pay-as-you-throw" (PAYT), whereby individual households are charged according to the amount of garbage they dispose. Under PAYT systems, the less garbage that a household disposes, the less money they pay. The more they recycle, the less trash they will generate and the less they will pay for trash disposal. These variable fee systems are based on the number and size of bags or barrels a household purchases to dispose of its trash, with the recycling cost built into the cost of each bag or barrel fee.

I started my research on PAYT programs by reading seminal reports produced for the US Environmental Protection Agency by solid waste consultant Lisa Skumatz. The Brookline SWAC, along with a supportive public works department, collaborated with the town administrator, finance committee, and select board members to conduct waste audits and hold multiple public

3. Cary Funk and Brian Kennedy, "Everyday Environmentalism," Pew Research Center (October 4, 2016), https://www.pewresearch.org/science/2016/10/04/everyday-environmentalism.

4. "Public Opinion Surrounding Plastic Consumption and Waste Management of Consumer Packaging: A Report to World Wildlife Fund," Corona Insights (2020), https://www.merkley.senate.gov/imo/media/doc/Public%20Opinion%20Research%20to%20WWF%202021.pdf.

5. US Environmental Protection Agency, "Recycling Economic Information (REI) Report," accessed March 23, 2023, https://www.epa.gov/smm/recycling-economic-information-rei-report.

hearings. I spent months estimating expected cost savings, waste reduction, and increased recycling from implementation of a PAYT program. On average, these programs can increase recycling rates by up to 59 percent and decrease waste generation by up to 27 percent within a few years.[6] We responded to community requests for increased household options by incorporating multiple bag and barrel choices into our proposal and addressed equity issues by adding low-income and large-family allowances. We handled every question asked at meetings and hearings, continually modifying the proposal.

After facing overwhelming initial opposition at every step, our SWAC managed to convince every committee and layer of government to strongly support the proposed new policy. All that was needed for the policy to become law was approval by a majority of the 240 town meeting members, Brookline's elected legislative body. At the town meeting, however, on the night that citizen representatives were slated to vote on our proposal, loud opposition was expressed by one university professor and one realtor association representative. Although both had numerous opportunities to speak with us either privately or in well-publicized public forums, we had not heard from them until the night of the vote. They posed a few technical questions for which we did not have ready answers. In a matter of minutes, these two opponents paralyzed the SWAC committee and town representatives, who decided to withdraw the warrant article from a vote. That one decision in 1992 took with it voluminous hours of research and recommendations.

It took another 25 years before I had the opportunity to personally stand in front of that same town meeting governance body and again advocate for a PAYT program for Brookline. I warned the group not to let the one loud opponent—there was, and often is, just one loud opponent—block the program again. We had seen this before and it cost Brookline millions of dollars of lost financial savings and significant waste reduction over the previous 25 years. This time the new SWAC chair, John Dempsey, a former Brookline school principal, did not relent. Our combined voices, mine echoing the past and his the present, were enough to gain overwhelming support from town meeting voters and catapult the new PAYT system into law, to the pleasant surprise of the town administrator and public works commissioner. With its passage came town savings of $1 million per year reaped through a financial incentive to use smaller garbage carts and the automation of large recycling cart handling.

My experience with the Brookline SWAC galvanized my interest in further developing local recycling and waste reduction policies. After being laid off from a recycling and waste management startup during the 1990 recession,

6. US Environmental Protection Agency, "Pay-As-You-Throw: A Cooling Effect on Climate Change" (March 2003), https://archive.epa.gov/wastes/conserve/tools/payt/web/pdf/climpayt.pdf.

I took on multiple consulting projects, including researching and writing a detailed manual on safe HHW management that officials in Westchester County, New York, used to provide instructions over the telephone to residents. I also developed a Brookline resident's guide for reusing and recycling packaging and products. In addition, I wrote locally syndicated newspaper columns on environmental issues, including recycling, composting, and waste management. The weekly columns forced me to learn quickly about many issues, understand multiple stakeholder perspectives, and offer opinions on relevant policy topics backed by research, facts, and reason.

STATE RECYCLING POLICY

In 1993, when the waste policy director position opened up at the Massachusetts Executive Office of Energy and Environmental Affairs (EEA), I used my columns and past consulting projects to highlight my research-based approach to developing policy. I was hired by Undersecretary Leo Roy and Secretary Trudy Coxe, both of whom were avid environmental advocates and staunch Republican political appointees. By contrast, I was a liberal who had never worked on a political campaign. I soon learned that personal conviction and vision, no matter what political party, was as important to environmental protection in Massachusetts as political affiliation.

Being in the executive office, which reported to Governor William Weld, a strong environmental advocate, buffered us from the waste industry. By contrast, the Department of Environmental Protection (MassDEP), one of about six departments under EEA, had close daily contact, through issuance of facility permits and enforcement actions, with those companies impacted by waste policies. To me, it appeared harder for MassDEP staff than it was for our office to refuse the waste industry's constant demands for permission to site more disposal facilities.

Together with my MassDEP colleagues, I designed, implemented, and evaluated policies to reduce, reuse, and recycle the roughly eight million tons of MSW generated each year in the state.[7] Our executive office had the authority to develop statewide recycling and waste management policies that would influence 351 municipalities, as well as state parks, offices, facilities, and other state activities. I was given few internal constraints. It was like being in a proverbial candy shop, with a sweet tooth. I didn't leave for seven years.

7. "Beyond 2000 Solid Waste Master Plan—A Policy Framework," Massachusetts Department of Environmental Protection (December 2000).

One of my initial assignments as waste policy director was to secure funding for municipal recycling programs. Fortunately, by the time I worked for the state, the Massachusetts courts had awarded the Commonwealth ownership of unclaimed bottle deposits in a lawsuit brought by bottlers.[8] In 1981, Massachusetts became the seventh state to pass a bottle bill, which places a five-cent deposit on the sale of a range of beverage containers, such as beer and soda.[9] Consumers redeem their five cents when they return an empty container to a retail store where they purchased the beverage or to a redemption center. In 1993, the number of unclaimed deposits totaled about $30 million each year, which was placed into a fund controlled by the Massachusetts State Legislature. My first task was to lobby the legislature to secure the $30 million per year for recycling programs.

My inclination was to draw on my mediation background, gained through graduate school studies, trainings, and mediating small claims cases in the local district court. I developed strong personal and professional relationships with two state leaders from opposing factions to forge a coalition to lobby the legislature for funding. Amy Perry of the Massachusetts Public Interest Research Group, a statewide environmental advocacy group, had led the effort in Massachusetts to place a moratorium on the construction of new waste-to-energy plants in the state. That policy was a cornerstone of the 1992 Massachusetts Solid Waste Master Plan that I inherited upon my arrival to the agency. Susan King worked for the Southeastern Massachusetts Resource Recovery Facility (SEMASS), now owned by Covanta, which produces energy from burning 1,500 tons of waste per day. Together, the three of us rallied support from local governments, recyclers, environmental groups, and others to lobby the legislature to appropriate the funding.

Our efforts succeeded in securing all $30 million, of which half was designated for recycling and the other half for other environmental programs.[10] My MassDEP colleagues and I now had $15 million each year to develop multiple programs to support municipal recycling. I worked closely with many at MassDEP, most notably Robin Ingenthron, Greg Cooper, and Brooke Nash. Together, we developed the Municipal Recycling Incentive

---

8. *Massachusetts Wholesalers of Malt Beverages, Inc. v. Commonwealth*, 414 Mass. 411, 609 N.E.2d 67 (Mass. 1993), https://casetext.com/case/mass-wholesalers-of-malt-beverages-inc-v-cwealth.

9. Provisions for Recycling of Beverage Containers (Bottle Bill). 301 CMR 4.00. (2013). https://www.mass.gov/regulations/301-CMR-400-provisions-for-recycling-of-beverage-containers-bottle-bill.

10. Apparently, a significant number of MassDEP staff were already funded through the state budget, and the legislature decided to designate half of the unclaimed bottle deposits ($15 million) to fund those staff positions.

Program, a potpourri of optional initiatives from which municipalities could choose. For example, if a municipality collected a greater-than-average number of material types, conducted above-average education and outreach, or implemented PAYT programs, they received more state grant funds. The more programs they implemented, the more money they were eligible to receive. The state funds did not fully cover the cost of municipal recycling programs. Instead, they acted as incentives to entice local governments to spend their own money to take steps that would earn them valuable equipment, outreach materials, or cash to enhance the efficiency of their unique local program.

From the executive office, we developed the state's first comprehensive plan for managing HHW that included grants for storage sheds and statewide collection and disposal service contracts to safely manage HHW such as paint, pesticides, and batteries. We funded, and helped develop, the state's first scrap electronics recycling program conducted at the University of Massachusetts by Lynn Rubinstein, former longtime executive director of the Northeast Recycling Council. We also developed the state's first program for developing and promoting PAYT programs, which now cover more than 153 municipalities (about 30 percent of the state's population).[11] The MassDEP also enacted state bans on the disposal of certain recyclable materials, and our office paid for enforcement staff to inspect loads. Jointly, we developed new regulations that required waste-to-energy plants to remove mercury products, such as thermostats and fluorescent lamps, from the waste stream prior to incineration. By 1999, our agency and MassDEP increased statewide waste diversion (recycling and composting) to 38 percent.[12]

Massachusetts was not unique. Initiatives to reduce toxics and increase recycling were taking place across the country. There was no federal leadership and no blueprint or single recipe for success. State by state, locality by locality, thousands of governments took it upon themselves to educate residents about recycling and waste reduction, provide funding to increase recycling, reduce toxics by collecting household hazardous waste, and hold citizens and haulers accountable for disposing of recyclable materials. Governments embraced this responsibility because their citizens wanted it. In turn, the opportunities for residents to recycle and reduce toxics vastly expanded in most states across the country.

---

11. "Massachusetts 2030 Solid Waste Master Plan," Massachusetts Department of Environmental Protection (October 2021), accessed February 20, 2023, https://www.mass.gov/doc/2030-solid-waste-master-plan-working-together-toward-zero-waste/download.

12. "Beyond 2000 Solid Waste Master Plan—A Policy Framework," Massachusetts Department of Environmental Protection (December 2000), accessed February 20, 2023, https://www.mass.gov/doc/beyond-2000-solid-waste-master-plan-a-policy-framework/download.

WASTE DISPOSAL POLICY

While the exciting part of my state policy job was to promote waste reduction, reuse, and recycling, the ramifications for failure were severe. Not meeting our recycling goals meant we needed to permit additional in-state disposal facilities. These facilities were always in somebody's backyard, so to speak, and would impact hundreds of people, lower neighborhood property values, and be a public health concern for nearby residents. By contrast, not having enough in-state capacity for waste disposal in landfills and incinerators could drive up disposal prices for households and businesses by having to export waste to other states.

Maintaining enough waste disposal capacity is a tricky balance. When the economy booms, consumers buy more products and more waste is created. When the economy hits a downturn, consumers buy fewer products and waste generation is reduced. Although economic cycles are typically about five years in duration,[13] adding disposal capacity can take anywhere from a few years to expand landfill "cells," to a decade or more to permit and build new landfills and waste-to-energy facilities. Waste capacity decisions are not nimble, flexible policy tools. In addition, increased reuse and recycling decreases the need for disposal capacity. Big decisions on adding capacity depended on whether our state could reduce, reuse, and recycle at a high enough rate to avoid inevitable disposal impacts.

When I took over as waste policy director, I inherited four bold policy decisions set forth in the 1990 solid waste master plan (SWMP) under the former administration of Governor Michael Dukakis. This work was led by former EEA secretary John DeVillars, assistant secretary Ralph Earle, MassDEP commissioner Dan Greenbaum, and MassDEP Solid Waste Division director Willa Kuh. First, the plan set an ambitious 46 percent recycling goal to be reached by 2000. The 1990 SWMP also committed the state to *remove* disposal capacity by closing more than 100 unlined, polluting landfills. It also placed a moratorium on the construction of new waste-to-energy plants. Finally, the SWMP included a policy of self-sufficiency that committed the state to maintain in-state disposal capacity for waste that could not be reused or recycled.[14] The 46 percent recycling target gave our agency a clear goal and strong incentive to achieve it. If we did not meet this goal, we would need to add in-state landfill disposal capacity. While these policies signaled strong

13. "What's an Average-Length Boom and Bust Cycle?" Investopedia, updated November 30, 2022, accessed March 23, 2023, https://www.investopedia.com/ask/answers/071315/what-average-length-boom-and-bust-cycle-us-economy.asp.

14. "Beyond 2000 Solid Waste Master PlanA Policy Framework," Massachusetts Department of Environmental Protection (December 2000), iii, accessed June 29, 2021, https://www.mass.gov/guides/solid-waste-master-plan.

support for recycling, they also increased pressure on the state to provide enough disposal capacity for materials not recycled.

I soon learned in my first year with the agency how tenacious the waste industry was in promoting the building of new waste disposal capacity, particularly after MassDEP and our office closed all unlined landfills. The National Waste and Recycling Association represents small and large companies in every facet of the business—hauling garbage and recycling, and operating transfer stations and landfills. The waste-to-energy industry is large and consolidated, and has its own association, the Waste-to-Energy Association. In the 1990s, and through to 2021, waste disposal was more lucrative than recycling for most waste management companies.[15] Although there was a moratorium on building waste-to-energy plants, there was extreme pressure to permit new landfills and add newly permitted cells (landfill units) to expand existing landfills. Our office also experimented heavily with "landfill mining" for recyclable materials by permitting companies to sift through garbage in existing landfills and permit the disposal of trash in space vacated by recovered materials.

Our agency was responsible for determining whether adding in-state capacity was needed and justified, which would open the door for more trash trucks to rumble through neighborhoods and dispose of waste. The alternative was to stand firm, bank on more recycling, and shut the door on adding capacity. We felt intense pressure—on one side from the waste management industry and their litigious lawyers, and on the other side from environmental groups and impacted communities ready to blow the media whistle. Our agency eventually permitted a number of new landfill cells to replace cells filled to capacity and taken out of service. Even so, we resisted adding unnecessary disposal capacity, particularly constructing new landfills. Governor Weld came to office with strong support from environmental groups in a state with nearly three times as many registered Democrats as Republicans.[16] These landfill capacity decisions were highly political, and a goal of our executive office was to keep these conflicts out of the governor's office.

AN EPIC BATTLE

Unbeknownst to me when I started at the Massachusetts EEA in 1993, our office was already in the midst of a multiyear battle with Douglas

15. Cole Rosengren, "Waste Management Q2: Landfills More Profitable, Recycling 'Fallen Off a Cliff,'" *WasteDive* (July 26, 2019), https://www.wastedive.com/news/waste-management-q2-2019-earnings-landfill-recycling-cliff/559546/.

16. "The Commonwealth of Massachusetts, Enrollment Breakdown as of October 29, 2022," https://www.sec.state.ma.us/divisions/elections/download/research-and-statistics/enrollment_count_20221029.pdf.

Environmental Associates, Inc. (DEA), a developer seeking to construct a landfill on a parcel in Douglas, Massachusetts, on behalf of the site owner, Vincent Barletta. DEA's first step had been to obtain a site assignment, which is a declaration by the state that the land is technically suitable to use as a site for a solid waste management facility. Although this is not a green light to start construction, it was a first major hurdle that DEA needed to pass in the long approval process.

As recorded by the appeals court in 1999: "DEA began its quest in 1987 with a site assignment proceeding before the Douglas Board of Health. [MassDEP] affirmed the site assignment[17] in March 1991. The town of Webster and several individuals appealed DEP's decision to the Superior Court. In September 1992, a Superior Court judge upheld DEP's decision, and the site assignment approval became final. Later that month, DEP issued a draft permit to DEA authorizing the landfill to accept 1,500 tons per day (tpd) of solid waste, of which 600 tpd could be municipal solid waste. In April 1993, after two public hearings and intense opposition to the proposed facility, DEP denied DEA a final permit for the project."[18]

Over six years, from 1987 to 1993, the saga had taken a major twist. Although MassDEP designated the parcel of land fit for a landfill in 1991, it reversed its decision in 1993 and denied the permit based on the newly found presence of an endangered salamander, which was detected during a study, ironically conducted by the landfill proponent. DEA and MassDEP became locked in negotiations to develop a mitigation plan to accommodate the salamander while also analyzing the impact that mitigation would have on the economic viability of the site. While the negotiations were taking place, our agency conducted an updated capacity assessment in 1995[19] that now showed *excess* in-state disposal capacity. Our finding meant that the site was *no longer needed*. The site was also located on a large geologic fault and included wetlands bordering Massachusetts, Connecticut, and Rhode Island.[20] Even so, landfill proponents persisted.

Landfill opponents persisted as well. One night I attended a community meeting in Douglas that was held by neighborhood activists who vehemently

17. According to [310 Mass. Reg. 19.006], a site assignment "designates an area of land for one or more solid waste uses subject to conditions with respect to the extent, character and nature of the facility that may be imposed by the assigning agency after a public hearing."

18. Massachusetts Cases, Published Opinions from Massachusetts Courts, accessed June 28, 2020, http://masscases.com/cases/sjc/429/429mass71.html. DOUGLAS ENVIRONMENTAL ASSOCIATES, INC., & another vs. DEPARTMENT OF ENVIRONMENTAL PROTECTION & others; TOWN Of WEBSTER & others, interveners. 429 Mass. 71. December 11, 1998 - March 1, 1999. Suffolk County [MA].

19. Our office conducted a subsequent capacity assessment in 1999, which was published in 2000, showing similar results of excess in-state capacity.

20. "Landfill Site Becomes State Forest," *SouthCoast Today* (January 11, 2011), https://www.southcoasttoday.com/article/19981225/news/312259976.

opposed the landfill. As the waste policy expert accompanying Secretary Trudy Coxe to the home of the lead activist, I listened to passionate pleas for help from dozens of residents not to permit the facility. They were keenly aware that the land on which they lived would be changed forever, or at least for their lifetimes, if the landfill was permitted. The house overflowed with people. I was nestled in the nook of a doorway adjacent to the living room as I watched Secretary Coxe give hope to people who were desperately fighting a power seemingly beyond their control. This was their last and final chance to stop a massive disruption to their lives and their environment. Their only hope lay in the state denying a final permit and a court backing up that finding.

The residents got one but not the other. While MassDEP stuck to its latest decision that the site was not needed, DEA continued to litigate, eventually receiving court approval upon appeal in 1999 to construct the landfill. The Massachusetts Supreme Judicial Court (SJC) agreed with the appeals court finding that "DEP's change of position on MSW disposal for DEA's facility was arbitrary, capricious, and an abuse of discretion."[21] The court further agreed with the appeals court that "DEP could not in fairness abandon the allocation made in DEA's draft permit"[22] and that the length of the proceeding was unfair to DEA. In other words, even if there was excess capacity at the time of the final permit denial, the court ruled that MassDEP took too long to get to that decision in reversing an earlier permit approval. The court ruling seemed to send a decisive blow to landfill opponents.

In one final twist, however, Governor Paul Cellucci, who assumed the post in 1999, signed legislation authorizing the taking of the land by eminent domain. As reported in the local press, "Three hundred acres once slated to become part of a giant landfill were purchased by the state and became part of Douglas State Forest, ending a 12-year battle between area residents and the proposed dump operators. What was once permitted to be the site of a new 1,500 ton per day landfill that would have operated for 20 years was eventually turned into a state forest."[23]

In the process to determine how much money the state owed DEA for the land, I was one of several agency staff subpoenaed to testify about agency deliberations and subsequent decisions not to permit the site. To bolster its hand, DEA claimed that our agency rigged the assessment to show excess disposal capacity. Although my MassDEP colleagues felt strong pressure

21. Massachusetts Cases, Published Opinions from Massachusetts Courts, accessed June 28, 2020, http://masscases.com/cases/sjc/429/429mass71.html. DOUGLAS ENVIRONMENTAL ASSOCIATES, INC., & another vs. DEPARTMENT OF ENVIRONMENTAL PROTECTION & others; TOWN Of WEBSTER & others, interveners. 429 Mass. 71. December 11, 1998 - March 1, 1999. Suffolk County [MA].

22. Massachusetts Cases.

23. Massachusetts Cases.

from our office to push back against the developer, I felt then, and even more strongly today, that Secretary Coxe's leadership and instincts produced a stunning result. She could have given up on our ambitious 46 percent recycling rate target and claimed that residents would be unlikely to reach that goal. Governor Cellucci could have accepted the appeals court ruling that MassDEP could not reverse its previous decision.

There were two distinctly different lenses through which this situation was viewed. One lens assumed that the current parameters would continue: waste generation would fluctuate with economic activity, and recycling would fluctuate with markets. This view doubted that the public would increase recycling enough to decrease disposal capacity need. There was justifiable concern about the significant financial investments of facility owners and about workers who relied on those jobs for their livelihoods. There was also concern about rising costs resulting from a constriction of in-state disposal capacity. But there was also interest in profiting from waste disposal. Through this lens, the Douglas landfill provided needed capacity for residential and business waste.

Another lens, however, saw a feasible future reality and a pathway to reach it. This was the path our office took. We knew that waste disposal results in unquantified health, environmental, and socioeconomic impacts that fall disproportionately on those living near, or on the road to, garbage disposal facilities. Instead of planning to permit the amount of disposal capacity that *might be needed*, we planned for a new reality of reduced waste. That meant projecting that the state would meet its 46 percent recycling goal and plan to permit only enough landfill capacity to make that vision a reality. Once the facility was built, there would be no turning back. We believed we could reach 46 percent recycling by 2000 because that was what was needed to avoid inevitable impacts, as well as to reap the economic benefits of recycling. Through our lens, disposal capacity offered by the Douglas landfill was not needed.

Our agency did all we could to resist what seemed to be an inevitable decision to turn a forest into a garbage dump. Secretary Coxe, with the support of Governors Weld and Cellucci, pushed to the brink until the Supreme Judicial Court ended proceedings with a ruling in favor of the landfill proponent. This forced the state's last hand—to take the land by eminent domain and make it part of the Commonwealth's heritage. It cost the state $1.5 million to assume possession of the land, in accordance with a subsequent court ruling.[24] Another payment, which I recall being around $30 million, was required to settle a lawsuit brought by DEA against the state. Those payments ended the Douglas landfill saga for Massachusetts, but the trash that evaded that landfill

24. Massachusetts Cases.

did not disappear.[25] While the export of waste to other states is a tough reality today, waste export was an even bigger issue back in the late 1980s when the Douglas landfill case was just beginning.

WASTE BARGE WITH NO HOME

In 1987, a privately owned garbage barge called *Mobro 4000* set out for North Carolina with 3,168 tons of New York City and Long Island trash from the Town of Islip's landfill. By the time the barge got to Morehead City, North Carolina, the public erroneously perceived the trash to be filled with medical waste and rejected the stinking barge of garbage, owned by Alabama businessman Lowell Harrelson.[26] As later reported by Charlie Rudoy of the Brooklyn Public Library, "For months the *Mobro* was rejected by every state between Florida and Texas, Mexico, the Bahamas, and Belize" and was the butt of jokes on late-night television.[27]

Eventually, the barge made its way back to New York City, where it was burned at Brooklyn's city-owned incinerator after the New York State Supreme Court ruled against Brooklyn Borough president Howard Golden, the New York Public Interest Research Group, and national environmental leader and professor Barry Commoner, all concerned with hazardous emissions being dispersed throughout the community. Although this incident was a defeat for the Brooklyn neighborhood, Golden highlighted an important point that goes to the heart of producer responsibility. He emphasized the inequity of city taxpayers having to "foot the bill for burning private garbage."[28]

Two additional lawsuits followed as a result of the *Mobro* incident, each related to responsibility for waste impacts. As Rudoy reported, "In September of 1987, New Jersey sued New York for a garbage slick that New Jersey officials alleged floated from Gravesend Bay and ended up on Jersey Shore beaches. Later, an additional case was brought by officials on Long Island due to the toxins found in the trash's ash before it was buried in Islip."[29] These impacts, which imposed real social, health, and environmental costs on the citizens of New York and New Jersey, were not remedied by the companies that sold the products and packaging that were the source of the waste problem. These lawsuits were early cases representing an awakening among the

---

25. "Massachusetts 2030 Solid Waste Master Plan," Massachusetts Department of Environmental Protection (October 2021), accessed February 19, 2023, https://www.mass.gov/guides/solid-waste-master-plan.

26. Charlie Rudoy, "If You Can Make It Here, They Won't Take It Anywhere," Brooklyn Public Library (May 1, 2019), https://www.bklynlibrary.org/blog/2019/05/01/if-you-can-make-it-here.

27. Rudoy.

28. Rudoy.

29. Rudoy.

public that garbage disposal causes real impacts that are externalized on those who do not cause the problems.

Although the *Mobro* garbage barge incident has, by now, been buried in our past, Rudoy viewed it as "a cautionary tale about the waste society produces."[30] No one wanted to take responsibility for the trash from New York City that was hauled around for months. The *Mobro* became a metaphor for the waste we all generate but don't want to face, no matter who we are or where we are. Our waste is filled with toxics, can become putrid, attracts disease-causing pests, and can drip gunk from trash trucks that stinks. That the *Mobro* trash was eventually burned at Brooklyn's city-owned incinerator in a low-income community was not without consequence.[31] Rudoy asserts that the incident "ultimately emboldened recycling and environmental activists."[32]

The *Mobro* incident was not only a New York City event; it became a national and international debacle, and it influenced decisions made in other locations across the country. In 1987, at the same time the *Mobro* was rejected by port after port, month after month, the residents of Douglas, Massachusetts, were rejecting a state-sanctioned fate to be the dumping ground for the region's trash. Just as the Douglas landfill was a turning point for Massachusetts, the *Mobro* was a turning point for the United States. No longer did people believe they needed to accept the inevitable dumping of waste in their communities and the pollution of their landscapes. They rejected the expectation that landfill after landfill would be permitted in communities to accept garbage that they could not stop from being generated. They forced government to challenge prior assumptions. Those of us in government knew we needed to come up with new strategies.

GOVERNMENT RECYCLING CHALLENGES

My MassDEP colleagues and I spent years designing, implementing, and evaluating policies and programs to remove toxic products from the waste stream, reduce waste, and increase recycling. Our job at the state was to make it possible for local governments to succeed. We were not unique. Similar initiatives were undertaken by state and local government officials across the country. The degree of program success depended on the level of priority given by local jurisdictions, which often related to the availability of

30. Rudoy.

31. Jake Mooney, "A Lesson in Civics: The Devil You Know . . . ," *New York Times* (September 17, 2006), https://www.nytimes.com/2006/09/17/nyregion/thecity/a-lesson-in-civics-the-devil-you-know.html.

32. Charlie Rudoy, "If You Can Make It Here, They Won't Take It Anywhere," Brooklyn Public Library (May 1, 2019), https://www.bklynlibrary.org/blog/2019/05/01/if-you-can-make-it-here.

funding, competing priorities, community demand, and other factors. Most importantly, through these efforts, thousands of governments took it upon themselves to educate residents about recycling and waste reduction, fund recycling and HHW infrastructure, and hold citizens and haulers accountable for disposing of recyclable materials. Governments largely assumed, even embraced, this responsibility because their community wanted it. In turn, the opportunities for residents to recycle materials and safely manage toxics vastly expanded.

For decades, government recycling programs for packaging materials have produced tremendous resource savings.[33] Over time, however, as these programs became more successful and residents recycled more materials, challenges emerged for governments and citizens alike. First, packaging became more costly to recycle if it could be recycled at all. Second, the cost to recycle is always compared to the alternative—the cost of disposal. In regions that hosted waste-to-energy plants, municipalities signed long-term contracts, some up to 20 years, that guaranteed they would send a minimum amount of waste to the facility. In exchange, facility owners guaranteed that the municipality's disposal fees would remain low—in fact, significantly lower than recycling—over the contract period.[34] Other regions, particularly those less populated, had an abundance of space for landfills, which also kept disposal fees much lower than recycling.

Nationally, recycling often lost out to disposal on a pure cost basis, and municipal officials were forced to justify why they should spend more to recycle than dispose of waste. They had an annual budget to balance and were often faced with the decision to either cut municipal staff or recycle. The financial calculation that is most important for municipalities is the cost to collect and recycle[35] materials compared to the cost to collect and dispose of waste. Those taking a pure market-based approach support recycling only if its cost is equal to, or lower than, the cost of disposal. Not covered in this approach are the full life-cycle environmental, economic, and social costs of pollution from mining, manufacturing, and waste disposal, as well as the value of recycling jobs and economic development. Although these economic costs and values are real, they are not quantified and, therefore, don't always factor into the recycling decisions of municipal officials desperately trying to balance a budget.

---

33. "National Overview: Facts and Figures on Materials, Wastes and Recycling," US Environmental Protection Agency, accessed February 19, 2023, https://www.epa.gov/facts-and-figures-about-materials-waste-and-recycling/national-overview-facts-and-figures-materials.

34. Charis Anderson and Doug Fraser, "SEMASS Costs to Spike for Cape Towns," *Cape Cod Times* (March 9, 2009), https://www.capecodtimes.com/article/20090309/news/903090311.

35. "Recycling" in the United States refers to a process where used materials are collected, processed, and sold to end markets that make new products and packaging. The meaning of the term may differ in other countries.

As waste policy director for Massachusetts, I calculated the recycling and disposal costs for all 351 municipalities in the Commonwealth, ranging from small rural communities to large cities like Boston, Worcester, and Springfield. Due to the prominence placed on a market-based approach to recycling, I reasoned that our agency's strongest argument to boost recycling rates was in proving the pure financial case for recycling. I found that, for one-third of all municipalities (representing two-thirds of the state population), the more their residents recycled, the more money the city or town saved. By contrast, for two-thirds of municipalities (representing one-third of the population), recycling was more costly.

A typical calculation is shown in table 2.1 for a hypothetical municipality. The cost to recycle materials in this municipality is $100 per ton—$75 per ton to collect and $25 per ton to recycle. By comparison, the cost to dispose of materials is $125 per ton—$25 per ton to collect and $100 per ton to dispose. Each ton of material recycled and not disposed *saves* this municipality $25. If that municipality generates 50,000 tons of waste each year, recycling half that amount will save it $625,000 each year (25,000 tons recycled × $25/ton).

In that group of municipalities for which recycling cost less than disposal was the City of Boston which, in 1995, was recycling at about a 12 percent rate. My calculations showed that, if Boston could recycle at the rate of Worcester, which was then at about 54 percent,[36] Boston could save $3 million each year. Since a key aspect of Worcester's system was its pay-as-you-throw program, I wanted to convince Boston to adopt PAYT to reap financial benefits. I held multiple meetings with Boston's public works commissioner and staff, as well as the city fiscal watchdog, the Boston Municipal Research Bureau. Unfortunately, the amount of projected savings of $3 million each year was not large enough to motivate them to adopt a PAYT program to boost recycling. The political risk was considered too great to implement a new system that charged residents for trash disposal, particularly when waste management was perceived as a "free" municipal service. The city's financial

Table 2.1 Calculation to Determine Recycling Cost or Savings—Idealized Municipality

Recycling	*Disposal*
$75/ton for collection	$25/ton for collection
$25/ton for recycling	$100/ton for disposal
$100/ton	$125/ton

36. "Worcester, Massachusetts 54% Residential Waste Reduction," US Environmental Protection Agency Solid Waste and Emergency Response (October 1999), https://nepis.epa.gov/Exe/tiff2png.exe/9100KCT4.PNG?-r+75+-g+7+D%3A%5CZYFILES%5CINDEX%20DATA%5C95THRU99%5CTIFF%5C00002302%5C9100KCT4.TIF.

pain was not deep enough, and the financial gain not great enough, for it to spend the time, effort, and political capital to significantly change its way of managing waste. The PAYT system was never proposed and, nearly 30 years later in 2023, Boston's recycling rate was still only 25 percent.[37]

The equation that compares recycling and disposal costs is, of course, more complicated. In fact, the cost of recycling varies by material type. If a company is willing to pay a high amount for a material after it is collected and processed, that material is considered to have stronger market value compared to another similarly processed material that has lower value. For example, since clear glass can be made into more products and packaging than colored glass, it tends to have a higher market value. White office paper has more value than cereal boxes. Aluminum cans, foil, and trays tend to out-value steel cans in the market.[38] The materials with a higher market value reflect stronger demand. Municipalities will be charged less, or might even get paid, to collect and process these higher-value materials as compared to materials with a lower market value.

Recycling, however, is rarely a money-making endeavor from a municipal perspective. There are costs to recycle just as there are costs to dispose of waste—except that recycling provides greater benefits, such as reduced environmental and health impacts and job creation. While it is possible to recycle a limited number of high-value packaging materials in some municipal programs under strong market conditions and make money, recycling even an average amount of packaging from a household in the United States costs money. The cost to collect and recycle materials will almost always be greater than the revenue derived from the sale of recyclable materials to end markets. The key is that, in many cases, recycling can be *less expensive than disposal.* In those municipalities, recycling will save money in comparison to garbage disposal.

China's recent restriction on accepting recyclable materials (known as the "National Sword" policy, discussed further in the packaging case study), however, has led even seasoned municipal recycling officials to erroneously claim that recycling used to make them money and now it is a cost. That is not accurate. What they mean is that, at one point in time, their municipal contract paid them for the sale of recyclables due to strong market demand from China, which accepted low-value, highly contaminated materials. These officials do not factor in the cost of recycling collection, education, and other

37. Zero Waste Boston, Recommendations of Boston's Zero Waste Advisory Committee (June 2019), 7, https://www.boston.gov/sites/default/files/embed/file/2019-06/zero_waste_bos_recs_final .pdf.

38. "Historical Recycled Commodity Values," US Environmental Protection Agency Office of Resource Conservation and Recovery (July 2020), https://www.epa.gov/sites/production/files/2020 -07/documents/historical_commodity_values_07-07-20_fnl_508.pdf.

program costs. In any case, recycling is not a reliable revenue stream for municipal governments, and deciding whether and how much to recycle is a complicated decision for every municipality.

First, they must know the relative price they are likely to pay *for each material collected*. Nearly all materials can technically be recycled. It is the cost to collect, transport, and recycle a specific material that becomes the deciding factor for municipalities as to whether they will choose to collect that material from residents, although contracts usually bundle materials together.

Recycling packaging in the proverbial "blue box" container is like going to the grocery store. Just as some goods are more expensive to buy, some packaging is more expensive to collect and recycle. For example, plastics such as PET (polyethylene terephthalate) and HDPE (high-density polyethylene)—also known as #1 and #2, respectively, for their chasing arrow recycling symbols stamped on the packaging—have greater market demand than PP (polypropylene, #5), PS (polystyrene, #6), and other plastics. For this reason, many governments have collected only #1s and #2s because they are less costly to recycle than other plastics. They might also collect white paper, newspaper, cardboard, aluminum, and clear glass as part of a basket of high-value materials, fating lower-value materials to be disposed of.

In cities and states whose residents value recycling to a high degree and where the cost to recycle (even if greater than waste disposal) is affordable and politically acceptable, then a municipality may collect those lower-value materials as a means of enhancing their recycling programs. These communities accept the lower-value materials because it makes it easier for residents to fill their recycling bins, increasing overall collections, including of high-value materials, which makes the higher cost justifiable—at least during periods of strong recycled commodity markets. By contrast, if a municipality is on a strict budget and/or doesn't have a high level of recycling support from residents, they might choose not to collect certain materials and, in turn, achieve a lower recycling rate.

Not only do municipalities need to understand the cost to recycle each packaging item; they also need to be prepared for changes in the commodity markets for each grade of paper, glass, metal, aluminum, and plastic. Although experienced officials have been around long enough to understand that markets fluctuate, they also know it is impossible to fully predict when changes will take place and this makes government funding unstable. These changes can be regional, national, or global, and they might impact each municipality differently depending on its recycling contract length and where in the contract cycle the price changes take place. Another variable that municipalities must consider is the degree to which the contract they sign with a waste management company passes cost increases onto their agency, rather than having it absorbed by the company. Given fluctuating markets, it behooves municipalities to advocate for a provision in their recycling contract that shares the revenue when markets are good and shares the risk when markets flatten.

Although a government might lock in a three- or five-year contract with a company to collect and process its recyclables, when that contract ends, a new contract can be vastly more expensive. If the market weakens for the basket of materials previously recycled, the municipality will be faced with higher costs. But if the market strengthens, the municipality might actually save money, sometimes contributing revenue that can allow it to accept a greater number of materials into their recycling program. These boom years can also send erroneous signals to government officials that recycling makes money. Again, it usually doesn't. It only saves money compared to the costs paid the year before.

The cyclical context of commodity markets is often forgotten in agencies with staff turnover and short-term, yearly budget memories. Years of high market demand for material are always followed by lower demand and higher prices. Municipal budgets don't typically account for potential swings in market prices, and governments are usually not prepared for cost increases during the next contract cycle. City and town officials are then faced with having to defend "higher" recycling costs, even if those costs were the same paid 10 years earlier. Those with an annual budgetary mindset do not recall recycling market fluctuations that are inevitable and add greatly to the complexity of planning.

Further complicating recycling costs was a transition starting in the late 1990s from using two small hand-carried bins into which residents source-separated paper and other recyclables (dual-stream recycling) to large carts into which residents could put all recyclable material (single-stream recycling). This transition, like accepting lower-value materials, was an attempt to increase overall participation and boost collection rates by making it easier for residents to recycle. Carts could be wheeled to the curb, providing greater convenience, and residents no longer had to sort material into two bins. This transition also had an added benefit in that carts could be mechanically emptied into a recycling truck without the driver leaving the vehicle, saving time and money. As a result, municipal recycling collection costs went down. Prior to China's National Sword policy, this system effectively brought in more of the high-value recyclables, albeit with low-value ones.

Unfortunately, commingling recyclable materials in one large-wheeled cart also resulted in greater contamination of material. For example, broken glass got mixed with paper, lowering processed paper value. In addition, a larger collection container led to "wish-cycling" among residents who want to recycle everything, even small plastic kiddie swimming pools, diapers, medical syringes, and plastic bags, all of which create operational problems in processing facilities, increasing cost. These facilities cannot create as clean a supply of processed plastic pellets, glass cullet, paper, and metal from commingled loads compared to materials that enter after being source separated. As a result, the average rate of contamination among materials recovery facilities that process recyclables has increased from 7 percent prior to the commingled revolution to 25 percent by 2019.[39]

If these changes are not challenging enough for municipal officials, consider what happened to the $30 million of unclaimed bottle deposits that my MassDEP colleagues and I used for recycling. To help balance the state budget, Governor Mitt Romney later diverted all those funds into the state budget. Recycling funding thus became reliant on the legislature's annual budget appropriation to the MassDEP, which then allocates funding across all its programs, including recycling. It wasn't until much later that a new recycling funding source was developed: energy credits through the burning of materials in waste-to-energy plants. Energy prices, however, fluctuate with the price of oil, natural gas, and renewables. Funding available for recycling in Massachusetts also fluctuates, creating ongoing uncertainty for municipal programs and the recycling markets that depend on that supply.

These are but a few of the many challenges municipalities face when deciding how to recycle. As the amount and type of materials have gotten more complex, and as global economic forces have made managing unpredictable fluctuations in commodity markets extremely difficult, municipalities have been forced to face a sober reckoning. Their costs to manage an increasingly complex waste stream have continually gone up. At the same time, the complexity of their task has also increased. Municipalities across the country now find that they are not in a position to contain steadily increasing costs that have spiraled out of their control.

How is it that our nation's municipalities got left holding the bag of waste problems, with little to no help from the brand owners who made the products and packaging that must be managed? Whether by reuse, recycling, composting, or disposal, municipalities stepped in, bore the costs, carried the burden, and added more cost and more burden onto their backs, until it became obvious that this should no longer be allowed to continue. The gap between what is possible for municipalities to manage and what is needed for effective management has resulted in a wide chasm. It is like watching a time-lapse video of the Colorado River slicing through sandstone layers over six millennia to produce the Grand Canyon. Slowly, a new landscape emerged. As recycling

39. Maggie Koerth, "The Era of Easy Recycling May Be Coming to an End," *FiveThirtyEight* (January 10, 2019), https://fivethirtyeight.com/features/the-era-of-easy-recycling-may-be-coming-to-an-end.

advanced from volunteer to municipal staffing, manufacturers began making products and packaging from more complex and often multilayered materials which, in turn, were difficult and more expensive to recycle. Materials that provided increased consumer benefits in product delivery also caused increased costs for municipal recycling programs. How did we get here?

THE LOCAL WASTE MANAGEMENT BURDEN

Although it took municipal governments many years before they accepted the responsibility of recycling, the way that garbage is managed has long been their domain, driven by public health concerns. Long ago, local officials removed manure from streets traveled by horse-drawn carriages and discarded these and other wastes on the outskirts of town. Governments assumed the function and cost of waste management as a service to the community. Most products were made of leather, wood, stone, and other organic materials that eventually decomposed. There were few items that could not be returned to the earth.

Over the years, manufacturers developed products that allowed our lives to be less burdened, healthier, and more enjoyable. Washing machines and dryers replaced the need to scrub clothes in the wash basin or hang them on the line. Garments made by machine obviated the need to sew your own. Electric mixers, blenders, and other appliances reduced the time to prepare foods that now come from mega farms and could be shipped all over the world. Packaged, processed food also became widely available and gradually accepted. These and other product developments brought positive societal changes, particularly for women in conventional households who were able to spend less time on household chores and had more opportunities for employment. The United States eventually became a global leader in production and innovation.

Economic growth and an increased standard of living also had less perceptible social, economic, and environmental costs. As manufacturing got more sophisticated, companies used cheaper material substitutes that lowered consumer costs so that more people could buy products and "enjoy a better life." Cheaper products democratized consumer purchases, allowing people with lower incomes to afford products previously purchased only by wealthier families. These products were often less durable, however, and needed to be replaced more often, resulting in more waste. To meet consumer demand, companies focused on making more, and less expensive, products. Of little concern was the postconsumer disposal of the unwanted remains. The manufacture of more products and packaging required more materials and more energy, resulting in even greater impacts.

As companies shifted their goals from product durability to production quantity and distribution, they had no incentive to repair older models and had every incentive to sell consumers a new product. Components became difficult to remove and replace and, in most cases, it was as expensive to repair a used product as it was to buy a new one. Some companies, such as Apple and John Deere, even opposed repair of their products by restricting information, tools, and parts for repairs.[40] As a result, product repair has largely become a lost art. Even new versions of many electronic products still require consumers to buy new battery chargers, adapters, cords, and other accessories while governments, taxpayers, and ratepayers bear the cost to dispose of obsolete and unfashionable items.

Another significant change in product and packaging development took place that increased the cost of managing the municipal solid waste stream: the introduction of toxins. Over time, some products that residential consumers purchased contained the same hazardous materials as those found in industrial products. Common household chemical products, including paint, motor oil, batteries, fluorescent lights, cleaning fluids, and solvents, also pollute our air, water, and land and pose a threat to health and safety. Retailers made it easy for customers to bring these hazardous products into their homes. In the early 1980s, Dave Galvin, a local government official from King County, Washington, referred to these products as "household hazardous waste" (HHW),[41] which started a national movement to divert these products from entering the household trash, sink and storm drains, and waterways. Municipalities began to collect and safely dispose of these products in hazardous waste management facilities due to their hazardous properties (e.g., toxic, corrosive, flammable, reactive).

While the public outcry to recycle bottles, cans, paper, and cardboard could be addressed by providing a bin to fill with material for curbside collection by a separate truck, managing HHW required yet another system because it could not safely be set on the curb for pickup. Residents stored these products in their homes and self-transported them to specified collection sites managed by trained personnel. Most regions in the United States began with periodic single-day HHW collection events, often twice a year. While useful, these events were rare and inconvenient for most people. To make collections more convenient and cost effective, a national trend toward permanent HHW collection facilities developed, enabling communities to hold more frequent collections and temporarily store and accumulate hazardous materials on-site for cost-effective material bulking and aggregation. Despite single-day

40. Editorial Board, "It's Your iPhone. Why Can't You Fix It Yourself?" *New York Times* (April 9, 2019), https://www.nytimes.com/2019/04/06/opinion/sunday/right-to-repair-elizabeth-warren-antitrust.html.

41. "Household hazardous waste" is a term coined by Dave Galvin in 1983 when working for the King County Local Hazardous Waste Management division: https://wasteadvantagemag.com/first-of-two-parts-luminary-interview-dave-galvi.

collections and the increasing number of permanent facilities in the United States, most communities did not have the funding to collect HHW at all, and so they were disposed in the trash destined for landfills and incinerators.

As government officials managed an ever-increasing waste stream, they began to further differentiate among types of waste. Bulky items, such as carpet, mattresses, and couches, took up valuable space in garbage trucks, required handling by multiple staff, and got tangled in equipment at transfer stations, landfills, and waste-to-energy plants. Since these bulky items required another truck to pick them up, they were more costly to manage than other trash. As more companies shifted to online sales and buying products was only a click away, consumer impulses translated easily into buying new stuff, shipped in greater quantities and different types of packaging. Packaging innovation reduced product spoilage and breakage in transport and increased efficiency, but often at the expense of packages that were not recyclable. The move toward more plastic packaging has been further driven by oil industry investments in new markets for oil.

Over time, as consumer products and packaging have gotten more complex, the cost of managing recyclables and garbage has continually increased. State and municipal officials routinely conduct waste audits to determine the composition of residential waste, which items are most wasted, and how much each item costs to manage at the end of the product's life. They prioritize their waste management problems, focusing first on products that are most toxic or dangerous (e.g., mercury thermostats, fluorescent lamps, batteries, pesticides, and pharmaceuticals), those costing the most to manage (e.g., household paint and electronics), and bulky items that are hard to handle and fill up landfill space (e.g., carpet, tires, furniture, and mattresses).

Enlightened local governments fund special programs to safely manage these products. Most agencies, however, cannot afford to, even as they recognize its importance. Municipal officials must balance funding for waste management with hiring staff to provide vital services like public safety, fire protection, and student education. As a result, many wastes that should be managed safely are not. The vast majority of HHW is still disposed of in the trash; down sinks, toilets, or storm drains; or dumped directly to land, and only 32 percent of the 300 million tons of municipal waste generated each year in the United States is recycled.[42] These statistics should be alarming to all of us. It is clear evidence that significant change is needed *right now*. We are throwing out almost all of the unwanted toxic products we buy and about two-thirds of all materials we use—that seems absurd and dangerous!

---

42. "National Overview: Facts and Figures on Materials, Wastes and Recycling," US Environmental Protection Agency, accessed February 20, 2023, https://www.epa.gov/facts-and-figures-about-materials-waste-and-recycling/national-overview-facts-and-figures-materials.

Local governments have long had the unenviable task of serving as a backstop for whatever manufacturers made and consumers disposed. They are required to collect more and more products that are being manufactured and consumed at a faster and faster rate. And, they do not have the funding to pay for it. This puts local officials in a bind. While they made great strides in managing waste, they are now strained to a breaking point. As municipal costs continue to mount due to an increasingly complex waste stream, governments have recently faced drastic fluctuations in global recycling markets that significantly increased their costs to recycle the packaging and paper products that comprise 40 percent of the municipal waste stream.[43]

It took decades to transition from a trash system run by municipalities to one that included recycling and HHW. And now, China's National Sword policy has imposed stricter standards on imported recyclables so that contaminated plastics and other materials are no longer accepted, uncovering a major weakness in the US recycling system infrastructure. Our country relies on local governments and their taxpayers and ratepayers to manage an increasingly complex and costly municipal solid waste stream over which they have no control. Increased costs resulting from recycling market restrictions have been compounded by the ubiquitous nature of plastic pollution and its cleanup cost. Through the work of activists, academics, organizations, and governments worldwide, a global consciousness has trumpeted vivid images of uncontrolled and grotesque plastic pollution in oceans and waterways and on land. The moment is screaming for change.

CLIMATE CHANGE AND PRODUCT MANUFACTURE

The manufacture of products in the United States contributes 29 percent of US greenhouse gas emissions, accounting for the extraction of natural resources; the production, transport, and disposal of goods; and the provision of services.[44] However, since the United States is a net importer of goods, it is responsible for 44 percent of total world greenhouse gas emissions based on the manufacture of products *used* in the United States.[45]

Fossil fuels are consumed to make energy, which is needed to mine or manufacture products. Product manufacture, waste management, and greenhouse

43. "National Overview: Facts and Figures on Materials, Wastes and Recycling."

44. "Opportunities to Reduce Greenhouse Gas Emissions through Materials and Land Management Practices," US Environmental Protection Agency Office of Solid Waste and Emergency Response (September 2009), 13.

45. "Reducing Greenhouse Gas Emissions from Recycling and Composting," US Environmental Protection Agency (May 2011), https://westcoastclimateforum.com/sites/westcoastclimateforum/files/related_documents/Reducing_GHGs_through_Recycling_and_Composting_0.pdf.

gas emissions are all directly linked. If we do not recover the materials used to make products and packaging and reuse those products or incorporate the materials into the production of new products, companies will mine more materials and manufacture more products. Typically, it takes more energy to mine or manufacture new materials for use in new products and packaging than it does to recover postconsumer product discards and remanufacture them into new products and packaging. Thus, reducing, reusing, and recycling materials that are used in products and packaging reduces energy use and leads to fewer greenhouse gases.

According to the US EPA, in "2018, about 94 million tons of MSW in the US were recycled and composted, saving over 193 million metric tons of carbon dioxide equivalent ($MTCO_2e$). This is comparable to the emissions that could be reduced from taking almost 42 million cars off the road in a year."[46] The potential for reduction of greenhouse gases varies by product and material. Again, citing US EPA, "although carpet comprises only 3 percent of the waste stream in terms of tonnage in California, Oregon, and Washington, it is a material with one of the highest emissions reduction potentials through recycling in all three states."[47] For the core recyclables that we normally associate with curbside recycling, the US EPA reported the following: "The use of recycled aluminum reduces emissions by reusing the material. The energy input of producing a ton of aluminum, which is directly linked to emissions output, is 96 percent lower when recycled aluminum is used. This is due to the elimination of the mining and smelting process required for virgin aluminum. Thus, end-of-life materials management strategies such as recycling can lead to significantly lower emissions from early stages in the material life cycle, including material extraction, manufacturing, and distribution."[48]

US researchers have a thorough understanding of the impact that US consumers have on the environment, and how to reduce that impact. Unfortunately, in 2014, the United States emitted 15 percent of global carbon dioxide emissions, the major greenhouse gas that contributes to climate change, lagging only behind China at 30 percent.[49] The tremendous wealth

---

46. "Advancing Sustainable Materials Management: 2018 Fact Sheet," US Environmental Protection Agency (December 2020), 14, accessed February 20, 2023, https://www.epa.gov/sites/default/files/2021-01/documents/2018_ff_fact_sheet_dec_2020_fnl_508.pdf.

47. "Reducing Greenhouse Gas Emissions from Recycling and Composting," US Environmental Protection Agency (May 2011), https://westcoastclimateforum.com/sites/westcoastclimateforum/files/related_documents/Reducing_GHGs_through_Recycling_and_Composting_0.pdf

48. "Reducing Greenhouse Gas Emissions from Recycling and Composting."

49. T. A. Boden, G. Marland, and R. J. Andres, "National CO_2 Emissions from Fossil-Fuel Burning, Cement Manufacture, and Gas Flaring: 1751–2014," Carbon Dioxide Information Analysis Center, Oak Ridge National Laboratory, US Department of Energy (2017), doi 10.3334/CDIAC/00001_V2017.

amassed by the United States in the last century and the technological progress we have enjoyed and shared with the rest of the world has also come at a great cost to ourselves and others.

Climate change, including the warming of our planet, has already caused billions of dollars of physical damage worldwide. With global warming, weather patterns have changed, and storms, flooding, fires, drought, and other calamities have gotten more severe.[50] Lives have been lost, particularly among those most vulnerable to the environmental shifts that have made some areas of the world uninhabitable. Billions more dollars are being spent to fortify coastlines and other fragile environments at risk from a rising sea level. This threat has spawned an entirely new job sector, called resiliency planning, which plans and implements protective measures to combat ongoing threats from the physical environment caused by climate change.

The costs of physical damage resulting from climate change and the cost of mitigation strategies to prevent further calamities are all massive externalities. These costs are not paid directly or proportionally by those causing these impacts—the companies and individuals profiting from the sale of products (e.g., cars, fuels, and everyday household items) that are the mainstay of our economy. Instead, these costs are passed onto society—governments, taxpayers, and ratepayers—who pay to rebuild homes and roads wiped out by storms, house and feed people displaced from floods and fires, and prop up shorelines to protect coastal properties.

Just as the mismanagement of product and packaging waste has come back to encroach on human populations and caused financial, environmental, and social impacts, greenhouse gases pumped into the atmosphere have reached beyond the natural limit of absorption and now cause a major threat to humanity. Microplastics in our food chain and our bodies, as well as severe weather pattern changes that we now experience more regularly, are examples of the boomerang effect that results from ignoring responsibility for our actions. Climate change is the manifestation of colossally combined externalities that are not accounted for in burning fossil fuels to manufacture products. In other words, the cost we pay for consumer products does not incorporate the cost of emission controls and other interventions needed to reduce greenhouse gas emissions to safe levels. And those most impacted—low-income communities and communities of color—are often those *least likely* to be the cause of the impacts to which they are subjected.

50. "The Effects of Climate Change," NASA, accessed February 20, 2023, https://climate.nasa.gov/effects.

THE NEED FOR A PARADIGM SHIFT

Despite volunteer, government, and industry efforts, the United States is now one of the worst countries in terms of product and packaging material consumption. Waste generation in the United States actually *increased* 16.5 percent from 2010 to 2018, with each person now generating about 4.9 pounds of waste every day. This is nearly one ton per person per year or about 292 million tons of MSW annually for the entire country. About 32 percent of the amount generated is recycled or composted, while nearly 12 percent is combusted with energy recovery and about 50 percent is landfilled.[51] Over the past decade, the US recycling and composting rate has stagnated, while the cost to process waste—either by recycling, composting, or disposal—continues to be a growing problem.

Our society has always managed to rise to the challenge that is created by an evolving waste stream. We started collecting waste to clear the streets, prevent disease, and maintain public health. We piled waste into landfills away from the majority of the population. Incineration technology burned our waste to reduce garbage piles. When landfills leached into our water supply, we enacted laws to contain the contamination through an expensive landfill lining and leachate collection system. When incinerators produced air pollution, we required air emission controls, which were further improved by a process to turn the waste into energy.

As we became conscious of the amount of waste we generated and how it impacted the environment we love, recycling took a strong hold. Volunteers started the recycling movement until we realized that the job required permanent, paid staff. We lobbied governments to provide us with recycling service. When products became more hazardous, governments held single-day HHW events to remove paint, mercury products, pesticides, and other toxics from the regular household trash and safely manage them. Some regions built permanent HHW facilities to consolidate waste and reach greater efficiency.

We began to understand that many wastes should be treated separately from the regular household trash, not only toxics but those that were difficult to manage, like carpet, tires, and mattresses. As pharmaceuticals showed up in our waterways and the opioid epidemic hit, leftover medications were added to the list of wastes requiring special collection. We now see solar panels, bicycle batteries, and vape pens in the waste stream. What will be

51. An additional 17.7 million tons of food (6 percent) were managed by other methods, including animal feed, bio-based materials/biochemical processing, co-digestion/anaerobic digestion, donation, land application and sewer/wastewater treatment; "National Overview: Facts and Figures on Materials, Wastes and Recycling," US Environmental Protection Agency, accessed February 20, 2023, https://www.epa.gov/facts-and-figures-about-materials-waste-and-recycling/national-overview-facts-and-figures-materials.

the next troubling waste that municipalities are called upon to manage? Our national recycling rate has stalled for a decade, and waste generation has gone up. Cash-strapped local governments appear to react like a yo-yo on a string, pulled without notice by the vagaries of international market forces. With a growing and evermore complex waste stream, we are at a crossroads.

The trail behind us of successful waste management decisions tells us we have come a long way on our journey. We have made people safer, protected our environment, and developed recycling into an economic force. We no longer have rivers on fire, like the 1969 Cuyahoga River disaster in Ohio, in which "a spark flared from the train tracks down to the river below, igniting industrial debris floating on the surface of the water," with flames "in some places reaching five stories high."[52] There are no longer tire piles burning out of control, like the 1983 tire fire in Rhinehart, Virginia, that burned seven million tires for nine months, polluting nearby water sources with lead and arsenic, and depositing polluted emissions in three states.[53] There are no longer garbage barges being rejected from port to port as they search for a place to dispose of trash.

But we have also polluted our oceans and created a global climate crisis that is, in large part, due to excessive material consumption. Once again, we are at an inflection point where we need to act. Resources are being consumed at an unsustainable level. Human health and environmental impacts from consumer products take place all along their life cycle, most of which we know little or nothing about. Municipal waste programs are breaking under the financial strain of international recycling market fluctuations and a complex and rapidly changing waste stream. The way we manage our garbage needs to change, now.

We have crossed the Rubicon in product management in the United States. The public now expects companies to make products and packaging without toxics, and to make them reusable and recyclable, and with recycled content. They expect companies to provide reuse and recycling opportunities for their packaging in public spaces, to educate citizens about how to recycle, and to fund and manage these systems. They expect manufacturers to take financial and managerial responsibility for their products and packaging after consumers are done using them. Basically, people want to be safe, protected, and secure against product impacts from material extraction to end-of-life management.

---

52. Tim Folger, "The Cuyahoga River Caught Fire 50 Years Ago. It Inspired a Movement," *National Geographic* (June 21, 2019), https://www.nationalgeographic.com/environment/2019/06/the-cuyahoga-river-caught-fire-it-inspired-a-movement/#close.

53. "Tire Fires," US Environmental Protection Agency, accessed February 23, 2023, https://archive.epa.gov/epawaste/conserve/materials/tires/web/html/fires.html. The tire storage facility, however, is listed by US EPA as still being cleaned up as a Superfund site, at significant taxpayer expense.

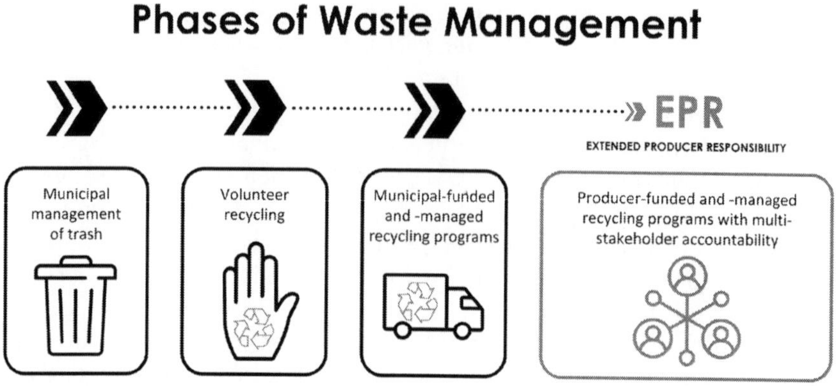

Figure 2.1 Phases of Waste Management
Source: © 2023 Product Stewardship Institute, Inc.

As with all systemic change, there will be some people who argue for continuing along the same treacherous path, but it is time to forge a new path for waste management (see figure 2.1). We can no longer rely on local governments alone to bear the responsibility for an ever-changing waste stream. The financial incentives to reduce waste need to change by placing financial and management responsibility on those who design, manufacture, and package the products we consume. Only through producer responsibility policies will product and packaging choices better reflect the true cost of goods, while also providing financial incentives to redesign them to reduce health and environmental impacts.

In the next chapter of this book, I will present the environmental policy concepts of product stewardship and extended producer responsibility (EPR), how they developed in the United States, and what those who support product stewardship and EPR seek to achieve, which is nothing less than a paradigm shift.

Chapter 3

Producer Responsibility in the United States: Definitions and Core Principles

Product stewardship is a policy that seeks to change the ways in which products and packaging are designed and waste is managed and financed. It is a paradigm shift and reorganization of relationships among manufacturers, retailers, consumers, recycling and waste management companies, and governments—those who make, sell, use, and manage leftover products and packaging. It is a system that responds to the need for these stakeholders to share responsibility for the costs and initiatives required to protect public health and the environment from the impacts of the products we use every day. It is about ensuring that the materials we use to enrich our lives have maximum value and minimal impact during their entire life cycle, from creation and use to their potential reuse and return to the earth. This chapter explores the principles and goals of the product stewardship and Extended Producer Responsibility (EPR) movement in the United States, as well as the roles of important stakeholders.

WHAT ARE PRODUCT STEWARDSHIP AND EXTENDED PRODUCER RESPONSIBILITY?

"Product stewardship" is the act of preventing the health, safety, environmental, and social impacts of products and packaging throughout all life-cycle stages, while also maximizing economic benefits.[1] The Product Stewardship Institute (PSI) developed the nation's first *Principles of Product Stewardship*

1. "Product Stewardship and Extended Producer Responsibility," Product Stewardship Institute (2011), accessed February 24, 2023, https://productstewardship.us/wp-content/uploads/2022/12/2022_update_product_stewards.pdf.

in 2001[2] and, along with the California Product Stewardship Council and the Product Policy Institute (now Upstream), updated them in 2011 to harmonize terminology in the United States and streamline the development of programs, policies, and legislation. "Extended producer responsibility" (EPR) is a mandatory type of product stewardship that includes, at a minimum, the requirement that the producer's responsibility for their product extends to postconsumer management of that product and its packaging[3] (see figure 3.1).

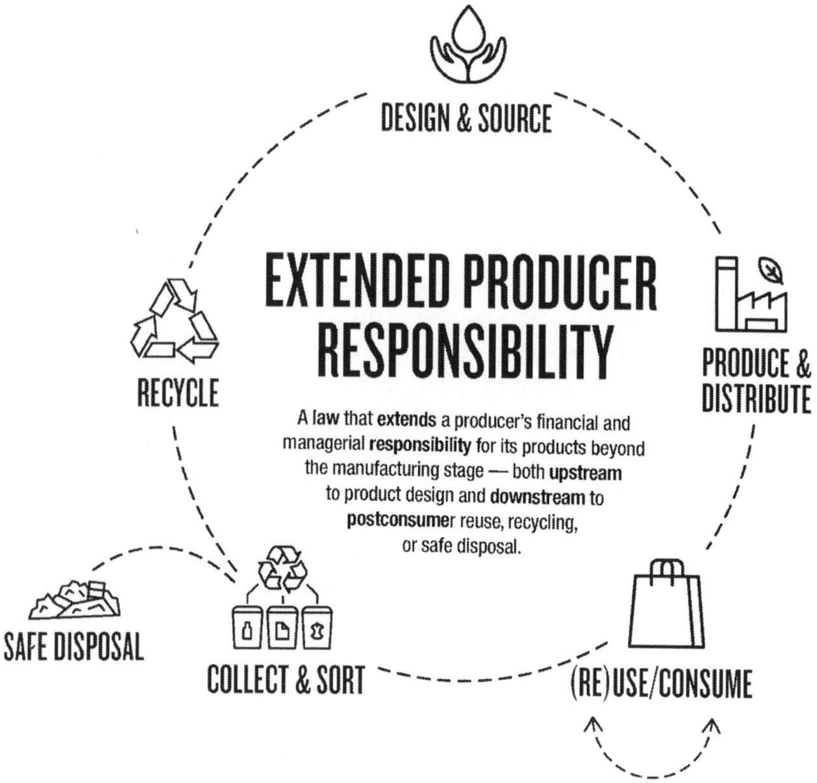

Figure 3.1 Extended Producer Responsibility (EPR)
Source: © 2023 Product Stewardship Institute, Inc.

2. "Principles of Product Stewardship," Product Stewardship Institute (2001), accessed February 24, 2023, https://productstewardship.us/wp-content/uploads/2023/02/Principles-of-Product-Stewardship.pdf.

3. "Product Stewardship and Extended Producer Responsibility," Product Stewardship Institute (2011), accessed February 24, 2023, https://productstewardship.us/wp-content/uploads/2022/12/2022_update_product_stewards.pdf.

There are two important features of EPR policy:

1. shifting primary financial and management responsibility upstream to the brand owner or product manufacturer (typically referred to as the producer) and away from the public sector, with government oversight; and
2. incentivizing producers to incorporate environmental considerations into the design of their products and packaging.

The core tenet of product stewardship is that producers take responsibility, either through government regulation or voluntarily, for reducing impacts all along a product's life cycle. It creates a thread of accountability for those engaged in mining through manufacturing and sale all the way to the ultimate fate of materials downstream—reusing and recycling ("cradle-to-cradle") or landfilling or incineration ("cradle-to-grave"). Since product stewardship encompasses both voluntary and regulatory initiatives (including EPR), it is the broadest term used to describe actions for which producers have a central role in the financing and/or management of their products and packaging when no longer wanted by consumers. Included in this broad term are voluntary initiatives that companies fund and manage, such as long-standing US rechargeable battery and mercury thermostat recycling programs or the take-back and recycling of water filters, toner cartridges, and other products. Product stewardship can also be undertaken by government regulation, such as a ban on the sale of new mercury thermostats, plastic bag bans and paper bag fees, and through EPR laws (see figure 3.2).

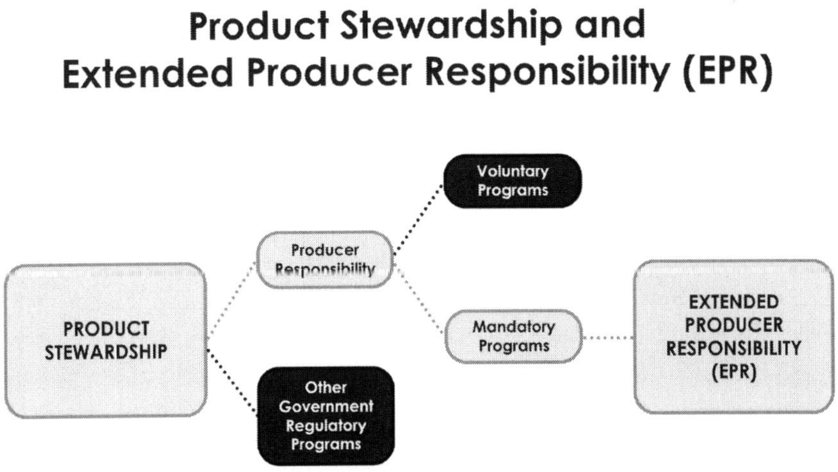

Figure 3.2 Product Stewardship vs. Extended Producer Responsibility (EPR)
Source: © 2023 Product Stewardship Institute, Inc.

CORE PRINCIPLES OF EPR

There are five core principles of all EPR legislation: (1) producer responsibility, (2) level playing field, (3) results based, (4) transparency and accountability, and (5) roles for key stakeholders. Together, they achieve maximum results and thus are considered a best practice to include in all EPR legislation. Although these principles have been applied differently by US jurisdictions, adhering to these principles provides the best opportunity to harmonize legislation nationally, through a unified set of state laws or with a federal law. Since these principles are informed by global norms, established through experience gained in European and Canadian EPR programs, they provide the opportunity to develop globally consistent EPR policies and programs. Based on these five principles, PSI developed a set of best practices that all EPR laws should contain (see chapter 6).

Principle 1: Producer Responsibility

The bedrock EPR principle is that producers have the greatest responsibility to reduce the financial, environmental, and social impacts caused by their products and packaging. Since they know what materials are used to manufacture their goods, they are in the best position to reduce those impacts and create postconsumer value to return used materials to the circular economy. This principle further clarifies that a producer's responsibility is to finance and provide end-of-life management of their products and packaging as a condition of sale. If a producer funds and manages take-back programs voluntarily, we refer to it as product stewardship. If performed as the result of a law, it is called EPR.

EPR laws hold producers responsible for funding and largely managing the steps needed for a material to go from consumer waste to a new product or package. Producers pay into a fund, which they usually manage, and use the funds to hire companies to collect materials at locations that are convenient for consumers to access. They also hire recyclers or waste management companies to make sure the collected material is managed in accordance with the jurisdiction's materials management hierarchy for that material.[4] Producers also develop consumer educational materials, or do so in conjunction with local and state governments, as well as with collectors and recyclers. These

4. The standard materials management hierarchy includes source reduction, reuse, recycling, waste to energy, and landfill. Most government agencies also place recycling material back into the same product or into a similar value product that can be recycled multiple times (closed-loop recycling) higher in the hierarchy than downcycling material into a product that is recycled only once or twice and then tossed.

materials are intended to guide consumers on proper product and packaging reuse, recycling, composting, or disposal.

Principle 2: Level Playing Field

EPR laws level the playing field for all producers within a particular product category by requiring that companies compete under the same requirements. If a company does not fulfill its legal responsibility to fund and manage EPR programs, the government oversight agency has the authority, and responsibility, to prohibit the company from selling products covered under the law in that jurisdiction. The agency can also enforce penalties against a noncompliant retailer for knowingly selling products covered under the law. Creating a level playing field through enforcement is a critical role and responsibility of oversight agencies. If they fail to uphold this responsibility, they provide an unfair advantage to noncompliant companies at the expense of those who play by the rules.

Under voluntary take-back programs, producers that take responsibility for their products and packaging pay for those who don't take responsibility and will always be at a competitive disadvantage. Voluntary take-back systems must contend with "free riders," companies whose waste products get collected and managed in the system into which the free rider does not financially contribute. That is, they benefit from the efforts of others and not due to their own efforts. For this reason, voluntary stewardship programs cannot achieve fairness among producers. EPR laws are fairer because they require all companies to finance and manage the system.

Principle 3: Results Based

EPR systems are established to prevent negative consequences. They are no different from laws that require citizens to stop at red lights to prevent an accident. Wasting resources results in the need to mine more minerals to manufacture more products and packaging, and the emission of more greenhouse gases (GHG). It means more waste going to landfills and combustion facilities (i.e., incinerators), more truck traffic, and more truck exhaust and stench in neighborhoods. It means fewer recycled materials are available for recycled product manufacturers to create new products and packaging and complete the circular economy, which creates new jobs and economic value.

EPR laws reflect our society's evolution of thought and understanding about how to reduce impacts that result from resource consumption. These laws also reflect an understanding that producers, not governments, should be responsible for meeting measurable performance targets that serve as a surrogate for broader environmental, social, and economic goals. Many

US federal, state, and local environmental laws already require companies to meet results-based targets. For example, governments issue permits that allow companies to emit specific levels of pollutants into the air and water and onto land from sources such as smokestacks and pipes. Companies that own and operate combustion facilities (also known as waste-to-energy facilities) and landfills must also meet air and water emission standards. What is different under an EPR system is that producers are held responsible for meeting reduction, reuse, and recycling targets for the product and packaging waste created by the materials (i.e., products) they put into the marketplace.

Meeting these targets used to be the domain of state and local governments. Some state agencies have tried to impose mandatory recycling requirements on local governments, only to be met with a backlash from local officials who argue that imposing requirements without funding is an "unfunded mandate."[5] Some local governments have attempted to fine residents for not recycling, but these acts have been politically perilous, particularly because recycling can be confusing to residents. Infractions often have more to do with unclear instructions, difficulty in distinguishing between recyclable and non-recyclable materials, or changes in materials collected from one town to another. Long ago, the federal government realized the difficulty of regulating millions of households under federal waste management law, even exempting the disposal of hazardous household products by residents. It's very challenging to monitor what households slip into garbage bags and barrels and enforce solid waste management compliance.

In some states, environmental agencies have banned specific materials from disposal, such as yard waste, lead-based cathode ray tubes,[6] mattresses, and recyclable packaging, then held companies responsible for collecting and recycling them. Instead of requiring households to recycle, governments often enact bans that hold collectors accountable. Noncompliant loads are then rejected by state officials at transfer stations, landfills, and waste-to-energy plants. While these bans send the clear signal that recycling and composting are state priorities, waste collectors find it unpopular to police their customers unless items are large, like television sets, which can be left on the curb. These challenges are exacerbated by the unreliability of government funding for compliance officers to inspect waste disposed at numerous facilities throughout the state.

---

5. Scott Cassel, "Product Stewardship: Shared Responsibility for Managing HHW," in *Handbook on Household Hazardous Waste*, Second Edition, ed. A. Cabaniss (Lanham, MD: Bernan Press, 2018), 159.

6. Cathode ray tubes, or CRTs, were used in older model televisions and computer monitors to display images on the screen. To block harmful radiation, CRTs contain lead that protects users from radiation.

EPR, by contrast, creates a chain of accountability that starts with the brand owner but must, by necessity, involve all others in the chain of responsibility, including collectors, material recovery facilities, state and local governments, and of course, the consumer. Since the brand owner chooses the materials for their products and packaging, they are on the hook for ensuring that a measurable amount of these materials are reused or recycled. Other stakeholders, though, also have a role to play to make sure that the brand owner is successful.

Performance Goals

Performance goals are needed to ensure that programs are effective and efficient, and that they achieve the policy intent. EPR systems require that all stakeholders be clear about what they are trying to achieve and who is responsible for which element of the system. Under EPR, performance goals can be set in statute, regulation, or through stewardship plans that producers submit for approval to the government oversight agency. In effective EPR laws, if goals are not met, the oversight agency is given authority to require the responsible party to expend more effort—for example, producers might need to provide additional convenient collection sites, more educational materials and outreach, and take other actions to meet program goals. EPR laws also include financial penalties for not meeting goals after repeated attempts.

Typical performance goals require brand owners to collect, reuse, and recycle a minimum volume or weight of a given material within a certain time period (e.g., one year, or by a certain date). For example, under Vermont's single-use battery recycling program, battery manufacturers were required to recycle 20 percent of single-use batteries they put on the market in 2020. The state oversight agency, in this case the Vermont Department of Environmental Conservation (VT DEC), required the producers to submit a plan detailing how they intend to meet the 20 percent collection rate goal throughout the term of the approved stewardship plan. Goals such as these are, not surprisingly, called "rates and dates." Government does not dictate how producers should meet the goal but seeks compliance assurance from producers through a detailed stewardship plan that outlines collection locations, a public education and outreach program, and other variables. If the battery collection rate performance goal is not reached in Vermont, DEC has authority under the law to require modifications of the plan and can issue penalties for repeated failure to meet the goals.

A key question is, who is responsible for meeting these performance goals? Early US EPR laws required companies or producer responsibility organizations (PROs) to allocate responsibility by company in accordance with the percentage of their own branded products *returned* for collection by consumers. If, for example, 1,000 pounds of computer equipment covered by an EPR law were collected in a region and 100 of those pounds came from one computer manufacturer, that company would be responsible for paying into the system 10 percent of the total cost.

This type of calculation, known as "return share," assigned each collected item to an individual company. While accurate, it was extremely resource intensive to administer. Over time, the calculation of this goal was replaced by a method based on the amount of equipment the company sold into the market, its "market share," which was determined through industry sales data. If a company sold 10 percent of the pounds of new electronics each year into the jurisdiction, they would be responsible for financing the collection and processing of the same amount of electronics through the program. Agencies also used this same market share percentage to assign responsibility for what I like to call "ownerless products," which are products from defunct companies (also called "orphan products" in the field). The company with 10 percent market share, for example, would also be responsible for funding and managing the collection and recycling of 10 percent of the ownerless products.

Holding producers responsible for meeting specific goals allows government to step back from micromanaging the process and provides producers with the flexibility to innovate. Even so, the experience in the United States over the past 22 years has shown the reluctance of producers to accept rate-based goals, largely due to the difficulty in developing a methodology that producers will accept. There will always be a degree of uncertainty in establishing goals that are based on a percentage of unwanted material that needs to be managed. For example, although it is possible to calculate the *actual* pounds of mercury thermostats *collected* in an EPR program in a year's time (the numerator in the equation to establish a rate-based goal), it is only possible to *estimate* the pounds of mercury thermostats *available* for collection (the equation's denominator). The denominator is determined by a mathematical calculation that factors in variables such as the average amount of time that mercury thermostats stay on the wall before being replaced with newer technologies, which can vary by geographic area, economic activity that drives home renovations, and other variables (see equation 3.1).

$$\text{goal} = \frac{\text{mercury thermostats collected (lbs)}}{\text{mercury thermostats available for collection (lbs)}} \tag{3.1}$$

Since the variables that influence the denominator can only be estimated, there will always be uncertainty about the amount of mercury available for collection, especially since mercury thermostats were no longer sold in the United States after about 2006[7] (although they can remain operable for decades). Uncertainty breeds opinions, not facts, which often leads to program delays caused by legal challenges. Since the ramifications for not meeting a target can be significant, producers often vigorously contest methodologies used to determine goals, as well as the data plugged into models and formulas that lead to the calculated measure of program performance to which they are held accountable. For a case in point, in California, the Thermostat Recycling Corporation (TRC) hired its own consultant to develop a methodology for calculating a recycling rate for mercury thermostats for which they were evaluated by CalRecycle, the state oversight agency. TRC later rejected their own consultant's methodology when the results showed they did not meet their statutory performance goal. Many agencies withstand these challenges and rely on best available data. Others defer to goals that are less reliant on specific performance data, such as convenience standards.

Convenience Standards

Over time, some US government officials have grown weary of constant challenges by brand owners to how performance goals are calculated and have shifted their emphasis to evaluating programs based on the convenience that programs provide to consumers. Convenience standards ensure there are adequate opportunities for consumers to reuse, recycle, or safely dispose of their unwanted products and packaging rather than evaluating how much they use these opportunities. Convenience standards are always coupled with EPR statutory requirements for producers to educate consumers about the importance of diverting used products and packaging from disposal, along with specific opportunities for collection.

A typical convenience standard might require pharmaceutical producers to place a permanent collection kiosk in at least one retail pharmacy per county with a population of 10,000 or more, and an extra kiosk in other retail pharmacy locations in that county for every additional 10,000 people. Another standard might be that paint producers ensure that 90 percent of residents have an opportunity to drop off their leftover paint at a permanent collection location within 15 miles of their home. Since rural areas have fewer pharmacies, paint collection locations, and other convenient product collection sites, periodic one-day collection events and mail-back options are written into many stewardship plans. Events require close communication

7. The Vermont and Maine thermostat EPR laws included a ban on the sale of mercury thermostats after July 1, 2006. The California thermostat EPR law prohibited the sale of mercury thermostats after January 1, 2006. Similar provisions were included in many of the other thermostat EPR laws that were passed around this time. These state bans, along with earlier state mercury product bans (see chapter 5), resulted in the cessation of the sale of mercury thermostats nationwide and encouraged thermostat manufacturers to develop alternative non-mercury thermostat technologies.

and coordination among producers, collection contractors, local government agencies, and the state oversight agency. For small items, like leftover medicine and used syringes, prepaid mail-back envelopes and containers that are available online or at convenient locations can also become part of a stewardship plan.

Convenience standards ensure that consumers have opportunities to return their products and packaging to be managed in accordance with the standard waste management hierarchy—reduction, reuse, recycling/composting, waste-to-energy and, finally, landfilling (see figure 3.3). Opportunities may include other beneficial uses authorized by the oversight agency. Since it is easier for an oversight agency to determine the number of collection sites open to the public at convenient times than to calculate recycling rate goals, convenience standards have become critical to evaluating the success of an EPR law. Although many governments still prefer performance targets based on a recycling percentage, convenience standards have become either a supplement to, or at times a replacement for, the more specific performance targets. The emphasis placed on meeting performance targets and convenience

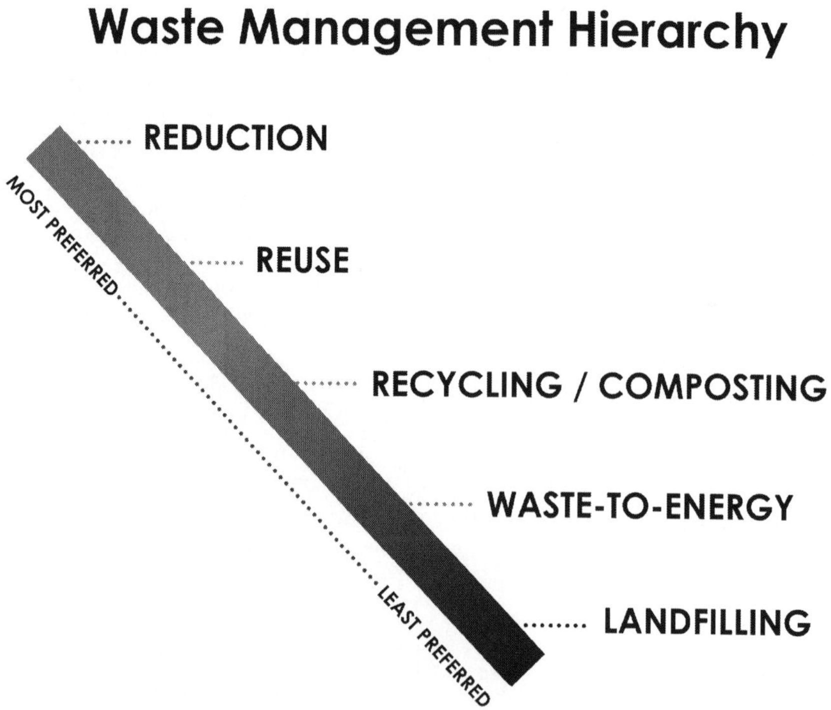

Figure 3.3 Standard Waste Management Hierarchy
Source: © 2023 Product Stewardship Institute, Inc.

standards highlights the need for accepted methods by which government oversight agencies can hold responsible entities legally accountable for the performance of EPR systems.

Principle 4: Transparency and Accountability

Government is responsible for ensuring that producer programs are transparent and accountable to the public, including providing opportunities for stakeholder input. The public has interests in many aspects of waste management, including where and how recycled materials are processed; the amount of collected material that is actually recycled versus the amount that is unusable due to contamination or a lack of markets; compliance with the law; and program costs. They also want to know where and how waste is disposed of when it is not recycled, the compliance record for those facilities, and associated health impacts.

The importance of transparency is tied to an oversight agency's need for data to evaluate its waste management programs. All stakeholders can agree that programs, policies, and laws should produce intended results. Otherwise, they should be changed so that they do. And the definition of success shifts over time, too, as stakeholders around the world gain EPR program experience and newly developed best practices are incorporated into existing EPR programs. Program improvements, however, are only possible if data is collected and available to those overseeing the program and those legally responsible for meeting the goals.

Over the past 22 years of the US EPR movement, the questions I am most often asked have shifted significantly from the simple (e.g., "What is EPR?") to the more complex (e.g., "Where do collected materials go for recycling and what products are they made into?"). As more people become aware of the concept of EPR and the brand owner's role in achieving a circular economy, questions will undoubtedly address how materials are sourced to make the product and packaging put on the market; the environmental, social, and economic impacts of mining and manufacture; the environmental and social impacts of recycling operations domestically and abroad; financial costs at each stage of the recycling and waste disposal process; amounts collected; and other variables and questions.

The public is also asking why so much waste is produced and what producers and governments are doing to reduce it through source reduction, reuse, and recycling. They are demanding reductions in greenhouse gas emissions that tie directly into product manufacture, waste, recycling, and the circular economy. They want to know what is downcycled to another use that only lasts one more product cycle (e.g., scrap carpet to decking board) and what is truly returned to the circular economy through multiple material cycles (e.g., scrap carpet into new carpet).

Businesses are already hyper-focused on cost efficiency and meeting their own publicly stated sustainability commitments, such as incorporation of recycled content into new products, which creates a demand for the material collected in EPR systems. Collectors want data to ensure efficiency and to identify system problems quickly and accurately. They want to be able to forecast how much material will come from each hauling route so they can develop more efficient trucking operations. Recyclers need to know the amount of material collected from multiple trucks and arrival times at their facilities. They track fluctuating commodity prices as well as the level of contamination of materials from each location, along with the cost of recycling each material. Governments need the full array of data to orchestrate an efficient and effective system that serves all citizens equitably—urban, suburban, and rural—as well as whether there are disproportionate impacts on specific communities, including air and noise pollution, litter, and related metrics.

Programs generate considerable amounts of data, and data management systems will continue to evolve to provide the data transparency and protection expected by, and of, each key stakeholder. These systems will also take into account a company's need to maintain business confidentiality. There will need to be a process to clearly determine what is considered confidential and whether the agency, the producer responsibility organization, or another entity is assigned the role of protecting confidential information from becoming public. The public's right to know how a program is managed must be balanced among multiple competing interests. Data management systems will be increasingly important tools to provide the information and insights needed for all stakeholders to make decisions that will ensure the attainment of program goals. These systems will provide the visibility into program performance that stakeholders need to track their own responsibilities, as well as the data transparency others need to ensure an EPR system is effective, efficient, sustainable, and publicly credible.

Principle 5: Roles for Producers, Government, Collectors, Recyclers, Retailers, and Consumers

Although EPR contains the bedrock principle of producer responsibility, this fifth fundamental principle acknowledges that EPR systems cannot be successful unless other stakeholders are held accountable for program aspects over which they have the most control. Reducing waste and hazardous ingredients (e.g., toxics), returning materials to the circular economy, and safely disposing of products that cannot be reused or recycled, will require the collective action of producers, governments, retailers, collectors, recyclers, and consumers. Without a system like EPR that holds each of these entities accountable for assuming specific roles in the context of a comprehensive system, we will not achieve shared sustainability goals.

The term *producer responsibility* took hold because we need those making products and packaging to take a leadership role and partner with state and local governments to set up a system that has the best chance of reducing waste. Another term that has been used to describe the multiplicity of relationships needed to reach these laudable goals is *shared responsibility* since, under EPR systems, multiple stakeholders share responsibility for managing postconsumer products and packaging. Each stakeholder—state government, local government, retailers, collectors, recyclers, and consumers—has a role to play in reducing waste, increasing material reuse and recycling, and returning materials to the circular economy, or in some cases safely disposing of them. These entities can only work in harmony through a comprehensive system that provides incentives for specific actions to be taken by each of these stakeholder groups so that measurable goals can be achieved. In essence, while both terms—*producer responsibility* and *shared responsibility*—are accurate, it is best to view EPR programs as producers taking *primary* responsibility while other stakeholders also assume essential roles.

Shared responsibility does *not* mean that governments, retailers, or others *must* accept shared financial responsibility in managing waste materials, although some may choose to do so. For example, producers might offer to pay half the costs of collection and recycling if local governments pay the other half. Or producers might offer to recycle materials if municipalities will collect them. That is *not* the intent of the EPR movement. Instead, the EPR movement seeks to provide a financial incentive to reduce impacts to those making and selling products and packaging, thus removing a financial and management burden from governments. In the case of packaging waste, however, some municipal and state governments have sought greater control over the EPR system because they do not have adequate assurance that producers will effectively assume the management of a well-established recycling system that has been run by municipalities and the state for decades. In these cases, a transition period might be warranted to build trust among producers, governments, and waste management companies, eventually allowing the shared responsibility system to transition to one more fully producer funded and managed.

EPR: A FRAMEWORK FOR STAKEHOLDER ACCOUNTABILITY

EPR systems provide a framework that ensures all stakeholders are accountable for their piece of the puzzle and impose consequences if they don't meet that responsibility. These systems provide the framework within which important, and often complicated, stakeholder relationships can be negotiated. For example, producers provide sustainable system funding. Retailers educate employees about new collection procedures for the public and often provide space for a collection kiosk or container. Recyclers handle materials accepted by the EPR program. Local governments provide space at their recycling and household hazardous waste (HHW) facilities for residents to return materials. And state governments oversee the entire system. Each stakeholder might contribute labor, store space, or other resources to make the entire system work.

Sharing responsibility is essential to achieving the shared outcomes sought. To develop a sustainable materials management system in the United States, each stakeholder group must be willing to accept that its specific interests are only one part of the equation. These stakeholders must also accept and value the unique interests of other groups critical to the success of the system. Carpet recycling, for example, requires the integration of interests among state and local governments, carpet manufacturers, recyclers, installers, waste management companies, environmental groups, and consumers. Each product has its own unique array of stakeholders, and each of those interests is important. Collectively, under EPR, they are melded into a comprehensive policy that can be implemented efficiently and effectively, with flexibility to address issues that no one can fully anticipate at program inception. Above all, we need to think of EPR in the context of a system that assigns important roles, responsibilities, and accountability to those needed for program success.

My work at the Product Stewardship Institute has been partly driven by an interest in harmonizing EPR policies and programs. Standardizing elements in EPR laws will, over time, result in greater familiarity with these laws and will reduce costs for producers who must comply. The more we can streamline these systems across the United States, either through model state legislation or a federal law, the more efficient these programs will be. Even so, no two US states will interpret the concepts of producer responsibility or shared responsibility exactly the same way. There will always be some variation, even as PSI seeks to narrow that band by setting joint expectations through negotiation.

During the legislative process, the political give-and-take will produce outcomes that reflect the values of each state, administrative and legislative cultures, and ways that existing EPR or other environmental laws have been enacted. Each industry group also has its own culture and way of doing business, which might vary based on the geographic area in which they operate, their market share, or their business model. Retailer interests vary by whether they are independent family-owned businesses, chains, big box outlets, or franchises. Consumer attitudes about recycling and environmental conservation also vary by the regions in which they live. These attitudes often influence the strategies taken by local environmental groups to advocate for community interests.

A key goal of the national product stewardship movement has been to provide a basic framework for all key stakeholders no matter where they live or operate in the country. Through the development of policy models, PSI and our state and local government members have developed principles, core policy elements and options, and best practices that have become common nationwide for products that had limited collection and recycling infrastructure 22 years ago, such as mattresses and other bulky items as well as paint, thermostats, and other toxic materials. These policies, programs, and laws are fairly uniform in their core elements. Even so, they reflect unique state and regional characteristics, often resulting in differences in program performance.

By contrast, household and commercial packaging recycling systems have developed over the past 50 years into complex and nuanced relationships among governments, taxpayers, ratepayers, and recycling companies. These systems are already much more variable than current US EPR systems for toxic and bulky wastes. Even so, those developing US packaging policies seek to craft EPR systems around common principles that align the unique aspects of the current US waste management system with global best practices.

THE ROLE OF PRODUCERS

Under EPR laws, brand owners of consumer goods (or "producers" as they are called globally) are required to finance and manage the collection and reuse, recycling, or safe disposal of their products after consumer use. Although most US EPR laws are written so that a company can choose to be individually or collectively responsible, most choose to join a producer responsibility organization (PRO) that takes on the obligation to meet legal requirements. By holding producers responsible for managing their products at the end of life (postconsumer), they have a direct financial incentive to ensure those products retain maximum value or cost as little as possible to manage. Requiring producers to "own" their products at the end of their useful life ensures they will consider the amount and type of materials they use during design and manufacture. By internalizing these costs, producers have incentive to reduce costs just as they would labor, capital, and other company expenses.

While many stakeholders have roles to play in EPR systems, those with the greatest ability to reduce a product's impact over its entire life cycle—brand owners—have the greatest degree of responsibility. Only producers know the materials contained in their products and have the power to reduce or change them at the source. These systems shift the cost of end-of-life management of products from taxpayers and municipalities to producers (which, in some cases, are retail brands like Target or Walmart). By shifting costs, policymakers create the funding base needed to sustain postconsumer management programs that reduce public risk and create environmental and economic benefits. In addition, shifting costs from governments and taxpayers to producers is a fairer system because those who consume more pay more, rather than all taxpayers covering those costs even if they do not create the waste. If your neighbor purchases and discards more products and packaging than you, do you want to pay for their waste?

The concept of EPR connects the responsibility that companies have long assumed in a manufacturing facility to the end of life (postconsumer) management of their products and packaging. EPR recognizes that the old way of handling residential waste—with the financial and management burden entirely on governments—no longer works. EPR sets out a new vision, a paradigm shift that changes the roles of those held responsible, by law, so that the burden is shared and the primary responsibility rests with the entity that is in the strongest position to reduce impacts: the producer. EPR extends a producer's responsibility for eliminating the health and environmental impacts from the manufacturing plant through to postconsumer waste collection, transportation, reuse, recycling, or disposal.

EPR is also a concept that explains how upstream mining and downstream recycling and disposal are directly connected. Under an EPR law, if a company manufactures batteries for example, which requires the extraction of manganese, copper, lithium, and other materials, it will have the responsibility to finance and manage the collection and recycling of those batteries, which significantly reduces impacts that occur in mining and manufacturing. That company will retain a financial connection to those batteries. Regardless of whether the metals extracted from a recovered battery type have high or low market value, EPR systems are designed to internalize the full cost to manage that postconsumer battery into the cost of doing business, no different than the cost of labor, fuel, materials, and other related expenses.

THE ROLE OF GOVERNMENT

Although the term *EPR* places producers in the policy crosshairs, governments have the political responsibility to develop that policy. Without government will and action, EPR systems will not be put in place and implemented effectively. All three branches of government—legislative, judicial, and administrative—have played key roles in advancing the concept of EPR in the United States and making it possible for societal views on resource management to change.

Most of US EPR legislative activity has taken place at the state level, although local EPR laws have been strategically passed in order to prompt state action. As of June 2023, there were 133 EPR laws enacted (106 state and 27 local) on 17 products in 33 states (including Washington, DC). By contrast, there have been no federal EPR laws passed, although federal EPR legislation has been introduced on packaging, pharmaceuticals, and electronics.

Judicially, one case of national significance to EPR laws was the 2014 case *Pharmaceutical Research and Manufacturers of America, et al. v. County of Alameda, et al.*, in which the pharmaceutical industry claimed that the Alameda County, California, drug take-back program violated the federal Commerce Clause because residents outside of Alameda County would be forced to pay for the costs of pharmaceutical take-back in Alameda County. The US Supreme Court refused to take up this case, however, which left standing the Ninth Circuit Appeals Court ruling that allowed the county's pharmaceuticals take-back program to proceed.[8] The standing decision that the EPR ordinance did not violate the Commerce Clause was significant, not only for pharmaceuticals take-back laws. It also upheld the central tenet of all EPR laws that require producers to fund and participate in the postconsumer management of their products and packaging.

In the United States, the main federal administrative agency overseeing waste management, the US Environmental Protection Agency (US EPA), has placed most of the responsibility for municipal solid waste management (MSW) on state and local governments. US EPA provides technical support to these agencies but has not typically participated actively in promoting EPR. Federal laws, such as the 1976 Resource Conservation and Recovery Act, have set legal guidelines for MSW disposal facilities—landfills and combustion facilities—but have not regulated household and most small business waste. As a result, in all states (except California) it is legal to dispose

8. *Pharmaceutical Research and Manufacturers of America; General Pharmaceutical Association, Biotechnology Industry Organization v. County of Alameda, Alameda County Department of Environmental Health*, https://cdn.ca9.uscourts.gov/datastore/opinions/2014/09/30/13-16833.pdf.

of all MSW (including household hazardous waste) in MSW landfills and combustion facilities as long as they meet US EPA guidelines, are operated within permit limits, and meet permit standards.

Even so, the US EPA has advocated for the voluntary reduction, reuse, and recycling of waste, reduced product toxicity, and efficient and effective waste management. This voluntary approach to MSW management has been interpreted differently among the agency's leaders over PSI's first 22 years. Under the administration of President George W. Bush (2001–2009), PSI received several resource conservation grants that enabled us to facilitate multi-stakeholder meetings and develop EPR policy models on mercury thermostats, fluorescent lamps, paint, pharmaceuticals, and other difficult-to-manage products. Under the Bush administration, EPA officials led an effort to seek a national consensus on electronics EPR legislation and were instrumental in the development of several PSI EPR models, including the national agreement that PSI mediated with the paint industry, which led to the passage of paint stewardship laws in 10 states and the District of Columbia by 2023.

Later, under the administration of President Barack Obama (2009–2017), the EPA eliminated "voluntary" programs, including the grants that previously funded PSI's multi-stakeholder dialogues that led to agreements on EPR models. Ironically, some Obama officials took a more voluntary approach to MSW management during their administration than under the Bush administration. As federal leadership dissipated on EPR, state and local agencies carried the torch forward.

State Legislative and Oversight Activity

Although there is often a dynamic political tension between state legislators and state agency officials (who represent the governor), these two branches of government play critical leadership roles in the passage of state EPR laws.

Legislator Role

Over the past 22 years, PSI has focused on building the capacity for state and local government agencies to develop and introduce EPR bills, as well as manage them once they become law. The strategy for educating legislators, however, is different. Legislators have little time to ramp up on an issue, and have multiple issues that vie for their attention. They must be quick learners but also be provided concise information about bills that they can easily use. Many legislators seek EPR product bills that have already passed or been introduced in other states. They have little time to refine a bill without assistance from outside policy experts.

One of PSI's roles has been to provide technical policy expertise to legislators and their staff. Educating legislators is a constant effort, especially with elections, term limits, and aides who frequently move on to other jobs. While educating agency staff often produces long-term benefits, educating legislators must be done strategically for resources to be used judiciously. Legislators with secure seats, along with the knowledge and interest to introduce EPR bills, are local stars in the EPR movement. Those who can forge bonds among their colleagues, as well as with agency staff, the regulated industry, environmental leaders, and their constituency are the most effective. Choosing the right legislative sponsor for a bill is a skill, and legislators can be reassigned to a new committee by leadership that changes a bill's political calculus from year to year.

Agency Role

Under current US EPR laws, legislators assign the role of oversight to a specific local or state government agency, such as a state department of environmental protection. Under a federal EPR law, which currently does not yet exist in the United States for any product or packaging, the US EPA would likely become the oversight agency. That agency's role is to manage the entire program and ensure that each stakeholder in the system is held accountable for performing its function as articulated in the law. Effective agency staff develop strong relationships with each stakeholder, including the public, to ensure that the overall system functions as intended. In essence, they are like the conductor in a symphony, blending various instruments to produce harmonized sounds that are made at specific times. The music (the law) must be well written so that each instrument (each stakeholder) knows their role. The conductor must see it all, hear it all, communicate clearly with each entity, and make corrections as needed. Each stakeholder in an EPR law is clearly an important and distinct instrument in the symphony that plays a complex piece of music that requires the steady hand of an able conductor—government.

Passing good EPR bills and preventing bad ones from passing are part of the significant role played by government. And they must play this role for effective EPR bills to pass and for material use to be "less unsustainable," as Dr. Bob Giegengack ("Gieg"), my undergraduate college professor from the University of Pennsylvania, liked to say. Government officials need to be educated on the composition of good laws, how programs work, challenges that can arise, and how to address problems. They need to understand the interests and positions of all key stakeholders. They must also be experts on the process by which stakeholder interests can be melded into a viable program that fits within the agency's current body of EPR and environmental laws and programs.

Half-hearted legislative efforts resulting in poor laws, as well good laws that are poorly implemented and enforced, only tarnish the potential of these powerful systems. Governments not only have to be proponents of good EPR laws, but they also need to protect against bad EPR laws. In New York, a poorly designed EPR law in 2017 would have placed the entire burden to fund and manage pharmaceuticals on retail pharmacies. PSI partnered with the New York Department of Environmental Conservation (Peter Pettit), New York Product Stewardship Council (Andrew Radin), and Citizens Campaign for the Environment (Brian Smith and Adrienne Esposito) to convince Governor Andrew Cuomo to veto the industry bill that had been passed nearly unanimously by an unwitting legislature. Our joint effort resulted in the passage of a strong EPR law the following year that placed the responsibility on producers with other stakeholders also playing key roles.

EPR laws with design or implementation flaws can have serious political ramifications. The implementation of a series of EPR laws in Ontario, Canada, for example, caused the ouster of the province's minister of the environment when multiple fees appeared on consumer receipts due to the confluence of program implementation timelines. EPR outcomes, even if significantly better than those of prior systems, can also have negative impacts if expectations are not managed. For example, the passage of New York's electronics EPR law in 2010 significantly increased the amount of scrap electronics recycled and secured millions of dollars from producers for recycling. But it also left local government programs with costs beyond their expectations because the performance goal in the legislation was set too low and did not increase over time. When producers reached their recycling goal, they stopped collecting, leaving local governments with no choice but to continue to receive scrap electronics from their residents and pay recyclers with funds for which they did not budget. Governments collected more electronics through the EPR law, but not all of it was covered financially by producers, as was intended, because of a flaw in the bill's design. The increased costs for local programs were not expected and soured the appetite of many local officials for other EPR programs for several years.

Packaging and paper products, which represent 40 percent of the waste stream, have represented a unique challenge for all stakeholders, but particularly governments. If either the bill design or law implementation is mishandled, a community's packaging recycling system, even if currently inadequate, could result in confusion and anger not unlike citizen anger over botched snow removal during a major storm. Governments, which are inherently risk averse, need to figure out how EPR programs can work technically as well as politically. Since packaging recycling systems have been in place for decades in many communities, packaging EPR programs have been the most significant change facing all stakeholders. For this reason, governments initially proceeded more cautiously on packaging EPR even in the face of significant recycling and waste management cost increases.

These important government oversight responsibilities account for the often slow, deliberative nature of legislative and administrative actions. While there are risks for governments, the payoff is huge, and awareness of EPR as a tool to reduce government waste management costs has grown. Successful EPR law implementation will save significant amounts of money for local governments, create jobs, protect the environment, and have enormous positive benefits. These results will reflect positively on government leaders who recognize the need to change waste management and take bold steps to implement a new system.

PSI identified these positive changes through a 2017 evaluation of four EPR programs in Connecticut (electronics, mercury thermostats, paint, and mattresses) based on available data from 2011 to 2015.[9] Together, these programs diverted more than 26 million pounds of materials from waste, yielded a cumulative cost savings of more than $2.6 million per year to Connecticut municipalities, and provided services worth another $6.7 million. They led to the creation of more than 100 jobs, reduced greenhouse gas emissions by more than 13 million kg of carbon equivalent, and nearly all Connecticut residents now have convenient access to recycling collection sites for the four products.

Laws are only effective, though, if they are enforced both fairly and consistently by government oversight agencies. All stakeholders rely on these agencies to enforce EPR laws to level the playing field for all companies. In standard EPR programs, the oversight agency requires producers individually or collectively to meet performance standards, such as collecting and recycling a certain percentage of the material they put on the market, at product end-of-life. The agency requires producers to submit a stewardship plan for its review and approval detailing how they will meet all requirements of the law. The state also requires producers to submit an annual report that evaluates program performance based on parameters articulated in the law. To ensure an effective EPR program, agencies need a wide range of staff expertise that includes policy, science, communications, and legislative skills.

Governments overseeing EPR laws also have the responsibility to enforce those laws against noncompliant companies to maintain a level playing field for a fair market-based program. Governments are also expected by their constituents to look out for their interests. They need to know what those interests are. In an EPR context, that translates to having a program that is convenient

9. Product Stewardship Institute, *Connecticut Extended Producer Responsibility Program Evaluation: Summary and Recommendations* (2017), accessed February 26, 2023, https://psi.wildapricot.org/resources/Other%20Resources/2017-01-Evaluation-Connecticut-EPR-Programs-and-Recommendations.pdf.

for them to use, education that is clear and informative, outreach that is readily accessible, and program elements and outcomes that are transparent.

Agency EPR Leaders

Government officials play pivotal roles in their agencies to advance the notion that change is needed. They must move a bureaucracy beyond the status quo by articulating a problem and providing a vision for a new direction. They must often take risks and push for change against stagnant and entrenched ideas, skeptical superiors, or political currents "from above" that send subtle messages to lie low and not stir up trouble. Above all, these officials need to form alliances and coalitions that leverage their ideas and create changes in how waste is managed through programs, policies, regulations, and laws. They lead the passage and implementation of EPR laws that create incentives for each stakeholder to perform the role we need them to take for the sustainable use of natural resources.

As CEO of the Product Stewardship Institute, I have participated in numerous government policy discussions and actions, from conceiving an idea, to developing and passing a policy, and evaluating its effectiveness. PSI's role is to catalyze government action, add fuel and direction as needed, help governments engage industry and environmental groups in collaborative problem-solving, build coalitions that support EPR bills, and advocate for bill passage. There is an abundance of state and local government officials whose leadership has resulted in the passage of EPR laws and who have significantly changed the course of waste management in the United States. They have protected their own environment; saved their government and other governments thousands, if not millions, of dollars; created jobs; and redirected the trajectory of US waste management toward a more sustainable path.

Governments rely on officials who retain knowledge and implement policies regardless of which political affiliation is in power. There are countless state and local, and a few federal, officials who paved the way and maintained momentum, building the capacity for US product stewardship and EPR, nationally and in their own geographic regions of the country. Though there are many more than I can name, I cannot overstate the tremendous importance of these government pioneers' passion and expertise. These have been some of the people with whom PSI staff have spent the most time and who have formed the backbone of the EPR movement in the United States, often becoming PSI board members and the EPR leads at their agencies. To understand the gravity of their contribution, keep in mind that when PSI and some of these government officials started to promote product stewardship and EPR policies in 2000, almost no one knew what those words meant, few governments or environmental groups were involved, and those who would eventually oppose these policies had not yet heard about them. Other officials who came to the movement later also forged new ground in their respective states.

At the risk of inadvertently omitting many deserved people, here are some of the most influential government product stewardship and EPR leaders over the past 22 years: Jennifer Holliday, Chittenden County, Vermont; Cathy Jamieson, Vermont Agency of Natural Resources; Ann Pistell, Carole Cifrino, Paula Clark, and Elena Bertocci, Maine Department of Environmental Protection; Peter Pettit, New York Department of Environmental Protection; Andrew Radin, Onondaga County, New York; Dawn Timm, Niagara County, New York; Kate Kitchener, New York City Department of Sanitation; Tom Metzner, Connecticut Department of Energy and Environmental Protection; Jen Heaton-Jones and Cheryl Reedy, Housatonic Resources Recovery Authority, Connecticut; Walter Willis, Lake County, Illinois; Marta Keane, Will County, Illinois; Garth Hickle, Minnesota Pollution Control Agency; Mallory Anderson, Hennepin County, Minnesota; Fenton Rood and Patrick Riley, Oklahoma Department of Environmental Quality; Steve Danahy, Nebraska Department of Environmental Quality; Dave Berger, St. Louis-Jefferson Solid Waste Management District, Missouri; Lisa McDaniel and Nadja Karpilow, MARC Solid Waste Management District, Missouri; Angie Snyder, Ozarks Headwaters Recycling and Materials Management District, Missouri; Scott Mouw, North Carolina Department of Environment and Natural Resources; Sego Jackson, City of Seattle, Washington; Dave Galvin and Lisa Sepanski, King County, Washington; Jay Shepard, Dave Nightingale, Janine Bogar, and Megan Warfield, Washington Ecology; Scott Klag and Jim Quinn, Metro, Oregon; Abby Boudouris, David Allaway, and Cheryl Grabham, Oregon Department of Environmental Quality; Shirley Willd-Wagner and Cynthia Dunn, CalRecycle; Rich Berman, Santa Monica, California; Bill Pollock, Alameda County; Jen Jackson and Maggie Johnson, City/County of San Francisco; Raoul Clarke and Jack Price, Florida Department of Environmental Protection; Becky Jane, IL Environmental Protection Agency; Teresa Stiner, IA Department of Natural Resources; Steve Brachman, University of Wisconsin-Extension; and Cynthia Moore, Barb Bickford, and Jennifer Semrau, Wisconsin Department of Natural Resources.

Some local government officials are registered lobbyists who are adept at introducing and passing state EPR laws. Jen Holliday, of Chittenden County Solid Waste District in Vermont, is one such person. She has used her technical knowledge as director of public policy and communications to lobby for EPR policies. And, as chair of the Vermont Product Stewardship Council, she does so with the full support of other solid waste district managers across the state. Walter Willis, of Lake County, Illinois, plays a similar role with the Illinois Product Stewardship Council, and Dawn Timm (Niagara County) and

Andrew Radin (formerly of Onondaga County) have played key roles with the New York Product Stewardship Council, as has Jen Heaton-Jones with the Connecticut Product Stewardship Council. These people work statewide to form broad coalitions with other local governments, state officials, legislators, corporate leaders, and environmental leaders. Since local governments are hit hardest by the financial and managerial burden of waste management, they often play key advocacy roles in their state legislatures to pass EPR laws. They provide a sincere voice regarding the specific waste problems they face and the benefits of an EPR approach.

While state officials also testify in support of bills, this has been far less common in my experience, although state officials are beginning to take a stronger role on packaging EPR bills. The process for seeking approval from the governor's office is often laborious. In my seven years as the liaison on waste policy issues with the governor's office in Massachusetts, our agency did not sponsor any environmental bills on which I was the lead. State officials will usually promote a bill only if it is a priority for the governor or the agency (with the governor's approval). It is more common for state officials to testify on a bill to provide basic information, protect the agency's interests regarding the need for funding to oversee the program, ensure the bill is not onerous to implement, and ensure that they have authority to adjust requirements if program goals are not met. While some state officials are the drivers of EPR legislation, their most common role is to provide extremely important technical assistance and to educate agency management internally to support bills if introduced by others.

The Changing Role for Government

Local governments assume roles in EPR systems in accordance with their abilities and interests. Many communities take pride in their recycling programs and want to continue to perform that function. They often have a personal connection to residents and don't want to lose that interface. They send recycling information to each household, answer questions by phone and online, staff collection facility events, and provide a community service of which they are proud. Local governments have come a long way from the days of volunteer collection programs and they do not want to relinquish their role so quickly, even as they are desperate for funding to maintain and enhance their programs. They built recycling and other waste management expertise over the past five decades. It will require conversation and negotiation to find the right role for each community to play in EPR programs.

Governments are no longer expected to be the lone entity funding and managing society's waste. It was too easy for companies and residents to leave waste cleanup to the government, taxpayers, and ratepayers. Many governments no longer want to act as the cleanup crew and pick up whatever consumers throw in the trash. They view EPR laws as marking the end of the assumption that the production and consumption of resources can continue unabated and uncontrolled.

THE ROLE OF RECYCLERS

A sustainable recycling system requires a profitable recycling industry. Public and private recyclers are the backbone of the circular economy. They produce on average 10 times as many jobs as those providing a disposal service.[10] Producers who wish to tout their achievements in recycling their products and packaging have an obligation to ensure a viable recycling industry. Producers are increasingly expected to create postconsumer markets for the products and packaging they sell. They not only rely on recyclers to recover and turn their materials into new products and packaging, but they also revel in the success of recyclers in meeting targets for recycled content, reuse, recycling, and zero landfilling. Recyclers, in turn, rely on producers to put onto the market the materials that they can reuse or recycle. In fact, we all rely on recyclers. These companies perform a vital service that combines technology and human labor to sort and process materials that must meet precise specifications set by end-markets.

Recyclers commit to a specific level of service amid a shifting commodities market. They know the amount of contamination allowed for each market and communicate that to their material suppliers—households and businesses—through their collectors. They invest in new equipment, train staff, and protect workers against accidents. These companies also take a business risk. If the equipment they purchase does not recover revenue in a reasonable period of time, they will lose money or market share, and/or go out of business. They must rely on residents to recycle only those items that can be recycled. Otherwise, the recyclable material they collect will have less value, which increases their cost of operation. Recyclers are challenged by packaging materials that residents put in recycling bins for which there are no markets. They contend with plastic swimming pools, plastic bags that get stuck in processing equipment and cause costly plant shutdowns, and other "wish-cycling" items consumers desperately want to recycle.

To keep contamination at a minimum, recyclers often rely on governments to educate their residents about what to recycle, how to recycle, and the importance of recycling right. In other cases, they may educate residents

10. Tellus Institute with Sound Resource Management, *More Jobs, Less Pollution: Growing the Recycling Economy in the US* (2011), https://www.nrdc.org/sites/default/files/glo_11111401a.pdf.

directly, working with the municipality to ensure their messages are consistent. Recyclers will benefit from EPR laws because they provide sustainable funding for their services that are not jeopardized by the vicissitudes of municipal finances. EPR laws also regulate the type of materials sent to recycling facilities and can stimulate the market for materials that recyclers turn into bales of recovered material. The circular economy relies on recyclers to turn discarded material into a viable commodity for its next use as a product or package. Recyclers, in turn, need a consistent system that provides strong financial incentives for producers to put on the market the materials that can be recycled. EPR provides such a system because it connects producers to collectors, recyclers, consumers, governments, and other key stakeholders in the chain of product supply and material recovery.

If we want to sustain a healthy recycling ecosystem, producers, municipalities, and state governments need to nurture companies that perform recycling services. EPR systems must account for existing investments that companies and governments have made in recycling infrastructure. Since EPR systems require producers to take a greater role in funding and managing the recycling system, those currently owning and managing recycling facilities are rightly concerned with losing those investments. EPR systems will need to incentivize the use of existing, efficient operations and also provide a degree of flexibility and management decision making for those paying the recycling bill—the producers—who want to ensure that the system efficiently and effectively processes the materials they put on the market. Navigating the transition from a current statewide recycling program to an EPR system hinges significantly on how the new EPR system uses existing recycling infrastructure while driving efficiency and expanding collections to new materials currently not yet recycled.

THE ROLE OF COLLECTORS

Just as truckers transport consumer products to our homes, collectors transport recyclables from homes, municipal depots, retail stores, and other locations to recycling facilities and on to end markets. These are the people we rely on to skillfully back into narrow alleys, turn sharp corners, maneuver around double-parked cars, and ensure the safe and efficient delivery of commodities to recycling facilities. They drive large trucks often under extreme pressure to meet schedules while seeking to prevent accidents. When operations work well, we often don't think about these people and take them for granted. It is only when there is an accident, or our recyclables are not picked up on time, that they get our attention. Collectors may be the most underrated aspect of the materials recovery supply chain.

There is currently an abundance of collectors who transport recyclables from our homes and businesses to processing facilities. Most collectors are small independent companies with fewer than 20 trucks[11] that developed contractual relationships with local governments and businesses that solidified over years. Over time, many small companies collected waste and recyclables from small geographic areas. In time, a few large regional or national companies emerged, consolidating the industry somewhat. However, there is still a predominance of small collectors in the business. As has occurred all across the country, collectors often pick up a different set of recyclables prepared to different specifications from each household, business, or municipal district. The current recycling system is fragmented, uncoordinated, and inefficient. This is not the fault of collectors, recyclers, or governments; it is the way our recycling system has evolved over time to meet community members' needs.

Changing to an EPR system will produce greater efficiency through consistent funding, collection of the same basket of recyclables statewide and prepared under the same instructions, and efficient hauler routes that are coordinated statewide. Even so, the transition to a more efficient EPR system with greater control given to producers must account for the livelihood of existing collectors and the people they employ. Just as our country has begun the transition away from coal, oil, and gas, we are transitioning away from an inefficient recycling system to one that will provide greater environmental, economic, and social benefits. However, it is also our societal responsibility in both cases to make that transition sustainable for workers and offer companies who employ them a transition period to the new frontier.

There are often a large number of collectors in one small geographic area serving individual households ("subscription service"), and the transition to EPR systems must account for the investments of these existing businesses. For that reason, there needs to be collector representation in the development of EPR policies, just as there needs to be recycler representation because of the investment they have made in recycling facilities. EPR systems seek to provide incentives for producers to use existing collectors and recycling facilities since that is often the least costly option. However, it is possible that some small collectors may need to be a subcontractor to larger service companies to compete in a less splintered and more comprehensive and cohesive EPR system. In any case, the move to an EPR system must minimize negative repercussions on collectors, even as it seeks greater efficiency, environmental gains, and financial savings for taxpayers, ratepayers, and governments.

11. "Over 96 percent of trucking companies are small businesses with fewer than 20 trucks; 87 percent have 6 or fewer trucks." Brian McGregor, Dr. Ken Casavant, "Truck Transportation," in *Study of Rural Transportation Issues* (USDA Agricultural Marketing Service), 404, https://www.ams.usda.gov/sites/default/files/media/RTIReportChapter13.pdf.

Companies that transport recyclables from the curb or drop-off locations are often the same ones that process recyclable materials. This vertical integration gives them an advantage over companies that solely transport materials to recycling and waste disposal facilities. In the United States, these companies typically contract with a municipality, business, or household. Collectors can also be part of a municipal workforce. Depending on the EPR system, private collectors might continue their current contractual arrangements, or instead contract with a PRO. Since these private collectors directly serve households and businesses, they are in a good position to communicate with those entities regarding the steps they need to take to reduce contamination and increase the value of recyclables.

Many waste management companies have generally supported EPR programs for difficult-to-manage and toxic products because EPR removes these pesky products from the waste stream. In addition, by creating a dedicated funding source from producers, EPR systems create new business opportunities for collectors, recyclers, and waste management companies. Many have set up new business units to manage mercury products, electronics, pharmaceuticals, syringes, and other products. However, many of these same companies perceive EPR for packaging as a potential threat, since it could change the nature of their core business.

These waste management companies often have leverage in negotiations with individual households, municipalities, and businesses that they believe could be weakened if they have to contract with producers. This is not the case, however, under three of the four packaging EPR laws and most of the proposed bills, all of which do not require producers to contract directly with collectors. Instead, collectors continue their contracts with municipalities, which then get reimbursed by producers, with the only exception being individual subscription households that opt into the EPR system. At first, when PSI promoted EPR for packaging in 2007, the initial reaction of waste management companies was to maintain the status quo by claiming EPR does not work and that the United States needs a different system. As stakeholders have learned more about EPR, they find ways to address the interests of collectors and recyclers in EPR systems. In fact, they are key stakeholders needed "at the table" to help create an equitable transition to a new EPR system that is desperately needed.

THE ROLE OF RETAILERS

Local retailers play a key role in defining the character of our communities. They have often become an essential part of the fabric of take-back programs by serving as a convenient place to drop off unwanted or scrap products for reuse, recycling, or safe disposal. When picking up new medication at a pharmacy, residents can drop off unused medication in an in-store kiosk, thereby reducing the risk of accidental poisoning or addiction if stored in the home, and water contamination if flushed or disposed in the trash (due to potential leachate). When buying new batteries, residents can drop off spent batteries in a countertop box, reducing waste toxicity and increasing recycling. When buying new paint, they can drop off leftover paint that had been sitting in their workshop, shed, closet, or basement (often for years), returning it to the circular economy to be recycled into new paint.

Many other products and packaging materials are routinely collected voluntarily by retailers—toner cartridges, fluorescent lamps, thermostats, plastic bags, and electronics, among others. The motivation for these retailers is to generate increased foot traffic. Take-back programs get customers into stores, which then become an important destination for residents who seek a familiar place to recycle their used products and packaging, as long as it is convenient for them. Convenience for the consumer is the most important aspect in maximizing the collection of products and packaging. Neighborhood retailers factor prominently in providing a convenient take-back experience for residents. If a retailer is part of a consumer's routine—picking up new batteries while shopping, for example—consumers will more quickly adopt a new habit of adding the return of old products to their routine.

Small community-based retailers have a significant incentive to provide excellent customer service, which is often lacking in online or big box retailers that compete more on price. Large retail outlets like Walmart, Target, Home Depot, and Lowe's have, to date, rejected being a location for the public to drop off old items. They fear that they will become a recycling or household hazardous waste collection depot for all products sold in their stores. Small retailers, therefore, often benefit from EPR programs that require producers to provide retailers with educational materials for consumers, as well as equipment and services to collect, reuse, recycle, or safely dispose of products. These retailers attract customers to their stores for a valued service at no extra charge, sealing their loyalty and often additional sales. Product take-back programs offer small retailers an opportunity to better compete with larger retailers by providing a community collection service that larger stores are ill equipped to match.

In 2004, PSI designed and implemented a five-week pilot computer take-back program with retail store Staples and US EPA. It was the first retail take-back of electronics in the country. As part of the program evaluation, I conducted customer interviews that are still vivid in my mind. Nearly all those with whom I spoke said that their sense of loyalty to Staples deepened because of the take-back program. Based on this pilot, Staples rolled out the program in Washington State for two years before taking it national, becoming a take-back model for Best Buy, Office Depot, Office Max, and other retailers. These retail sites became part of the required collection infrastructure when EPR laws were later passed and continue to play a major role in the collection and recycling of electronics in the United States.[12]

PSI evaluations of paint take-back programs in several states with EPR laws also yielded evidence that retailers voluntarily and enthusiastically collect leftover paint from consumers, as long as the costs of collection bins, transportation, recycling, and educational materials are covered. According to PSI program evaluations in 2016, 88 percent of retailers in California[13] and 78 percent of retailers in Connecticut[14] indicated that it was easy or very easy to participate in the paint EPR take-back program. More than half of the retail drop-off sites in both California and Connecticut indicated that foot traffic in their stores increased on account of participation in the program, and 80 percent of retail respondents in Connecticut indicated that they were satisfied or very satisfied with the paint EPR program. According to one retailer surveyed in Oregon in 2013, "It's a great program! We get more foot traffic in our store (which is sorely needed these days) and our customers (as well as potential new customers) feel great about recycling their old paint, in turn helping our environment."[15]

Retailers also play an essential role in producer compliance with EPR laws. The most effective EPR laws require producers to reach a specified level of consumer convenience that is detailed either in the law or in the stewardship plan submitted by producers to the state oversight agency for approval. Not meeting a "convenience standard" will make take-back less convenient for consumers, which typically translates into a lower recovery rate of the materials targeted under the law. Lower rates of return might also trigger producer noncompliance with meeting a certain collection rate by a specific date (i.e., performance targets).

Producers, therefore, must often partner with retail storefronts because they are an essential part of the EPR collection infrastructure. Although each retail location might receive a small amount of material, these locations should

12. Product Stewardship Institute, *The Collection and Recycling of Used Computers Using a Reverse Distribution System: A Pilot Project With Staples, Inc., Final Report to the US Environmental Agency* (June 2005), accessed February 26, 2023, https://psi.wildapricot.org/resources/Electronics/2005-06-Report-Electronics-Recycling-Via-Retail-Reverse-Distro-Staples-Pilot.pdf.

13. Product Stewardship Institute, California Paint Stewardship Program Evaluation (January 7, 2016), accessed February 26, 2023, https://psi.wildapricot.org/resources/Paint/2015-12-Evaluation-Paint-California-Paint-Stewardship-Program-Evaluation.pdf.

14. Product Stewardship Institute, Connecticut Paint Stewardship Program Evaluation (November 21, 2016), accessed February 26, 2023, https://psi.wildapricot.org/resources/Paint/2016-11-Evaluation-Paint-Connecticut-Paint-Stewardship-Program-Evaluation.pdf.

15. Product Stewardship Institute, Oregon Paint Stewardship Program Evaluation (March 13, 2013), accessed February 26, 2023, https://psi.wildapricot.org/resources/Paint/2013-03-Evaluation-Paint-Oregon-PaintCare-Program-Evaluation.pdf.

be viewed in the context of the entire infrastructure needed for producers to comply with performance targets and convenience standards. In the case of leftover paint, for example, three-quarters of the voluntary paint take-back locations in states with EPR programs are owned by small retailers (e.g., paint dealers, hardware stores, and lumberyards), while the other one-quarter of sites are municipal-run locations that collect larger volumes per site. Retailers and local governments are thus both critical to producer compliance with EPR laws. Compliance, in turn, ensures program effectiveness and efficiency.

When producers cover the costs of logistics, collection bins, transportation, recycling, and educational materials—as they have under EPR programs—many small and independent retailers have seized the opportunity to become part of the product take-back infrastructure. While some retailers prefer not to train staff to accept products, or don't have the floor space for a collection kiosk, or have other current priorities, an increasing number of retailers have taken on this responsibility and turned it into a business opportunity. Even if they don't collect materials themselves, retailers can provide a valuable service by educating their customers about what to do with products and packaging when they no longer need them.

With the rise of online sales, local retailers have been further challenged to compete with the convenience of the internet. EPR laws now include provisions that hold online sellers, including Amazon, responsible for paying into the system that, in turn, funds take-back programs. Although local brick-and-mortar retail stores have struggled against online sellers, they remain the most convenient location to return used products. Mailing items back from individuals is almost always more expensive than asking individuals to self-transport to a retail collection location where products and packaging are consolidated. In addition, requiring someone to request a pre-paid mailer online or find a box and mail it is often not as convenient as dropping a product off the next time you go shopping at a familiar location. Even pharmaceutical mail-back envelopes distributed at retail pharmacies are often not as convenient as dropping off items in a pharmacy kiosk, unless you are homebound or in a less populated location without a pharmacy.

Some retailers collect and recycle used products outside of EPR programs and cover the costs. However, unless they are part of a producer-funded EPR program network, most retail collection sites will not collect enough material to make an environmentally beneficial difference. For example, while grocery take-back programs for plastic bags were a step in the right direction and gained the public's attention, they failed to quell the backlash against plastics production, use, litter, and trash disposal. Collectively, these programs did not provide a sustainable solution to problems created by the proliferation of single-use plastics. As a result, many view these programs as marketing efforts to do just enough to avoid government regulation.

Other retailers might get too much material returned by consumers, which is what I learned firsthand from staff at one large food retail chain near my home. That company cut back on in-store collection of various packaging materials because the store space was too valuable for them to collect the large amount of material that customers returned. In addition, meeting the demand for collection would have significantly increased the store's costs. In this case, a successful collection, which was convenient for customers, was a financial threat to the store. This food retailer paid to recycle materials that other retailers did not, thus putting them at a competitive disadvantage.

Retailers might not think of themselves as anything other than sellers of goods to consumers, but they can play a meaningful role in collecting products and packaging from those same customers. They are one of the few stakeholders, in addition to recycled product manufacturers, that play a dual role of selling new products and receiving scrap products and packaging when consumers no longer want them. Retailers need to be integrated into the collection and recycling network. They can provide a valuable service by providing real solutions to environmental problems.

Being the nexus between producers, government, and consumers is a powerful place for retailers. This power also comes with a responsibility. In effective EPR programs, retailers are required to inform their customers about available product take-back locations using educational materials paid for and provided by producers. Under EPR laws, retailers are also typically prohibited from selling products from brand owners not in compliance with the EPR law, usually because the producer did not pay their fair share into the system. The payoff is marketing, customer loyalty, and sales.

Retailers are an important part of the EPR collection infrastructure. Certainly, non-retail collection options still play a significant role in take-back infrastructure in the United States. Curbside recycling of packaging like bottles, cans, containers, and paper is still more convenient than traveling to a retail location. And less populated areas often do not have conveniently located retailers. These locations require drop-off and mail-back options, along with periodic collection events to provide convenience matching that of product purchase. But in many circumstances, EPR laws provide brick-and-mortar retail stores with a purpose that tightly binds them into the community fabric at a time when online sellers have gutted downtowns. One California retailer participating in the paint EPR program operated by PaintCare said, "It's helped us generate more retail business—because people are satisfied that we take back paint."[16] In a circular economy where materials

16. Product Stewardship Institute, *California Paint Stewardship Program Evaluation* (January 7, 2016), accessed February 26, 2023, https://psi.wildapricot.org/resources/Paint/2015-12-Evaluation-Paint-California-Paint-Stewardship-Program-Evaluation.pdf.

flow into a community through sales and flow out of the community to reuse and recycling facilities, or for safe disposal, retailers will play an increasingly important role.

THE ROLE OF CONSUMERS

All stakeholders who share responsibility for providing effective EPR programs have one key stakeholder in mind: the consumer. EPR laws create a comprehensive system that ties all entities together and holds them accountable for sustainable production and consumption of natural resources. If done well, these stakeholders provide important ways for community members to "do good for the environment." To get to this level, the system needs to take human nature into account so that all consumers can participate. It must be easy to use, have convenient collection locations, and provide educational materials that make it clear why people should care, where to bring used materials, and how to prepare them for collection. Even better is the inclusion of a motivational message to encourage desirable consumer behavior. And, as beverage container deposit return systems have shown, financial incentives work wonders.

Consumers cannot be expected to drive 30 minutes to drop off cardboard and other packaging delivered conveniently by Amazon's army of trucks. Collection needs to be convenient in their weekly routine. Community members cannot be expected to understand that they might not be able to recycle their polypropylene yogurt container even though it has the chasing recycling arrows on it with the number 5 in the center. First, producers need to stop making packaging from materials that cannot be reused or recycled. They also need to provide systems to collect and reuse or recycle the packaging they do create. Then, consumers need to be continually educated about which materials are accepted through which recycling channel to minimize material contamination.

EPR programs put the most emphasis on producers to take responsibility and finance and manage take-back programs. Governments, collectors, recyclers, and retailers also play key roles. But once an EPR program is in place, including adequate education and a convenient collection system, consumers have the responsibility to act. When convenient collection opportunities are provided, all of us as consumers have a responsibility to use them. We, too, play a key role in helping producers meet their legal obligations under EPR systems. We touch hundreds of products each week as we live our lives. Although we cannot possibly understand the impacts that each product has on the environment, if there are convenient opportunities to reuse, recycle, or safely dispose of items, we need to use them. Eventually, consumers must also be held accountable for protecting their neighborhoods and the wider environment in which we all live. Once comprehensive EPR systems are in place, we will expect more of consumers.

IN SUMMARY: STAKEHOLDER ROLES

Consumers have long been blamed for the problem of overconsumption. They do have a major role. But we cannot place full blame on the consumer when they are only one part of a broken system. Why should a person care if all taxpayers in their city or town subsidize their purchase of multiple computers, home furnishings, cars, or other goods if they are not required to pay for the reuse, recycling, or disposal of what they bought and no longer want? Why should a producer care if municipalities, taxpayers, or ratepayers keep paying for the reuse, recycling, or disposal of whatever they put on the market?

Materials are extracted, manufactured, and consumed at rates that are out of balance with our global ability to recover, reuse, and recycle those materials. EPR laws create a comprehensive system of accountability that requires all key players to step up and do their part to reduce resource overconsumption. Our global environment is in distress. The climate has changed. The world has warmed. The solution for a significant part of this is clear. We all have a role—perhaps not in equal amounts, but we all need to act. And government leaders have the ultimate responsibility to find the right balance of interests that suits their state's political and cultural characteristics.

An ethic of recycling has taken hold in the United States over the past 50 years. That was good enough for that period of time, but no longer. We are still generating huge quantities of waste and reusing and recycling far too little. Companies and governments are star performers that need to change their orientation to better manage the global problem of resource overconsumption and disposal. Waste rumbles through our neighborhoods, fills landfills and incinerators near our homes, and causes the generation of GHG emissions by requiring the additional manufacture of products and packaging we toss in the trash. The only way to change this reality and create a more sustainable one is for all those who produce and consume materials to embrace their unique responsibility. We all have a part to play, and now we need to meld those parts into a cohesive program. EPR systems provide a framework that can hold us all accountable for what we are best able to do, in synchrony with others.

Chapter 4

Origins of Producer Responsibility

PROTECTING BODY AND SOUL

The growth of the US product stewardship movement cannot be separated from the beginning of the modern American environmental movement of the 1960s and the pioneering work of the naturalists who came before. John Muir and other nature lovers founded the Sierra Club in 1892 and grew it into the country's first prominent national environmental organization. Muir, President Theodore Roosevelt, and other early preservationists and conservationists created and nurtured the growth of the national parks system, national monuments, national forests, bird sanctuaries, and other unique places of splendor. Their appreciation of the awe-inspiring natural world was coupled with a will to protect special vistas in Yosemite Valley, the Grand Canyon, and other sites of breathtaking beauty. David Brower, Ansel Adams, and numerous others fostered a reverence for our nation's natural treasures by encouraging personal experience, through hiking, climbing, photography, and other forms of recreation that integrated a love of the environment with a political voice to protect it.

The human spirit is stirred by nature, and people fight to protect places that touch their soul. This growing sentiment nurtured the environmental movement of the 1960s, which rose up against the brazenness of the industrial revolution that sought to control the natural environment. People began to demand that government act in the face of rising industrial pollution. Rachel Carson's publication of *Silent Spring* in 1962 tied the production and use of pesticides to large-scale environmental degradation. Carson's powerful blend of science and writing communicated to the American public the tremendous impact our actions can have on nature and human health. *Silent Spring*, written in peace and solitude on the Maine coast, resounded like a cymbal in the public ear.

Silent Spring, and Carson's subsequent convincing testimony before a US Senate subcommittee, woke up government officials and the public to the dangers of pesticides and the impacts from unfettered scientific and technological experimentation. Her written and spoken words put the pesticide industry on trial, and her scientific exposé linking the pesticide DDT to human peril was groundbreaking. Equally apparent was the pesticide industry's furious attempts to defend itself and discredit Carson, which laid bare to the public a paradigm of corporate behavior that would be repeated by tobacco executives who hid nicotine addiction,[1] pharmaceutical executives who masked opioid addiction,[2] and fossil fuel executives who denied their awareness of climate change in the 1980s.[3] This pattern began even before Carson launched her career. The case against lead in gasoline was made in the 1920s by Alice Hamilton, a pioneering chemist and medical doctor, among others. Their work did not result in the laws needed to address the problem until the passage of the Clean Air Act in 1970 and the phaseout of lead in gasoline in the late 1980s, 50 to 60 years later.[4]

These cases are all instances in which companies knew of problems their products created and concealed them, even against overwhelming evidence. The US conservation and environmental movements emerged because a vast number of people sensed that the ways in which they were living their lives, and the ways that companies were conducting business, were out of balance with nature and were unsustainable. Many people felt a sense of responsibility to protect truly awesome places from degradation. Many also became frightened about the impacts of technology on their own health and well-being.

PROTECTING AND CONTROLLING NATURE: A FLEETING BALANCE

The modern-day environmental struggle between protecting and controlling nature is epic: on one hand we might feel *apart from* nature, able and willing

1. Alison Kodjak, "In Ads, Tobacco Companies Admit They Made Cigarettes More Addictive," NPR (November 27, 2017), https://www.npr.org/sections/health-shots/2017/11/27/566014966/in-ads-tobacco-companies-admit-they-made-cigarettes-more-addictive.

2. Barry Meier, "Origins of an Epidemic: Purdue Pharma Knew Its Opioids Were Widely Abused," *New York Times* (May 29, 2018), accessed February 27, 2023, https://www.nytimes.com/2018/05/29/health/purdue-opioids-oxycontin.html.

3. Suzanne Goldenberg, "Exxon knew of climate change in 1981, email says—but it funded deniers for 27 more years," *Guardian* (US) (July 8, 2015), accessed February 27, 2023, https://www.theguardian.com/environment/2015/jul/08/exxon-climate-change-1981-climate-denier-funding.

4. David Rosner and Gerald Markowitz, "A 'Gift of God'? The Public Health Controversy over Leaded Gasoline during the 1920s," *American Journal of Public Health* 75, no. 4 (April 1985), accessed February 27, 2023, https://doi.org/10.2105/ajph.75.4.344; This Lead is Killing Us—A History of Citizens Fighting Lead in their Communities," US National Library of Medicine, accessed February 27, 2023, https://www.nlm.nih.gov/exhibition/thisleadiskillingus/index.html.

to manipulate resources for our own needs and desires. We are capable of tapping the tremendous potential of human ingenuity to manifest a reality only achievable through a combination of vision and action. And yet, we can also feel *part of* nature, integral to the ecosystem on which we all rely for survival. Both approaches can result in human experiences that stir the soul. Standing at the rim of the Grand Canyon as a vast curtain of clouds is yanked aside revealing magnificence is just as brilliant as watching a rocket launch from Cape Canaveral on its way to the moon.

These two attitudes seem to reside in each of us to varying degrees. Some people have an abundance of reverence for nature, others have an affinity to control nature, and yet others are more balanced in their perspective. The two contrasting forces—control and protection—can also continually ebb and flow within each of us, creating difficult environmental choices about products to purchase, foods to eat, modes of transportation, and employment. As a society, our challenge is to balance these forces so that our collective lives align with earth's natural rhythm and carrying capacity. At present, there is a significant need for correction and realignment.

The concept of responsibility is ingrained in our lives. We might take *personal responsibility* in raising a family or obeying traffic laws, while others might feel that they have moral responsibility to care for the welfare of others and not only oneself or one's family. Many cultures seek to create norms that pertain to responsibility by establishing laws that require one person to compensate another if, for example, their animal wandered on another's property and caused damage. It might require that a person build a secure railing around the perimeter of a roof on their home to avoid injury. Different compensation levels might be set if an action was intentional or accidental.

These laws also seek to find a cultural balance in the desire to control and protect natural resources. If waste is a byproduct of exploited resources, those responsible for exploitation might be obligated to compensate those injured and to prevent future adverse impacts from taking place. While the paradox of controlling and protecting nature is part of ancient human civilizations, our goal is to find the right balance for us today. Unfortunately, knowing when our actions cause impacts on others has become more complex in today's society than in ancient times.

ENVIRONMENTAL EXTERNALITIES

I love the alabaster light fixture in my living room. The natural swirls in the stone provide a beautiful design when illuminated, giving light and warmth. The brass and wire accoutrements add to its décor. The price I paid for the fixture includes the cost to design, manufacture, package, transport, and sell the product to me, and also includes the cost of regulations to meet pollution control standards in the country in which it was made.

What I do not know is whether other costs of getting that product into my hands were included in my invoice. Was the rock mined in a manner that caused water pollution? Were toxic chemicals used in the mining process that caused air pollution? Were workers injured as a result of the mining operation? How much greenhouse gas was emitted as a result of energy consumed to extract the rock, ship it to the facility, make the product, and ship it to me?

The economic, environmental, and social costs of products that are external to a company's business costs are known as "externalities." Any impacts, even ones of which I am unaware, are externalities. The company does not pay for them, and I do not specifically pay for them. Essentially, society pays those costs—for example, workers and communities may suffer health impacts from pollution, and taxpayers who did not get a benefit from the products and packaging we purchase must still pay the cost of recycling or disposal. Clearly, the full cost of the lifestyle we lead, no matter who we are or where we live in the world, is not fully incorporated in what we pay for the product.

The management of postconsumer products and packaging results in an array of externalities. The costs to reuse, recycle, compost, and dispose of these items are picked up by governments, taxpayers, and ratepayers who have little control over the materials they buy. Government recycling programs have to react to changes in products put on the market, with no notice and no compensation for extra costs these changes impose. For governments, this lack of control means that their costs can rise at any time and by any amount, and they will always be counted on to pick up this cost or to decrease services provided to the public that has grown to expect them. Governments, taxpayers, and others thus absorb externalities imposed by product brands.

In the 1990s, HP Hood LLC, a local Massachusetts company that sells milk and other products, switched from using clear high-density polyethylene (HDPE) plastic containers to a white opaque plastic. Its goal was to entice consumers to buy their milk based on the presentation of the product, claiming it kept the product fresher (i.e., less likely to degrade nutritional value and flavor) by blocking light. To recyclers, however, opaque plastic has a lower commodity value on the recycling market, which meant they got paid less for the same level of effort. Those recyclers lost money or passed the cost on to the municipalities they serviced—all because the milk company, the product's brand owner in this case, switched packaging material. This externality was absorbed by recyclers and governments (and their taxpayers) and not by the manufacturer that caused the increased cost for others. Under the law, companies like Hood have every right to choose their packaging material. The problem is that their choice creates costs borne by others.

Every day, manufacturers make millions of products and packages. Their goal is to make high-quality, low-cost items that are readily available, attractive, and safe for consumers. Unfortunately, there is still a startling disconnect between the materials that a manufacturer uses to make its products and packaging and the cost that municipalities bear as a result of those design choices. The externalities from consumer product waste have grown steadily to the point that they harm our health and environment to a degree never before experienced. This includes plastic bags caught in neighborhood trees, floating bits of plastic waste in the ocean, and millions of tons of waste disposed each year in landfills and incinerators.

Pollution is not new, but it has grown exponentially and has now boomeranged on us in the form of human health and environmental impacts. Our garbage has come back to envelop us and invade our bodies and those of other animals (e.g., fish, sea turtles, and other aquatic organisms). Plastic polymers and microplastics, pharmaceutical compounds, mercury, and other elements and materials sold in the marketplace have been detected in our bodies at alarming rates. Most alarming is that pollution is not limited to a community, a nation, or a continent. It is global. And the biggest global externality in our lifetime—climate change—is caused in large measure by burning fossil fuels to mine and manufacture raw materials and products to replace those we waste by not reducing, reusing, and recycling.

These costs are externalities because they are external, or outside, the monetary costs we pay to purchase products. External does not mean nonexistent. They are very real to those impacted by pollution. For example, bauxite ore mining for aluminum might pollute a community's water supply and leave behind real human health and environmental costs that are not fully incorporated into the company's ledger sheet. Additional mining will be needed due to the lack of aluminum container recovery and subsequent reliance on virgin material.

Externalities like these emanate from all industries and all materials. External costs are difficult to quantify because they depend on assumptions and models. They often include taxpayer subsidies to mine and manufacture the products we consume and the packaging we use, creating unfair competitive advantages for virgin materials at the expense of reusable and recycled alternatives. Our goal is to internalize the cost to *prevent* externalities. Let's take a look at the emergence of a key concept that seeks to account for externalities and that spurred the modern environmental movement.

POLLUTER-PAYS PRINCIPLE

The concept that a producer should be responsible for postconsumer product waste when governments have taken on this function for centuries has been slow to permeate US culture. Some people in the United States have long adhered to an ethic of "rugged individualism,"[5] which celebrates the spirit and strength of the individual. As governments worked to manage the impacts of industrial pollution, however, costly and complicated cleanups led to the governing concept that the entity that causes pollution should be responsible for the costs of mitigating and further preventing those impacts. In other words, the individual entity has a responsibility to avoid harming others. No matter how innovative an individual or company may be, they have a responsibility to not negatively impact others.

This concept, which aligns with foundational religious and spiritual concepts of personal and moral responsibility, held companies causing pollution responsible for paying for, and preventing, damage. This concept of responsibility became known as the "polluter-pays principle." Delving into how this principle was interpreted, however, reveals two starkly different, but related, perspectives. According to Jean-Baptiste Fressoz, a French historian of the environment, industrialists in the nineteenth century proposed to authorities that companies pay to be allowed to pollute a certain amount rather than be shut down completely by police.[6] This arrangement provided industry with certainty to operate their businesses while they paid a fee in exchange for *permission to pollute.*

Over time, the principle was also applied to pollution prevention. In 1972, the Organization for Economic Cooperation and Development (OECD) adopted the polluter-pays principle "as an economic principle for allocating the costs of pollution control."[7] Under this principle, the polluter was expected to internalize the costs of pollution prevention and control measures that ensured that the environment would be "in an acceptable state" according to public authorities. The polluter was thus legally bound to cover all costs to protect the environment, "such as measures to reduce the pollutant emissions

5. President Herbert Hoover coined the phrase "rugged individualism" in 1928 to refer to the Wild West can-do attitude of the US spirit. David Davenport and Gordon Lloyd, *Rugged Individualism: Dead Or Alive?* (Stanford: Hoover Press, 2017), https://www.hoover.org/research/rugged-individualism-dead-or-alive-0.

6. Nic Ulmi, "Aux origines de la crise écologique [The Origins of the Ecological Crisis]," *Le Temps* (October 18, 2016), accessed February 27, 2023, https://www.letemps.ch/sciences/aux-origines-crise-ecologique.

7. Organization for Economic Cooperation and Development (OECD), *The Polluter-Pays Principle: OECD Analyses and Recommendations* (Paris: OECD Environment Committee and Environment Directorate, 1992), 5, accessed February 26, 2023, https://www.oecd.org/officialdocuments/publicdisplaydocumentpdf/?cote=OCDE/GD(92)81&docLanguage=En.

at source and measures to avoid pollution by collective treatment of effluent from a polluting installation and other sources of pollution."[8]

In addition, aside from narrow exceptions, companies "should not receive assistance of any kind to control pollution (grants, subsidies or tax allowances for pollution control equipment, below-cost charges for public services, etc.)."[9] According to the London School of Economics and Political Science, the polluter-pays principle is now "the commonly accepted practice that those who produce pollution should bear the costs of managing it to prevent damage to human health or the environment."[10] Thus, the basis for US and international environmental law, including extended producer responsibility (EPR) laws, emanates directly from this principle that requires companies, by law, to internalize costs to prevent pollution and keep people and the environment safe.

CORPORATE RESPONSIBILITY UNDER US ENVIRONMENTAL LAWS

The scientific integrity of Rachel Carson and other environmental scientists garnered overwhelming public support for the US environmental movement, leading to a string of federal laws that recognized a corporation's responsibility to take action to protect the public and the environment. The 1970 Clean Air Act regulates air emissions from stationary and mobile sources. In 1972, the Federal Insecticide, Fungicide, and Rodenticide Act gave the US Environmental Protection Agency the power to prohibit the sale, distribution, or use of pesticides; to require that pesticide users register; and to approve and license all pesticides used in the United States.[11] The Clean Water Act, passed in 1977, required companies to obtain a permit before discharging pollutants into navigable waters.

The concept of corporate responsibility is central to all of these laws since the federal government and, over time, state agencies came to regulate their activities. With the advent of environmental protection laws, companies could no longer release pollutants into the air, water, or land without a scientific and technical analysis to determine that the discharges were considered safe for public health and the environment. Although we can argue whether these

8. OECD, *The Polluter-Pays Principle*.

9. OECD, *The Polluter-Pays Principle*.

10. "What Is the Polluter Pays Principle?" London School of Economics and Political Science, accessed February 26, 2023, http://www.lse.ac.uk/GranthamInstitute/faqs/what-is-the-polluter-pays-principle.

11. "Overview: Key Federal Environmental Laws," FindLaw, accessed February 27, 2023, https://smallbusiness.findlaw.com/business-laws-and-regulations/overview-key-federal-environmental-laws.html.

laws are protective enough, particularly with the emergence of the science of cumulative impacts, they do establish federal science-based limits to commercial and industrial activity.

While these federal laws focus on pesticide use and application, water discharges, and air emissions, another federal law takes a more comprehensive view of pollution and addresses the entire system from waste generation to disposal, or from "cradle to grave." The 1976 Resource Conservation and Recovery Act (RCRA) gives the US Environmental Protection Agency (US EPA) the authority to control the generation, transportation, treatment, storage, and disposal of hazardous waste. Under this law, US EPA develops regulations, guidance, and policies that "ensure the safe management and cleanup of solid and hazardous waste, and programs that encourage source reduction and beneficial reuse."[12] Although RCRA "establishes the framework for a national system of solid waste control, states play the lead role in implementing non-hazardous waste programs."[13]

RCRA is the federal law that is most relevant to product stewardship programs. RCRA covers waste generation, including policies that encourage the reduction of waste at the source of generation. Under RCRA, US EPA develops regulations that set minimum national technical standards for disposal facility design and operation. States issue permits to ensure compliance with US EPA and state regulations, which are sometimes more stringent than the federal standard. My role as waste policy director in Massachusetts was part of the RCRA regulatory framework, since our agency's Department of Environmental Protection (MassDEP) issued permits for the operation of landfills and waste-to-energy facilities. One of the very first initiatives I addressed as a state policymaker in 1993 was to work with MassDEP staff to close more than 100 unlined landfills that did not meet state standards. A few years later, we succeeded in closing a city-owned incinerator in Fall River that, while providing revenue for the impoverished city, polluted its air and land due to ineffective pollution control technology. The authority to close polluting landfills and incinerators, as well as to design and operate existing facilities, emanates from RCRA.

Products with hazardous components that are used and disposed by most businesses are regulated under strict hazardous waste rules under RCRA, whereas the same products sold to residents are unregulated when disposed. Even toxic household waste—"household hazardous waste" (HHW)—which is generated by millions of individual households, is not regulated by the federal government. California is the only state that requires HHW to be

12. "Resource Conservation and Recovery Act (RCRA)," US Environmental Protection Agency, accessed February 27, 2023, https://www.epa.gov/rcra/resource-conservation-and-recovery-act-rcra-overview.

13. "Resource Conservation and Recovery Act (RCRA)."

managed as hazardous waste, although they put the onus on local governments to comply.

US EPA's main role in reducing, reusing, and recycling municipal solid waste (MSW),[14] including HHW, has been to promote best practices among state and local governments charged with protecting their residents. Although some local governments levy financial penalties on residents who dispose of recyclable waste, these measures are rarely enforced due to concerns about political backlash. In Massachusetts, our agency's policy was to enforce bans on the disposal of a standard set of recyclable materials at the point of disposal (e.g., at either a transfer station, landfill, or waste-to-energy facility). While this approach did not make our agency popular with waste haulers who dumped their loads at these facilities, it avoided a major political hurdle. Since haulers did not want to have their loads rejected, they, in turn, refused to pick up certain materials from residents, such as leaves in plastic bags and lead-filled computer monitors. We found that residents often changed their behavior once they understood why these materials were not picked up.

While these and other federal laws cover activities of the present, another federal law—the Comprehensive Environmental Response, Compensation, and Liability Act (CERCLA)—seeks to address mistakes of the past. Passed in 1980, CERCLA was sparked by the health, environmental, and financial disasters of Love Canal, New York; Times Beach, Missouri; and the Valley of the Drums, Kentucky. CERCLA then created a federal "Superfund" to clean up, contain, or remove pollutants and hazardous materials from abandoned hazardous waste sites and other toxic dumps. Under CERCLA, US EPA has authority to mandate that those responsible for creating the pollution pay for its cleanup. If whoever created the pollution cannot be found, US EPA can hold those within the chain of ownership and operation, and which are financially solvent, responsible to pay not only for their share of the costs but also for the total costs of the cleanup. In many cases, the agency cleans up the site and later seeks reimbursement for cleanup costs from the companies.

While RCRA authorizes US EPA to regulate toxics and other ongoing emissions and discharges from companies, Superfund (CERCLA) gives US EPA the authority to reach back to past actions that caused impacts on human health and the environment. This federal program seeks to redress the economic impacts from polluted land. According to the US EPA, $8.2 billion dollars have been paid by companies responsible for causing the pollution,

---

14. Municipal solid waste (MSW) is waste typically handled by municipalities, including household and small business waste. The definition of small businesses is most often determined by state law, although some states defer to the base level definition developed by US EPA. State solid waste planners traditionally focus on this part of the waste stream, usually leaving the majority of institutional, commercial, and industrial waste to corporate and large institutional generators to manage since they pay directly for their waste services.

with approximately $4.7 billion of that amount being spent on Superfund site cleanups and the balance of $3.5 billion planned to be used for ongoing or future Superfund cleanup work."[15]

As a result, hundreds of hazardous sites have been returned to productive use, creating economic value. While these results are significant, agency costs have not been fully reimbursed. According to the *Washington Post*, government expenditures for staff and cleanup costs that were not reimbursed add up to more than $21 billion since 1995, "while hundreds of companies responsible for contaminating water paid little to nothing."[16] These costs are externalities paid for by US taxpayers. CERCLA established the legal precedent that requires existing (solvent) companies to pay not only to manage their waste products at Superfund sites, but also those of companies no longer in business. This legal mechanism, known as "joint and several liability," is also applied under current EPR laws to producers of paint, mercury thermostats, electronics, pharmaceuticals, and numerous other products. In this case, companies still in business cover the costs of reuse, recycling, or disposal of products from companies no longer operating.

CORPORATE SOCIAL RESPONSIBILITY

As the polluter-pays principle became prominent in Europe in the 1970s, it quickly established roots in multinational corporations operating in the United States. Coupled with the rise of environmentalism, companies became increasingly aware of their legal liability and the potential for public outcry over environmental impacts for which they could be implicated. Being in compliance was paramount, but exceeding compliance was aspirational for leading companies that sought to further reduce impacts to air, water, and land during their manufacturing processes, as well as boost worker safety. Environmental engineers and scientists working to reduce facility impacts began to be called by new titles, such as "director of product stewardship," to indicate the dual role of environmental stewards and product manufacturers. In the 1980s, 3M and other companies emphasized the polluter-pays principle to underscore their significant investments in manufacturing processes that reduced or eliminated pollution compared to other companies that sold competing products at lower cost but with fewer environmental controls.

15. "Superfund Special Accounts," US Environmental Protection Agency, accessed February 28, 2023, https://www.epa.gov/enforcement/superfund-special-accounts.

16. Bryan Anderson, "Taxpayer dollars fund most oversight and cleanup costs at Superfund sites," *Washington Post* (September 20, 2017), accessed February 28, 2023, https://www.washingtonpost.com/national/taxpayer-dollars-fund-most-oversight-and-cleanup-costs-at-superfund-sites/2017/09/20/aedcd426-8209-11e7-902a-2a9f2d808496_story.html.

Companies that invested in technologies and approaches to reduce pollution from their factories below their legally permitted limits were heralded. They showed what is possible to achieve and tended to move entire industries forward toward greater sustainability. These industry leaders spawned a fast-growing global movement called corporate social responsibility (CSR),[17] which acknowledged that companies have a responsibility to the public that includes integrating environmental protection, philanthropy, racial equity, gender equality, and other social goods into their business operations. Many companies took action to save water and energy, reduce landfilling and materials use, and change product ingredients to make them less harmful for the environment. Companies taking these steps perceive a competitive advantage over other companies that do not make the leap to this more progressive business approach. They also seek to decrease their exposure to business risk, and thus take precautions against increased costs that could arise from new laws or a negative public perception of their brand. These companies go beyond basic compliance with environmental laws and make financial investments in company operations that have social value. These efforts seek to provide a return on investment through reduced company risk, greater market share, and/or enhanced customer loyalty.

CSR approaches, which became prevalent in the United States throughout the 1990s and are still popular today, are considered a form of industry self-regulation[18] based on the voluntary action of companies. While this movement established new social norms for corporate environmental behavior, a company is not required to participate. The need for companies to justify expenses makes them vulnerable to shareholder and investor criticism that money spent on CSR is unnecessary and cuts into company profits. This dynamic inevitably leads some companies to decide not to spend money on added environmental protection that goes beyond compliance.

CSR actions thus present a dilemma. On one hand, they can create social value, brand loyalty, and decreased risk of regulation. By contrast, they can also create a competitive disadvantage for a company if the extra money it spends is not perceived to equal the value it seeks to gain. In addition, CSR programs managed by individual companies can lead to a patchwork of programs, leaving consumers to discern whether one company's environmental

17. According to CSRwire, "CSR is the integration of business operations and values, whereby the interests of all stakeholders including investors, customers, employees, the community and the environment are reflected in the company's policies and actions." "CSRwire–Corporate Social Responsibility Newswire," Environmental XPRT, accessed February 28, 2023, https://www.environmental-expert.com/companies/csrwire-corporate-social-responsibility-newswire-6511.

18. Benedict Sheehy, "Understanding CSR: An Empirical Study of Private Regulation," *Monash University Law Review* 38, no. 2 (November 2012): 103–127, accessed February 28, 2023, https://www.researchgate.net/publication/228246181_Understanding_CSR_An_Empirical_Study_of_Private_Self-Regulation.

claim is superior to another's claim. CSR programs highlight both the benefit and vulnerability of voluntary actions. As pressure mounted on companies to manage their products and packaging after consumer use, many tapped into the concept of CSR as a first step.

VOLUNTARY STEWARDSHIP

Voluntary industry take-back programs fall under the wide umbrella of product stewardship. Solving a problem through voluntary measures—if it really can be solved—is sometimes preferable to risking additional cost through policy or regulation. Voluntary product stewardship is a good solution for products and packaging that have considerable value after consumer use. Many products circulating in our economy have secondary value (e.g., used cars, vintage clothes, heirloom jewelry, and millions of items sold in thrift shops and on craigslist, eBay, and other resale platforms). There is an assumption that, if a product has value, it will continue to be used through resale or donation, rather than being thrown away. These product reuse efforts avoid waste, at least temporarily, and create value in the economy. They can also function without an EPR policy, although they do need regulation to protect consumers against fraudulent sales and other infractions.

In general, voluntary stewardship efforts will work well for products and packaging with a resale value that is greater than the cost to collect, transport, and process that material. We say that these materials have "positive economic value." One such example is the voluntary collection of printer toner cartridges in retail stores, which are transported to facilities to be refurbished, refilled, and resold. There is little cost to consolidate these products when consumers bring them into a store for collection. In this case, the cost to transport, refill, and refurbish a cartridge is considerably less than the resale value. Even if the company refurbishing the product pays for the consumer to mail them their spent cartridge, the product will likely have positive economic value. Such a voluntary model is profitable without EPR legislation. Automobile batteries are another product that has positive postconsumer value due to the high commodity value of lead and acids in the batteries. The cost to collect, transport, and sell those materials to a secondary market is lower than the price paid for the recovered lead and acids.

However, for most other products and packaging, there is a net cost to recover those materials and prepare them for the secondary market. We refer to these items as having "negative economic value." Without sustainable financing to cover the net cost of recovering and processing materials to specifications acceptable to end markets, there is no financial incentive for those materials to be diverted from the dump.

Someone needs to pay the costs for these materials to be collected and reused or recycled. Having read this far, you know that this funding has largely come from local governments, taxpayers, and ratepayers, all of whom have little control over the material composition of the products and packaging that become municipal waste. More recently, the rise of CSR and corporate concern over liability have led more companies to fund and manage voluntary take-back programs. Increasingly, companies have felt inspired, or compelled by public pressure, to reduce their impact on the environment. Without legislation obligating them to cover costs, though, each company decides for itself if it wants to voluntarily pay the extra cost to manage their materials to avoid harm to others, even when their competitors do not. As a result, most companies don't add this cost to their business model because doing so might appear to put them at a competitive disadvantage. Why pay more when your competitor does not?

Company voluntary product take-back efforts do have benefits, however. They reduce waste and increase recycling. They can also show, by vivid example, that waste is a resource to be turned into new high-quality products and packaging. TerraCycle, Inc., a company that specializes in collecting and recycling hard-to-recycle materials, offers many voluntary take-back programs funded by individual manufacturers and retail brands. The company has proven that a dizzying array of materials, including plastic packaging, used diapers, cigarette butts, toothbrushes, and chewing gum can be recycled into new products. TerraCycle's programs show what is technically feasible, and they blow up the notion that certain materials cannot be recycled. They have also created recycling solutions for brand owners to recover and recycle their plastic bottles, flexible packaging, paperboard, and other packaging materials. And they have developed reusable (closed-loop) packaging systems for an increasing number of product brands.

Companies like TerraCycle show that reputable companies can develop a take-back system for the recycling of everyday packaging and products. According to Earth911, however, TerraCycle was the target of a 2021 lawsuit by The Last Beach Cleanup, which alleged that TerraCycle misled customers of eight companies into thinking they could send back products for recycling for free, even though only some customers could do so. The parties settled, but not without TerraCycle agreeing to changes in its business operations.[19] What motivates these companies to hire TerraCycle is an interest in showing the public that they care about the environment, even if the amount recovered is small compared to the size of the problem. The

19. Sarah Lozanova, "Is TerraCycle Greenwashing the Waste Crisis?" Earth911 (June 28, 2022), accessed February 26, 2023, https://earth911.com/business-policy/is-terracycle-greenwashing-the-waste-crisis/#:~:text=TerraCycle%20Lawsuit,products%20for%20recycling%20for%20free.

funds for these efforts usually come from company marketing budgets. Companies need to see a return on investment for spending money on an issue like the environment when their competitors do not. They want to differentiate their brand as one that reduces the impact their products and packaging have on the environment. Taking action with other companies, collectively, blurs an individual company's efforts to stand out from the pack. Brand recognition is their goal.

Voluntary stewardship programs can play a valuable role in the circular economy. They are easier to enact, fewer people need to be involved, and there are few regulatory hurdles to overcome. A retailer or manufacturer recognizing its responsibility to reduce product impacts can huddle internally, devise a plan, and roll it out. As long as the plan does not violate existing laws, make false claims, or create adverse impacts, a company has the freedom to act and seek credit for its actions.

In 2004, PSI developed a voluntary retail computer take-back program with Mark Buckley, former vice president of environmental affairs at Staples, and Christine Beling of the US EPA. Working across Staples' internal distribution network, we developed a reverse distribution solution whereby Staples' trucks delivered new products to the store from their distribution and fulfillment centers and backhauled used computers to a central location. The voluntary pilot program showed how computer recycling could be accomplished effectively and efficiently, and it was the first program of its kind in the country at the time.

As the industry leader, Staples' program prompted its competitors—Best Buy, Office Depot, Office Max, and others—to develop their own voluntary consumer electronics take-back programs. Eventually, these collections created an extensive infrastructure for the convenient collection of electronics in the United States. Later, as EPR electronics laws were passed in more than 25 states, these sites helped producers meet their legal requirement under the law to make their take-back programs convenient to the public. Not only did these voluntary collections provide retail brand recognition, but they later helped producers comply with state laws, with producers providing the funding for retailers to collect.

Some retail pharmacies have also stepped up to collect waste pharmaceuticals on a voluntary basis. Bartell Drugs, a regional pharmacy chain in the state of Washington, voluntarily collected consumers' unwanted medications in all of its locations across the state. Bartell's showed that pharmacy take-back programs can work, and that retailers can provide a valuable community service to residents. As with Staples, Bartell's voluntary collections provided initial locations that eventually became part of the infrastructure required under Washington's EPR pharmaceuticals take-back program, known as the Secure Drug Take-Back Act. Bartell's now has most of its costs covered by the pharmaceutical industry under the state EPR law, including collection kiosks, educational materials, transportation, and disposal, but not staff time. Some retail pharmacies in states without EPR laws also voluntarily collect waste pharmaceuticals as a community service and to increase store foot traffic. PSI's pilot drug take-back programs have shown the cost of collection to be relatively low (about $3,700 per year on average for small pharmacies).[20] As a result, some pharmacies continue to cover the cost at the conclusion of our pilot programs. The vast majority of pharmacies, however, do not have the funds to do so.

Although retailers sometimes initiate voluntary product take-back programs, most are developed by producers, and sometimes developed collectively. The longest-running industry-wide voluntary stewardship program in the United States has been operated by Call2Recycle, a consumer battery recycling organization, to collect and recycle rechargeable batteries that contain lithium, cadmium, and other metals. In operation for about 30 years, it has been one of the most effective voluntary programs in the country. One of my very first press events as Massachusetts waste policy director was with the governor, the environmental secretary, and "Al the Toolman," Call2Recycle's former brand ambassador. I watched as the crew filmed the governor call into a toll-free hotline and punch in a zip code to identify free collection sites for rechargeable batteries located in the vicinity of the State House.

The Call2Recycle program was not an act of pure philanthropy. It was developed in the early 1990s in response to legislative action by Minnesota, Florida, and several other states requiring the industry to collect and recycle rechargeable batteries due to their toxic content. To avoid a statewide patchwork of laws, the battery industry supported federal legislation that allowed it to voluntarily collect and transport batteries across state lines for recycling. In 1996, when the Mercury-Containing and Rechargeable Battery Management Act[21] was passed, it became easier and less costly to recycle rechargeable batteries and other designated products, paving the way for the national voluntary battery take-back effort that has endured to this day.

Call2Recycle's voluntary effort temporarily stopped the passage of additional state legislation by showing that the industry was addressing the problem. Although it avoided a patchwork of state laws by developing a uniform national program, the lack of requirements that would typically exist under EPR legislation has undercut the effectiveness of the program.

---

20. Product Stewardship Institute, "How-to Guide for Drug Take-Back: Managing a Pharmacy-Based Collection Program for Leftover Household Pharmaceuticals" (September 2016), 28, accessed February 26, 2023, https://productstewardship.us/wp-content/uploads/2022/11/160920_PSI_Pharmacy_Guide_vS.pdf.

21. For additional information, see Call2Recycle's website, https://www.call2recycle.org/1996-battery-act-2/, accessed June 15, 2021.

Call2Recycle recycles about 15 percent[22] of rechargeable batteries through a unified national program, extensive use of educational materials, and thousands of collection locations at retail and municipal sites. Unfortunately, the remaining 85 percent of non-recycled rechargeable batteries are landfilled or incinerated. The low collection rate, along with the problem of free riders, has prompted Call2Recycle to support EPR legislation, which is discussed further in chapter 5 and in the battery case study.

The same industry interest in avoiding a patchwork of state laws prompted the manufacturers of thermostats to create the Thermostat Recycling Corporation (TRC), which set up a nationwide free collection program for mercury-containing thermostats in 2006. As I'll cover in chapter 5, although the industry worked with PSI to develop a model EPR bill and supported passage of the nation's first thermostat EPR law in 2006 in Maine, it has opposed EPR legislation since that time. Instead, TRC promotes its voluntary national program that has now been in operation since 1998. The two top performing states, Maine and Vermont, have EPR laws that include a bounty of $5 per thermostat returned by contractors and residents.

What lurks behind producer-led, voluntary efforts is subtle industry acceptance of responsibility to manage their products at the end of their usefulness. Voluntary product stewardship efforts exemplify a willingness by individual companies to internalize at least some of the true costs of their products and packaging following consumer use. In so doing, these companies acknowledge that their actions result in external environmental, social, and economic impacts. Voluntary efforts, however, are also often an attempt to tame the will of government to regulate their activity. Just as companies in eighteenth-century England paid a fee to pollute, companies in the twenty-first-century United States pay for voluntary programs that prevent a small amount of pollution but still allow continued pollution from their products.

Even so, voluntary product take-back programs can begin to develop the collection and recycling infrastructure needed to bring materials back into the circular economy. They can show companies the benefits of recovering their products and, as done by Staples and Bartell Drugs, tie the program into an eventual EPR solution. A decade ago, I became intrigued by the potential for voluntary company take-back programs. Sensing a PSI business opportunity, I created a consulting service for the voluntary take-back of consumer products. During this period, the PSI team spoke with multiple brand owners about the take-back and reuse or recycling of products they sold on the market, including a national brand water filter manufacturer. I proposed a stepped up voluntary take-back program to this company, which at the time was

22. Call2Recycle, personal communication to author, August 2021.

recovering a tiny percentage of their product from customers and recycling it into reusable plastic cutlery and other plastic products. The recycling program was heavily promoted on the company website and marketed widely, but the steps to take back the product were not convenient.

During my presentation to the water filter manufacturer, company executives expressed a sincere interest in recovering their own products but were cool to my proposal to collect enough of their products to make an environmentally beneficial difference. They also had no interest in taking a leadership role in bringing other companies together to form a collective voluntary program that would solve their industry's problem. They saw others as competitors and since they already had a take-back program, even if it was anemic, they felt that they were still ahead. It quickly became apparent that this company was more interested in receiving a marketing benefit for its efforts than actually providing a solution that reduced the waste they created. They needed to justify their expense, and the value to the company—the public perception that they cared about the environment—could be purchased for the cost of recycling a tiny percentage of their product.

This company was one of about a dozen companies we tried to convince to voluntarily take back their products. We were not successful. After less than one year, I had to let go the employee I had hired for this new business venture. There are many motivations driving companies to spend money on social and environmental issues for which they are not required by law. It is often good business and fits well within the realm of corporate social responsibility. These companies have moved sustainability forward through voluntary actions. They make a difference. But unless all companies making similar products take similar actions, we will not solve our materials management and resource consumption problems. Without full-scale, industry-wide cooperation, we will continue to experience piecemeal and inadequate efforts. My costly business venture taught me a valuable lesson: EPR is needed to hold companies accountable for actions that will *meaningfully* reduce waste and return valuable commodities to the circular economy. In fact, this conclusion was already reached by environmental luminaries who have come before.

TRAGEDY OF THE COMMONS

In 1968, Garrett Hardin, a professor at the University of California at Santa Barbara, popularized a concept called "the tragedy of the commons" that was originally from the published lectures (1833) of British economist William Forster Lloyd. The term was used to describe the effects of a hypothetical case of unregulated grazing on common land in Great Britain and Ireland.[23] In his classic essay by the same name, Hardin uses herders as his example when he writes, "Each [herds]man is locked into a system that compels him to increase his herd without limit—in a world that is limited. Ruin is the destination toward which all men rush, each pursuing his own best interest in a society that believes in the freedom of the commons. Freedom in a commons brings ruin to all."[24]

By using the word *freedom*, Hardin taps into a visceral American ideal upon which our country was founded. Freedom from oppression. Freedom of religion. Freedom of speech, press, and assembly. Freedom to bear arms. Individual freedoms give each of us a unique identity. But Hardin makes a strong statement: some freedoms must be limited to protect other individuals, particularly related to our common environment. Put succinctly, "the tragedy of the commons is a situation in a shared-resource system where individual users, acting independently according to their own self-interest, behave contrary to the common good of all users by depleting or spoiling the shared resource through their collective action."[25] Under Hardin's example, each herder has a personal choice whether or not to add more of their own cattle to the commons. By leaving the decision to each individual, Hardin argues that this benefits those with greater self-interest and disproportionately harms those who exhibit behavior for the greater social good.[26] Inevitably, the sum of the individual voluntary choices leads to resource depletion and ruin of the commons.

The tragedy of the commons is a universal concept that has been applied to the depletion of all natural resources we hold in common—air, lakes and streams, oceans, and government land. Hardin's point is that these common resources must be managed through regulation and should not be left to the goodwill of individuals or companies acting in their own self-interest. Hardin's essay illustrates why we cannot protect our common resources through voluntary action. But to convince corporations and society that regulation is necessary, harm must be tangible and often proven.

PROVING CORPORATE HARM

US environmental regulations have long sought to balance the rights of business to operate and the rights of the public to enjoy clean air, water, and

23. Garrett Hardin, "The Tragedy of the Commons," *Science* 162, no. 3859 (December 1968): 1243–1248, https://doi.org/10.1126/science.162.3859.1243.
24. Garrett Hardin, "The Tragedy of the Commons."
25. Garrett Hardin, "The Tragedy of the Commons."
26. Garrett Hardin, "The Tragedy of the Commons."

land. In fact, all countries face the challenge of finding the right balance in regulation that prevents pollution but does not overregulate and unduly thwart economic activity. Regulations allow companies to emit a certain amount of pollution into the environment as long as they do not injure others. But how "injury" is proven is an enormous burden on a society with thousands of potential, and cumulative, pollution sources. Neither paying for environmental damage, nor preventing it, is an exact science.

Companies are more likely to accept responsibility and pay the cost if it is proven that they caused the negative impact. In accepting (or being compelled to accept) this responsibility, companies seek evidence of environmental and health impacts directly traced to their deeds—birds killed or injured from plastic products they manufactured, patients with leukemia caused by their dumping of hazardous waste that leached into the water supply of those contracting the disease, or asthma linked to emissions from a manufacturing plant in the neighborhood. Although these impacts are real to those injured, companies often fight being held responsible and seek exoneration by casting doubt on the sources of the harm and probable causation.

As a graduate student in 1986, I explored the complexities of holding companies responsible for pollution. As an MIT student, I was able to enroll in an environmental law class at Harvard Law School with Professor Zygmunt Plater, who was an experienced trial lawyer. Plater had previously brought a case before the US Supreme Court in defense of the tiny snail darter, an endangered fish whose habitat was jeopardized by the building of a dam in the Tennessee Valley. Plater subsequently introduced me to environmental lawyer Jan Schlichtmann, who at that time was representing eight families from Woburn, Massachusetts, who claimed they were injured by water that was tainted by W. R. Grace, Beatrice Foods, and Unifirst Corporation.[27] Schlichtmann was smack in the middle of major litigation attempting to prove the responsibility of these three corporate giants for allegedly poisoning Woburn's drinking water with trichloroethylene and other compounds dumped on several adjacent properties. The case fascinated me because it intertwined science and law to determine whether, and to what extent, the companies would be held responsible for leukemia deaths and other health impacts purportedly linked to past waste dumping practices on their properties.

I read stacks of scientific papers on the health effects of trichloroethylene exposure, studied the geology of the site around the impacted Woburn community, read about the dumping histories of past and present businesses

27. This case is covered in the excellent nonfiction book by Jonathan Harr, *A Civil Action*, and in the movie with the same title that featured John Travolta (as Jan Schlichtmann) and Robert Duvall (a lawyer defending W. R. Grace).

on the sites from which the contamination emanated, and pored through epidemiological studies that attempted to link the illnesses at Woburn with chemical compounds dumped on-site. I interviewed Jan Schlichtmann several times while he was litigating the case, and I eventually wrote an article that was published in *Environmental Impact Assessment Review*.[28] Although the illnesses were real and the evidence tying them to company actions compelling, the tenacity of the corporate fight against culpability was as epic as the pesticide industry's fight against Rachel Carson.

Schlichtmann's goal in the Woburn case was to prove to the jury that the companies were responsible for causing illness and death of the residents. Children had died from a terrible disease. The odds that it was a coincidence that so many of them had contracted leukemia in a small geographic area (a "cluster"), around which significant pollution occurred, were overwhelmingly slim. The legal system in which this case was tried did not allow for all parties to jointly seek the truth, then settle on the degree of culpability and appropriate compensation. There was no joint fact-finding and collaboration. This process instead pitted two opposing sides against one another in a fight to convince a jury of their version of the truth. The burden fell to the families to show they were harmed by the companies' actions. The companies, in response, spent huge sums of money to show they were not responsible for causing harm.

The financial resources, scientific and legal acumen, and personal stamina needed to sustain the case on behalf of the families were tremendous, and the settlements reached were a major victory in advancing the legal principle of causation ("the causal connection between an original cause and its subsequent effects").[29] Although the outcome reached—two settlements allegedly totaling $9 million and one acquittal—was not considered a resounding victory for the plaintiffs, it marked a major turning point for establishing corporate responsibility regarding environmental pollution, particularly chemical causation.[30] When my parents later moved to Toms River, New Jersey, I learned about another case in that town brought by Schlichtmann on behalf of children who contracted cancer allegedly due to pollution caused by Ciba Specialty Chemicals, Union Carbide, and United Water Resources. In December 2001, 69 families whose children were stricken with cancer,

---

28. Scott Cassel, "Woburn Revisited: An Interview with Jan Schlichtmann," *Environmental Impact Assessment Review* 7, no. 3 (September 1987): 259, accessed February 27, 2023, https://doi.org/10.1016/0195-9255(87)90015-1.

29. "Chain of Causation," *Merriam-Webster's Dictionary of Law,* accessed through FindLaw Legal Dictionary, accessed December 6, 2020.

30. Unifirst settled for an alleged $1 million, and W. R. Grace settled for an alleged $8 million. The jury did not find Beatrice liable within the segmented case structure dictated by the judge.

which they blamed on water pollution caused by these three entities, reportedly settled for at least $13.27 million.[31]

The Woburn and Toms River cases helped build the foundation on which the public has come to expect greater corporate responsibility. These cases had to be won in the courtroom before companies realized they had to take responsibility for their actions. Our society depends on legal teams representing those reportedly harmed by corporate giants. The first cases won by plaintiffs result in a corporate wake-up call. But as more of these cases are litigated and won by plaintiffs, they create a trend, and public opinion shifts along with the victories on behalf of those harmed. As cases like Woburn and Toms River are won or settled for large sums of money, companies begin to make the calculation that it might save time, money, and public embarrassment to prevent harm compared to significant payments to those injured, along with the negative publicity that impacts their brand.

Woburn, Toms River, Love Canal, Times Beach, and Valley of the Drums are all locations where pollution was so egregious that community members rose up to hold companies responsible for environmental pollution that caused sickness, death, financial catastrophe, and human suffering. In each of these cases, it took a battle within our legal system to move the needle of public perception regarding corporate behavior. These and other legal cases, along with the intense media attention they drew, were necessary to educate the public about human behaviors within companies that can cause sickness, even death, to innocent people. These cases are not as clear as a hand pulling the trigger of a gun. They required complex science, historical research, epidemiology, and pushing beyond the current limits of the law. But they all cried out for accountability no matter how complex the cause and effect might be.

We don't need to look too hard to find other cases in which companies knew that their actions were causing significant harm but covered it up. Tobacco industry executives lied under oath that they did not know nicotine in cigarettes was addictive. The courts found them guilty, requiring payouts of billions of dollars, while many lives were ruined from cancer linked to cigarettes. Pharmaceutical companies lied about their knowledge that opioids are addictive, and even lobbied the federal government to make pain a key health indicator, then convinced the medical community to prescribe the drugs they manufactured to alleviate pain. Drug companies have already been forced to pay out billions of dollars in the wake of a catastrophic public health epidemic they helped to cause. Yet, as of June 2021, members of the

31. "N.J. Cancer Suit Settled for $13.2M," *Huron Daily Tribune* (January 22, 2002), https://www.michigansthumb.com/news/article/N-J-Cancer-Suit-Settled-for-13-2M-7324187.php.

Sackler family, which leads OxyContin maker Purdue Pharma, continued to deny responsibility.[32]

Purdue Pharma became a partner of PSI in 2009 and supported our work to safely dispose of old medications, but then unexpectedly severed their affiliation with our organization in 2014. Around that time, King Pharmaceuticals, an opioid manufacturer, funded a section of the PSI website on drug take-back, as well as a large federal stakeholder meeting that resulted in near unanimous support for drug take-back. Representatives for the company told me that since their company made opioids, they believed they needed to provide safe take-back options for unwanted medications. Then, in 2011, Pfizer, another PSI partner at the time, acquired King Pharmaceuticals and stopped all association with our organization.

Currently, nearly all federal and state governments in the United States support drug take-back programs as the best and safest solution. Still, pharmaceutical companies continue to oppose take-back programs, even after the US Supreme Court in 2015 refused to hear the industry's final challenge to court rulings, thereby upholding the 2012 pharmaceutical take-back law in Alameda County, California, the nation's first.[33] Although held legally responsible for financing drug take-back programs in eight states and 23 local jurisdictions by March 2023, MED-Project, the industry's first and primary take-back implementation organization, still carries this disclaimer on its website: "This material has been provided for the purpose of compliance with legislation and does not necessarily reflect the views of MED-Project or the companies participating in the MED-Project Product Stewardship Program."[34] Competing take-back implementation organizations have since entered the market.

Even when companies assume responsibility for their actions that cause environmental and health impacts, there is undoubtedly a negotiation over how much money they need to pay to compensate for the pollution they caused. For example, hundreds of millions of gallons of oil were spilled by both the 2010 BP oil spill from an explosion on the *Deepwater Horizon* drilling rig in the Gulf of Mexico and the earlier 1989 *Exxon Valdez* spill in Alaska from a tanker that ran aground on rocks. The resulting environmental,

32. Meryl Kornfield, "Members of Family that Led Maker of OxyContin Deny Responsibility for Opioid Crisis in Congressional Hearing," *Washington Post* (December 17, 2020), accessed February 27, 2023, https://www.washingtonpost.com/health/2020/12/17/sackler-family-hearing-opioids/; Brian Mann, "Sackler Family Empire Poised To Win Immunity From Opioid Lawsuits," NPR (June 2, 2021), accessed February 27, 2023, https://www.npr.org/2021/06/02/1002085031/sackler-family-empire-poised-to-win-immunity-from-opioid-lawsuits.

33. *Pharmaceutical Research and Manufacturers of America, et al., Petitioners v. County of Alameda, California, et al.*, United States Supreme Court, accessed February 27, 2023, https://www.supremecourt.gov/search.aspx?filename=/docketfiles/14-751.htm.

34. "MED-Project," MED-Project, accessed February 27, 2023, https://med-project.org.

human health, and economic impacts were enormous. BP and Exxon did not deny their responsibility to clean up the pollution they caused. The key question became how to apportion fair compensation to those harmed.

About a year after the BP oil spill, my wife Susan and I rode the City of New Orleans Amtrak train down from Memphis to the Louisiana Bayou and toured the Gulf Coast to witness the spill's impact on the environment and its inhabitants. The handwritten wooden signs of distress were everywhere, deploring the conditions under which they were now living, even as the oil-soaked birds had, for the most part, been shoveled and disposed of. After years of legal wrangling, BP eventually paid out an estimated $65 billion[35] to remedy the pollution and pay for lost wages to owners of hundreds of fishing boats, lobster traps, restaurants, and other small businesses disrupted by the spill.

Unfortunately, we will never know if this amount of money was enough to fully cover the damage caused. We are, however, left with many questions. What will be the length of time it will take for the ecosystem to return to its pre-spill condition? How much of the cost was internalized by BP and its shareholders and how much was externalized on workers and the environment? Could the cost paid ever really cover the disruption and stress experienced by those who lost their livelihoods, whether temporarily or indefinitely? One thing is for sure, though—the 11 lives lost from the rig explosion were worth more than the cost of extra protection to prevent such a disaster. Indeed, a January 2011 government report to President Obama noted that the oil spill could have been prevented entirely had BP invested the time and money into correcting systematic failures in risk management that "place in doubt the safety culture of the entire industry."[36]

THE RISE OF EXTENDED PRODUCER RESPONSIBILITY IN THE UNITED STATES

Long gone are the days when US companies routinely poured their toxic wastes into the environment. This culture of contamination was halted by strong environmental laws that set up clear rules and penalties for noncompliance. Those laws focused on major sources of pollution to air, water, and

35. Adam Vaughan, "BP's Deepwater Horizon bill tops $65bn," *Guardian* (January 16, 2018), accessed February 27, 2023, https://www.theguardian.com/business/2018/jan/16/bps-deepwater-horizon-bill-tops-65bn.

36. *Deep Water: The Gulf Oil Disaster and the Future of Offshore Drilling–Report to the President*, National Commissions on the BP *Deepwater Horizon* Oil Spill and Offshore Drilling (January 2011), accessed February 27, 2023, https://www.govinfo.gov/content/pkg/GPO-OILCOMMISSION/pdf/GPO-OILCOMMISSION.pdf.

land, but they did not tackle the millions of American households that generate nearly 300 million tons of waste each year. In the United States, our early waste management efforts largely relied on a fragmented patchwork of local government recycling and household hazardous waste programs and company voluntary take-back efforts. In the late 1980s, however, a new policy framework started to emerge in Europe and Asia, and would eventually be imported to the United States. The concept that corporations should take responsibility for the postconsumer management of their products and packaging was first outlined in 1990 by Swedish professor Thomas Lindhqvist, who sought to extend a producer's responsibility beyond traditional facility operations both in upstream mining and downstream reuse, recycling, and disposal of a product and its package.[37]

It is not surprising that EPR originated in Europe, where there is generally more willingness to restrict an individual's behavior for the societal good than in the United States. EPR takes a completely different approach to regulation than previous major US environmental laws regarding who we hold responsible for preventing pollution. Under the Clean Air Act, the law regulates point sources of air pollution emitted from facility stacks. Likewise, the Clean Water Act regulates effluent that is discharged from facility pipes into waterways. Superfund holds companies "jointly and severally" liable for creating hazardous waste sites through past dumping practices. These and most other major federal environmental laws seek to regulate waste generators.

Municipal solid waste management, however, involves millions of household waste generators, who discharge their trash and recycling at curbside and drop-off locations, and release liquid waste to septic and sewer systems. Regulators long ago decided that these generators of MSW could not be federally regulated like larger point- and non-point pollution sources. Instead, federal regulators delegated that task to state and local governments, which have been challenged by a product and packaging stream that has become increasingly complex and costly to manage. Under EPR, the focal point of regulation is no longer the *generator* of the waste but the *producer* of the product or package that eventually becomes waste.

Our current US environmental regulatory system cannot adequately manage products and packaging across their life cycle. In fact, decisions made by producers, consumers, and others in the supply chain have contributed to pollution that exceeds the earth's natural capacity to absorb the impacts. Although US environmental laws address large point sources of pollution, residences and very small businesses have gone largely unregulated for waste

37. T. Lindhqvist, "Extended Producer Responsibility in Cleaner Production: Policy Principle to Promote Environmental Improvements of Product Systems," [Doctoral Thesis (monograph), The International Institute for Industrial Environmental Economics, 2000], IIIEE, Lund University, accessed February 26, 2023, https://lucris.lub.lu.se/ws/portalfiles/portal/4433708/1002025.pdf.

management. Once materials are used by consumers, only a small portion of them in the United States and globally are returned into the circular economy. Doing so on a greater scale through reuse and recycling will lower greenhouse gas emissions and overall environmental impacts.

EPR systems have potential to address the life-cycle impacts from manufacturing billions of products and packages each year. Preventing pollution from hundreds of millions of tiny household and small business sources of waste in our country is a monumental task. EPR is an elegant, if complex, solution to link these tiny sources so that our society can prevent pollution from the overconsumption of natural resources. Although we have always understood that waste creates local and even regional pollution problems, regulators are only now starting to acknowledge the connection between product manufacture and global climate change. Part of the solution to the climate change problem is solving the waste management problem. By continuing to make products and packages that become waste, especially single-use items, we extract more energy and minerals for the manufacture of more goods.

EPR recognizes that the producer has the most control over what they make, the materials they use (including postconsumer recycled content), how they are manufactured, and the degree of postconsumer value they will retain (e.g., using materials that have value to a recycler). Since producers have the most control, they have the greatest responsibility to reduce impacts. The concept of producers being held responsible for waste is a relatively new concept for the United States. As such, EPR policies should be viewed as the latest wave of rules that governments have placed on corporate behavior to temper their impacts on people and the environment. It is another step in our evolving global environmental consciousness that connects resource consumption with a threat to a healthy human existence. Just as burning rivers and toxic dumps spawned the US environmental movement, global environmental and human health impacts from overconsumption of natural resources have hatched the global EPR movement.

The "precautionary principle" warns us to take precaution in the face of uncertainty to avoid greater impacts and the associated costs. "The precautionary principle traces its origins to the early 1970s in the German principle *Vorsorge*, or foresight, based on the belief that the society should seek to avoid environmental damage by careful forward planning."[38] While permitting pollution will always be questioned because we can never know the exact impact of our actions, prevention will equally be questioned because we are seeking to prevent impacts that have costs we will never know. Once

38. "Precautionary Principle," Environmental Justice Organisations, Liabilities and Trade," accessed February 26, 2023, http://www.ejolt.org/2015/02/precautionary-principle.

again, we need to seek a balance. Significant environmental impacts in the United States and globally show that our calculations to prevent pollution are "off-base" and our environment is out of balance. The precautionary principle seems to have been flipped on its head, leading to environmental leaders trumpeting the phrase, "Companies privatize profits and externalize costs."

THE EVOLUTION OF CORPORATE ACCEPTANCE OF EPR

Throughout my product stewardship journey, my Canadian government colleagues and I would occasionally joke about the reaction from producers asked to take responsibility for the postconsumer management of their products and packaging. Industry response was so predictable that we could map the progression of their slow but inevitable transformation to finally accepting responsibility as governments became more committed to legislation. Their response was the same in both countries. Confronted initially with the problems that their products and packaging create, companies often react with denial of responsibility: "It is not our fault! We make products because consumers want them. After they buy our products, consumers should use them up, recycle the packaging, or dispose of them properly." This actually was the starting place for some paint manufacturers that later came around to jointly develop with PSI one of the most successful EPR programs in the United States.

Companies that are willing to engage with government officials and other stakeholders take a first big step because they must listen to other viewpoints. During PSI dialogues, agency officials present data regarding product impacts, which typically motivates companies to move to the next phase: they acknowledge the impacts but claim that government taxes should be used for waste cleanup. "Government should do more education and enforce more penalties against citizens and waste management companies to reduce pollution. Isn't this why we pay taxes?" Producers at this stage in the development of corporate consciousness will typically seek more government funding for education, but also trucks, bins, equipment, and technology to fix, strengthen, and expand the existing recycling system. After more than 15 years of PSI efforts to engage the packaging industry in EPR discussions, many manufacturers are still lobbying Congress for millions of dollars of taxpayer subsidies to better manage their postconsumer packaging.

Over time, however, companies seeking more government financial aid to manage their postconsumer waste will usually acknowledge that this solution would only lead to increased taxes or cuts in other essential services. They pivot and seek a viable solution with other companies. For example, in response to the crescendo of criticism of packaging pollution from governments, environmental activists, and the public, US packaging producers created a vast fund under the umbrella of a nonprofit organization to fortify themselves from claims of inaction. "Some local governments have already reached rates of 70 percent recycling. All the others need is technical assistance, equipment, and information to reach these same rates." This initiative poured millions of dollars into private-driven efforts to educate local governments and the public about recycling and provided collection equipment to make recycling more convenient for residents. At the same time, while taking on voluntary responsibility for the first time, these companies simultaneously opposed EPR bills introduced in multiple states.

Voluntary producer responsibility initiatives do help companies forge relationships with one another, recycling vendors, and consultants who guide them on the path toward greater responsibility and sustainability. Voluntary initiatives also demonstrate possibilities and allow company executives to experience tangible outcomes resulting from their own efforts. Most importantly, these initiatives educate corporations about the waste management challenges that their products and packaging create and engage them substantively, and safely, in discussions about solutions. Corporate staff get to understand how recycling carts help curbside municipal programs collect more recyclables. They are shown how consistent consumer education results in the collection of a cleaner stream of material.

Through these efforts, companies quietly learn how they can help solve the problem. They also begin to narrow the huge waste management learning gap between themselves and government and environmental groups. To some in industry, this gap might seem unfathomable and likely results in an initial unwillingness to engage with others who have decades of experience actively solving waste management problems. Voluntary producer responsibility initiatives help move industry along the path toward EPR, setting the groundwork for real solutions. But they are not quite there yet.

Ultimately, companies realize that voluntary efforts, no matter how robust, are just not up to the task of solving the pollution problems resulting from products and packaging. Companies come face to face with the tragedy of the commons. They realize that the recycling and waste reduction goals they announced publicly will not be met by just doing more of what has already been tried. Even so, they still want to figure it out themselves and achieve their own goals through their own solutions. "We are having discussions as an industry on the best approach to fix the recycling system, and we are all committed to finding the most effective solution."

Producers have taken another big step by acknowledging the problem and seeking ways to address it, but still need to be willing to let go of total control and instead find a solution with others. To return materials to the circular economy, prevent pollution, and safely manage waste, producers, haulers, recyclers, state governments, local governments, retailers, and consumers must act in concert. This necessary coordination does not happen through voluntary initiatives and not by acting alone. The burden to act shifts to government to step up and announce their interest in legislating an EPR solution with multi-stakeholder input that seeks to balance responsibilities among stakeholders, with producers playing a primary role.

As governments muster up enough political will to introduce legislation, some companies will take the next big step and engage with government and other stakeholders on how an EPR system might work. Others will develop their own version of EPR they are willing to accept. Companies and their associations might develop broad principles of EPR to signal an acceptance of regulation, although they still need time to determine what system they prefer. Miraculously, companies that vehemently opposed EPR for years begin to engage in discussions, internally at first, then publicly, about their concept of EPR. "We are actively figuring out the system that will work best for our members. If we are being asked to pay, we don't want to fund the status quo but instead fund new technologies that can recycle a greater number of materials."

Without the expression of strong political will, however, producers will gladly sit in mildly engaged EPR limbo for years. But when enough risk is created for producers, often as a result of negative publicity or the potential for passage of a bill they don't like, they tend to show a greater willingness to negotiate. To create this political momentum, multiple key stakeholders must advocate for passage of a bill, often led by local and state governments and environmental advocates. When the product is bulky and contains toxic materials, the coalition is often joined by recyclers and waste management companies seeking new business and to solve operational issues these products cause. When packaging waste is at issue, waste management companies have taken much longer to be willing to negotiate due to their longtime investments in curbside and drop-off recycling systems. As companies sense the political will rising to pass an EPR bill, more companies finally want a seat at the table to negotiate their interests, led by their trade association. By engaging in the legislative process, companies seek greater certainty to reduce their risk.

At this stage, though, many producers still cling to the hope of sharing financial responsibility with governments and other stakeholders in the supply chain. In Canada, several provincial EPR programs for packaging stayed at the shared financial responsibility stage for many years before packaging brands wanted greater management control of the system to reduce costs and, in exchange, fund the system in full. It takes companies time to understand the problem they are trying to solve and why EPR is the best solution. Each company is confronted with a different set of circumstances that they need to navigate. The transformation of US producer positions to first consider, and ultimately embrace, EPR can take a short time, as in the case of multinational corporations headquartered in Europe with experience in EPR systems, or a long time for US-based companies with less EPR knowledge. Brand owner views of EPR can be remarkably different when comparing companies operating in Europe and Canada to their US counterparts. The same brand company can be diehard supporters of EPR in Europe and Canada and dead set against EPR in the United States. Reasons could range from differences in public expectations of corporate responsibility, corporate acceptance of regulation, and other deep cultural values.

Once passage of a bill is inevitable, many producers will eventually seek a solution that protects their interests, and they try to cut a deal. This was the case with the International Sleep Products Association (ISPA), which resisted taking responsibility for several years for the estimated 40 million mattresses and box springs sold in the United States each year for use in residential and institutional settings.[39] In 2011, faced with a $300,000 per year expenditure to dispose of mattresses, the mayor of Hartford, Connecticut, Pedro Segarra, along with Marilyn Cruz-Aponte of the department of public works, championed a statewide EPR solution to cover the city's costs, increase recycling, and create jobs. The mayor was not alone in his disgust for mattresses strewn in vacant lots and gouging the budget.

Hartford and other local government members of the Connecticut Product Stewardship Council partnered with Covanta (owner of two Connecticut waste-to-energy plants) and others to hire PSI to set the EPR wheels in motion. Joined by the Connecticut Department of Energy and Environmental Protection, PSI facilitated multiple meetings to identify the problems with mattresses and the goals of the initiative. We conducted research, developed a detailed background briefing document filled with information, issues, and solutions, and engaged ISPA in finding a joint solution. When we got resistance, we offered options and pilot projects. There was more resistance, delay, and obfuscation.

Finally, our coalition decided it was time to act even if it meant doing so without the mattress industry. Working with an influential legislator, State Representative Patricia Widlitz, the coalition gained momentum in the legislature and was near to securing the votes needed for bill passage. At that time, ISPA's lobbyist told their client that negotiation was the only option because the strong political will of legislators made passage of an EPR law inevitable.

---

39. Product Stewardship Institute, *Mattress Stewardship Briefing Document* (July 25, 2011), accessed February 26, 2023, https://psi.wildapricot.org/resources/Mattresses/2011-07-Dialogue-Mattresses-Mattress-Stewardship-Briefing-Document.pdf. Mattress generation data was derived from stakeholder interviews.

ISPA was lucky that the political environment in Connecticut, Rhode Island, and California allowed for a compromise. They got the visible consumer fee they sought, even as they assumed responsibility for managing the mattress recycling program. Unfortunately, though, as I write this section of the book after spending part of my day testifying for a refined mattress EPR bill in Massachusetts, ISPA still clings to what is by now an outdated model that is in need of an upgrade based on over a decade of experience and best practices.

Companies take a big risk by resisting engagement with government to discuss and negotiate an EPR bill. Each year they delay passage of a law results in less work for producers and less money paid out if the financing system requires it to be internalized by the industry. During this delay, though, problems only mount. With delay comes frustration and distrust of industry. If companies wait too long, some governments will move on without them. Such was the case with packaging EPR legislation in 2021.

For 15 years, brand owners, packaging producers, and the waste management industry opposed any serious conversation about a packaging EPR bill in the United States, even as recycling stagnated and municipal recycling costs skyrocketed. This inaction caused significant distrust with many state and local government officials, as well as the environmental community. Year after year, producers refused to engage in *any* conversation that seriously explored the challenges and opportunities for EPR systems for packaging and paper products. I know because PSI offered numerous times to work together with packaging brand owners to explore recycling problems and potential solutions, including EPR. Producers refused the invitations except to present their viewpoint and did not stick around for the conversation that followed. I also participated in several dialogues conducted by other organizations, but—like our attempts—none resulted in serious discussions about whether, and how, EPR might work to address the packaging problems we faced.

Packaging problems continued to mount—recycling stagnated and the costs imposed on municipalities grew more apparent. Then, over a period of about four years, China restricted recycling imports and municipal costs skyrocketed. The recycling situation in the United States was a powder keg waiting to explode, and then it did. As state legislators scrambled to develop EPR bills, packaging brands scrambled to develop their own EPR concept around which they could rally. Since PSI had already developed a model packaging EPR bill with our government members, we reached out to the major brand associations to invite them into conversation. In 2019, only the Flexible Packaging Association (FPA), a subset of the packaging industry, was ready to engage. Although PSI ended up mediating an EPR agreement with FPA that included over 35 key stakeholders, it took other companies longer to coalesce around an EPR solution.

For some states, like Maine and Oregon, it was too late by the time producers coordinated themselves and were ready to talk. In Maine, the state agency, local governments, and the environmental community had already developed their own version of a system that suited them. In Oregon, the state and local governments leaned toward existing relationships with collectors and recyclers to develop a bill. In both states, without packaging brand owners willing to negotiate and travel the road to EPR together, the coalition of EPR supporters became averse to trusting an industry that waited too long to engage while municipal troubles mounted. By summer of 2021, Maine and Oregon became the first two states in the nation to pass a packaging EPR law, and many producers got stuck with systems they did not particularly like.

In each case, and in other states considering EPR bills, local governments, state governments, collectors, and recyclers grappled with how much control of the recycling system they were willing to relinquish to producers. The basic concept of EPR seeks to have producers take responsibility for financing and managing the reuse and recycling systems for products and packaging put on the market. The transition to these new systems requires trust. If there is little existing collection and processing infrastructure in place for a product, producer management is often welcomed. This has been the case for all previous non-packaging EPR laws passed in the United States on products that are bulky, toxic, and difficult to manage. For these products, the reuse and recycling infrastructure was in a nascent state before EPR bills were introduced. In these cases, there were relatively few existing collectors, processors, and customers.

By contrast, the packaging recycling infrastructure has evolved over 50 years. Companies and local governments have put in place nuanced and often complicated systems for the collection and recycling of packaging materials, as well as collection and disposal service for trash that often operates in tandem. Governments are entrusted by their residents to keep these systems functioning. Enhancing them is an added bonus. Screwups are political peril. For governments and others to willingly transfer this heavy responsibility, even if they desperately want to do so, they need a partner they can trust. Since most local governments already have deep relationships with recycling collectors and processors, these are the relationships they often fall back to when transitioning to an EPR system. By contrast, new relationships need to be built with producers, and this takes time. In the case of Maine and Oregon, the unwillingness of packaging brands to engage with governments did not play to their advantage.

Over time, however, with continued producer engagement in packaging EPR discussions, I expect US governments will be willing to relinquish more responsibility to producers to manage the packaging recycling system, as long as there are ambitious goals, best practice systems, serious opportunities for stakeholder input, and strong oversight. If producers use this opportunity to demonstrate competence in managing the recycling system, incorporate input from others, and continually adopt more sustainable practices, I believe that distrust and skepticism of producer control will dissipate. The United States is fortunate to be able to learn from our neighbors in Canada and colleagues in Europe. These systems have evolved considerably over time. Where they started is not where they are now. Over time, these packaging EPR systems have migrated to full producer financing and management.

The last stage in the long and winding evolution of producer engagement in EPR systems is reached when producers take pride in their EPR achievements. In the United States, the American Coatings Association (ACA), which represents the US paint industry, tops the charts in embracing their responsibility. Staff are engaged with stakeholders, and the industry has, for the most part, continually sought ways to improve its performance. ACA also sponsors the EPR model legislation developed with PSI and other stakeholders and works collaboratively to pass new EPR bills. Although paint manufacturers also traveled the typical EPR trajectory described in this section—from resistance to acceptance—its transformation was much quicker than others. This is partly due to its history of having been entangled in regulations related to lead paint and volatile organic compounds. PSI's national dialogue on postconsumer paint management offered the industry an opportunity to get ahead of a future problem. By engaging early with stakeholders, the agreed-upon model EPR bill incorporated ACA interests equal to those of other stakeholders.

The battery and mattress industries have, to varying degrees, also embraced their roles in collecting their used products and in recycling the recovered materials into new products. Both, though, still often advocate for their own bills. As another example, the electronics industry generally supported EPR bills from 2004 to 2010, then opposed bills for the next decade. Recently, pressurized gas cylinder brands have recognized that engaging with governments and other stakeholders is a preferred approach. Most other industries still resist taking responsibility by actively opposing EPR bills.

The carpet industry, led by the Carpet and Rug Institute (CRI), is a notable example. Since the late 1990s, CRI has adamantly supported voluntary agreements with government to avoid EPR legislation. Unfortunately, these agreements failed to divert scrap carpet from disposal. Then, in 2010, CRI created and passed its own weak EPR law in California. Due to poor performance, the law required two subsequent amendments, pushed by governments and environmental advocates, that eventually increased the carpet recycling rate to nearly 28 percent.[40] CRI continues to oppose EPR bills in other states that have had to develop bills without CRI input. For a time, CRI made recycling subsidy payments to carpet recyclers in exchange for their signatures on contracts that prohibited them from supporting EPR bills. This strategy worked for a time, as fearful recyclers hung on by a thread to receive funds that barely kept them in business. But as recycling markets fluctuated once again, the carpet industry reneged on its commitment and stopped their voluntary payments. In 2022, PSI worked in collaboration with other advocates and a carpet recycler to enact the country's second carpet EPR law, in New York, that is expected to significantly increase carpet recycling through cost internalization, performance goals, and other standard EPR elements.

PRODUCER RISK IN NOT EMBRACING RESPONSIBILITY

Producer responsibility is fast becoming an irrefutable environmental principle embedded in US and global culture. It is the heart of a social contract between corporations and the public regarding how they manage the products and packaging they put on the market. Products are *goods* that enrich our lives. They also make profits for those who sell them to consumers. Access to the market comes with a responsibility to manage the materials used in making the products and packaging sold into the marketplace. Companies are allowed to operate their businesses within the confines of local, state, and federal laws that seek to prevent harm to others. Pollution from waste is an impact that impedes basic human rights.

Under EPR, a producer's responsibility no longer ends with product sale or use. It extends downstream to postconsumer reuse, recycling, or safe disposal. The responsibility also extends upstream to the origin of the materials used in the product or package, like the cotton grown in the fields, pesticides used on the cotton, labor used to pick the cotton, and facility safety in countries far from our own. It has taken a constant drumbeat of education, persistence, and determination to create the paradigm shift for change in the United States over the past two decades.

With packaging EPR, the United States has reached a turning point. There will now likely be an EPR policy tsunami that signifies a new stage of maturity in the nation regarding producer responsibility for managing products and

40. "Carpet Stewardship Program Goals," CalRecycle website, accessed February 27, 2023, https://calrecycle.ca.gov/carpet/goals.

packaging. US packaging brands are now being required to reduce, and ultimately prevent, harm from their products and packaging. They must budget for these expenses, justify them to management and shareholders, and figure out how to control costs. These new obligations transform the role of a producer and subsequently change the roles of other stakeholders as well. This paradigm shift takes time; it does not happen overnight once an EPR law is passed. But it will happen, and when it does, we'll see which industries will be prepared, and which will not. Which will protect themselves by engaging with other stakeholders to develop EPR laws that work for their business model and which will increase their business risk by evading regulation and permitting their products and packaging to cause harm?

Recent US packaging EPR laws have already shown the risk to producers of engaging too late in the bill development process. But will that liability extend beyond having to comply with a law with parameters that run contrary to business interests? Will it involve punitive damages for willingly ignoring the harm that waste can cause to others? If we know that fish contain mercury and a company's product contributes to that pollution because mercury is released in emissions during waste combustion, and the company making mercury products willingly blocks EPR legislation that seeks to capture those products after consumer use (prior to disposal), does that company have some legal liability for the harm caused? If we know that lithium-ion batteries and gas cylinders cause repeated fires and explosions in recycling facilities, landfills, transfer stations, and waste-to-energy plants, what liability do these product manufacturers have related to those impacts? Should they pay for facility repairs? Sponsor EPR legislation to establish effective collection and recycling systems that eliminate the risk of explosions? What happens if a worker is injured or dies as a result of one of these explosions? Clearly, these companies should not wait for this to happen.

Let's go one step further. If we know that raw earth metals and other materials need to be mined, and more energy is used to remanufacture new electronic equipment, what level of responsibility do electronics manufacturers have for blocking the potential for equipment to be repaired and reused? Should these companies be forced to design their products so they can be more easily repaired? What liability does the electronics industry have for impeding the development of EPR legislation that seeks to recover the millions of pounds of equipment generated each year? If we know that 30 percent of MSW is solely single-use packaging used to transport products to consumers, what is the liability of companies that fail to do everything they can to reduce, reuse, and recover these materials? It will be hard to find fault with a paint industry that engaged with others when asked, developed an agreement, and continues to advocate for taking responsibility for leftover paint through EPR laws. If a state legislature fails to pass a paint EPR bill over seven legislative sessions, as has unbelievably happened in multiple states, it is time to blame legislators instead of the industry.

The Clean Air Act, Clean Water Act, RCRA, and Superfund have laid out the law of the land and the consequences for waste generators for not protecting human health and the environment. Lawsuits in Woburn and Toms River show the liability that companies face for past harm that they attempted to conceal. Exxon and BP illustrate a company's financial liability for massive cleanups due to a lack of safety precautions. Consumer product companies can be held responsible, through product liability laws, for direct harm caused by the use of their products. Regarding product and packaging waste, EPR shifts potential liability to the producers. Where companies block and stall on EPR, will they be held liable for harm caused by their inaction?

We know the companies that produce the packaging that causes impacts. Their logos proudly appear on the boxes, pouches, cans, bottles, jars, and tubs of products we buy every day. It is only a matter of time before researchers put a price on the cost to clean up and prevent plastic and other litter. They will estimate the cost of lost tourism, aquatic impacts, and harm caused by marine debris, including cigarette butts, abandoned fishing nets, and plastics. They will figure out the cost of other externalities that result from waste—localized impacts on communities living near landfills, transfer stations, and waste-to-energy plants.

We must also consider the portion of impacts related to climate change. We know from research study estimates that the manufacture of products (excluding food) in the United States contributes 29 percent of US greenhouse gas emissions, associated with the extraction of natural resources and the production, transport, and disposal of goods.[41] If our country is already spending billions of dollars at the local, state, and federal levels to remedy the impacts of a rising sea level,[42] what degree of responsibility do consumer product companies have to remedy and prevent this pollution and associated taxpayer cost? These impacts may not be as direct as workers dying from an oil rig explosion or birds dying after being soaked in oil from the spill. But the more we research, the more we will understand and the more culpable companies may be in contributing to impacts related to climate change. Will these companies be required to pay for these indirect impacts and to prevent

41. "Opportunities to Reduce Greenhouse Gas Emissions through Materials and Land Management Practices," US Environmental Protection Agency Office of Solid Waste and Emergency Response (September 2009), 13, accessed February 27, 2023, https://www.epa.gov/sites/default/files/documents/ghg-land-materials-management.pdf.

42. Maggie Koerth, "How Much Is The Government Spending On Climate Change? We Don't Know, And Neither Do They," *FiveThirtyEight* (February 8, 2021), accessed February 27, 2023, https://fivethirtyeight.com/features/how-much-is-the-government-spending-on-climate-change-we-dont-know-and-neither-do-they.

them in the future? I doubt it will be enough for a company to voluntarily collect a small amount of their products.

What will companies take away from the legacy-tarnishing lawsuits faced by pesticide, tobacco, and pharmaceutical companies that were dragged into court to pay for the trillions of dollars of social, economic, and environmental impacts caused by their negligence? Are consumer product goods companies at risk against lawsuits brought by public interest lawyers for not managing their postconsumer products and packaging to be safe for the environment and public health? What is the liability of companies that continue to oppose, thwart, and block the development and implementation of effective EPR laws that seek to prevent impacts? If a company impedes government action that seeks to prevent harm, can those companies be accused of causing that harm? Do companies that engage with governments and others to pass EPR laws and seek to prevent harm reduce their liability even for past impacts?

Many of these questions will begin to be answered in the coming years and decades. To bring our environment back into balance, we will need not only laws, but also attitudinal and behavioral changes, at the societal level, about waste and its impacts. The public has already taken a hard line against plastic pollution, not only forcing companies to take responsibility for plastics in the oceans, lakes, rivers, and land, but also seeking a halt on plastics production. Attitudinal changes will need to coincide with changes in the law. Otherwise, we run the risk of greater social polarization, with those living in states with EPR laws seeking to enact new EPR laws in states currently without them. Pollution knows no boundaries and such national cooperation will be needed.

EPR policy clearly states that producers are responsible for the impacts of their postconsumer products, both upstream and downstream. By now holding producers responsible for preventing harm from waste, they are accountable for not preventing harm. How industries will respond to this new challenge, and the roles played by governments, small businesses, environmental groups, and others, will be the subject of many policies, programs, papers, legal briefs, and books for years to come. How will we solve the problems created by overconsumption of natural resources? One thing is for sure: if we ignore humanity's principle of responsibility, natural systems will not let us forget.

Chapter 5

The Global Evolution of Producer Responsibility Policies

Globally, product stewardship and extended producer responsibility (EPR) programs seek to reduce product and packaging impacts *upstream* during material extraction, throughout the manufacturing process, and *downstream* to reuse, recycling, composting, and disposal. In practice, EPR programs have paid the most attention to the downstream portion of this equation. As effective EPR programs are put in place and our economic systems begin providing financial incentives for producers and other companies in the supply chain to internalize environmental costs, externalities such as ocean debris and air pollution should subside. This same pattern took place in the past as new policies effectively stopped rivers from catching fire, reduced the proliferation of massive hazardous waste sites, and eliminated huge scrap tire piles. With EPR programs in place, we will be better able to address new aspects of pollution, of which we were aware but unable to address. Our focus can then begin to extend upstream to eliminate impacts from mining and throughout the entire product life cycle to ensure companies uniformly meet environmental standards and uphold economic, social, and equity principles for each product and package—no matter the country. But first, let's take a look at where we are now and how we got here.

EPR IN EUROPE

The term *extended producer responsibility* (EPR) was first used in Europe to extend a producer's obligation for reducing environmental impacts both upstream and downstream of a manufacturing facility. ("Producer" identifies those who place a product for the first time on the market, which can be a manufacturer, importer, brand owner, or retailer that imports products directly.) Swedish professor Thomas Lindhqvist, who coined the term EPR, published his first paper in 1990 for the Swedish Ministry of the Environment describing this new concept. In a subsequent paper written for a 1992 Lund University seminar, Lindhqvist traced the concept that manufacturers should be responsible for their product waste back to 1975 in official statements made by Swedish government officials. However, he could not locate legal requirements for companies to do so.[1] In this paper, as well as in his 1992 doctoral dissertation, Lindhqvist describes EPR as a principle that must be implemented through various strategies to reduce environmental impacts across the full product life cycle, both upstream and downstream.[2] In other words, EPR was never intended to be one piece of legislation. Instead, it is a concept to be implemented through many policies and approaches, with the goal to reduce environmental impacts across all product life-cycle stages.

The leap from Lindhqvist's EPR theory to practice first took place in Germany in 1991 to reduce packaging waste. The EPR concept quickly spread from Germany to multiple European countries before the European Commission (EC), the administrative branch of the European Union (EU) that implements the law, issued a directive in 1994 to all EU member nations that required them to meet mandatory recycling targets. While EPR policy was mentioned as a possible tool to attain these goals, member states could choose how to reach the targets. By 2021, nearly all EU countries put EPR systems in place. Several still derive part of their funding through government-managed fees and taxes, which are not recognized as EPR principles in the United States. Best practice EPR systems with full producer cost coverage will become mandatory for all EU nations by 2023 for existing laws and by 2024 for new systems.[3]

Three other EC directives applied the concept of EPR to industry sectors beyond packaging: vehicles (2000), electronics (2003), and batteries (2006). The four directives, all of which were updated over time, thus laid the foundation for the concept of EPR to be applied throughout Europe.

1. Thomas Lindhqvist, "Extended Producer Responsibility as a Strategy to Promote Cleaner Products" (June 1992), Invitational Expert Seminar, Trolleholm Castle, Lund University, Sweden.

2. Thomas Lindhqvist, "Extended Producer Responsibility in Cleaner Production," Doctoral Dissertation (May 2000), Lund University.

3. "2. Member States shall ensure that, by 31 December of 2024, extended producer responsibility schemes are established for all packaging in accordance with Articles 8 and 8a of Directive 2008/98/EC." European Parliament and Council of the European Union, "Directive (EU) 2018/852 of the European Parliament and of the Council of 30 May 2018 amending Directive 94/62/EC on packaging and packaging waste," *Office Journal of the European Union* (2018): 10; "a) WFD 2018, deadline for existing EPR systems is 05. January 2023 following Art 8a No 7 WFD 2018." European Parliament and Council of the European Union, "Directive (EU) 2018/851 of the European Parliament and of the Council of 30 May 2018 amending Directive 2008/98/EC on waste," *Office Journal of the European Union* (2018), 10.

The Organisation for Economic Co-operation and Development (OECD),[4] of which the United States is a member, issued a seminal guidance manual on EPR in 2001, and it was updated in 2016.[5] The information provided by OECD on EPR theory and practice has been a main conduit for spreading the concept of EPR to the United States and around the world.

1991 GERMAN PACKAGING ORDINANCE[6]

The 1991 "German Packaging Ordinance"[7] laid the groundwork for the worldwide EPR revolution. By all measures, this initiative unleashed a tidal wave of opinion in which one side supported its waste reduction incentive for producers and the other opposed its strong regulatory approach and complex implementation. With its enactment, the concept of producer responsibility took hold in Europe.

The legislation included a unique provision: If the beverage container refillable rate dropped below 80 percent of all beverages put on the market, consumers would be required to pay a deposit on certain nonrefillable containers for which they would receive a refund when they returned the empty containers. In 2003, with the refillable rate still below 80 percent, this provision became law. It also obligated retailers to take back packaging from consumers and required producers to meet specific collection and recycling targets. Producers could meet these targets by either joining a government-run program or by running their own private system. The industry chose a private company, Duales System Deutschland GmbH (DSD), to set up Der Grüne Punkt (Green Dot), a program that put a symbol on packages covered by the law to indicate to consumers that producers were meeting their legal obligations. DSD, which was founded in September 1990 in anticipation of the

4. The OECD is a "forum where governments work together to address the economic, social, and environmental challenges of globalization." OECD, "Extended Producer Responsibility: Updated Guidance for Efficient Waste Management" (2016), 289, http://dx.doi.org/10.1787/9789264256385-en.

5. OECD, "Extended Producer Responsibility: Updated Guidance for Efficient Waste Management" (2016), http://dx.doi.org/10.1787/9789264256385-en.

6. This section on the German Packaging Ordinance is an updated and expanded version of the chapter: Scott Cassel, "Product Stewardship: Shared Responsibility for Managing HHW," in *Handbook on Household Hazardous Waste,* 2nd ed., ed. A. Cabaniss (Lanham, MD: Bernan Press, 2018), 167–222. The section here is written by Scott Cassel with significant input from Joachim Quoden, executive director of the Extended Producer Responsibility Alliance (EXPRA). Many case studies on the German Packaging Ordinance exist, including one written by Joachim Quoden, EXPRA: https://www.iswa.org/media/publications/iswa-extended-producer-responsibility-library/#countries/.

7. The Ordinance on the Avoidance and Recovery of Packaging Waste came into force in Germany on June 12, 1991. Federal Ministry for the Environment, Nature Conservation and Nuclear Safety, "Ordinance on the avoidance and recovery of packaging wastes [Packaging ordinance—Verpackungsverordnung—VerpackV1]," *Federal Law Gazette* (1998), 2379.

ordinance, subcontracted the collection, sorting, and recycling of packaging materials to waste management companies and municipalities.

At the start of the program in Germany, DSD was the only certified compliance scheme. It owned the Green Dot trademarked symbol and licensed its use to packaging producers to finance the recycling system. To avoid EU packaging trade barriers, DSD founded Packaging Recovery Organization Europe (PRO EUROPE) in 1995 and gave a license for the Green Dot to PRO EUROPE, which awarded the trademark to national collection and recovery systems within the EU, the European Economic Area, and other candidate countries in accordance with uniform rules and regulations.

In the late 1990s, some producers in Germany became dissatisfied with the service and attitude of DSD because it was perceived to be acting as a monopoly. Following lawsuits against DSD proving anticompetitive behavior, the European Commission, in 2001, forced open the packaging take-back market to allow for multiple competitive compliance schemes. DSD was thus required to allow the Green Dot symbol to be used by packaging companies that contracted with compliance services of DSD competitors. In a separate 2001 decision by German antitrust authorities, DSD was also required to allow their competitors to use the collection infrastructure they set up to comply with the packaging ordinance. The German packaging ordinance was finally stabilized in 2003 by awarding contracts through competition among collectors, sorters, and processors. DSD was later acquired by the financial investor Kohlberg Kravis Roberts & Co. L.P. in 2005.

While competing systems lowered costs, there was also a lack of accountability in reporting the materials collected. For example, in 2012, the 10 compliance systems collected around 2.4 million tons of lightweight packaging (plastic and metal packaging and beverage cartons) but only reported 1.2 million tons, a 50 percent gap. By reporting half of what was actually collected, the compliance systems overcharged some companies, which unlawfully allowed other companies not to pay into the system but still have their materials collected and processed. In the first quarter of 2014, the reporting gap had increased to more than 60 percent of what was actually collected, leading the Bundesrat, a legislative body in the second chamber of the German parliament, to seek an emergency revision of the packaging ordinance to close the most obvious legal loopholes and discrepancies.[8]

The two biggest lessons that emerged from this first EPR implementation were the need for competition among collectors, sorters, and processors and

8. Joachim Quoden, "EXPRA: The Operation of the German Packaging Recovery Organization," *International Solid Waste Association* (2019), https://productstewardship.knack.com/iswa-epr-library#menu/kn-asset/142-76-39-5d962e85b7e346001038580d/epringermany.pdf. Joachim Quoden, EXPRA, email communication, February 10, 2021.

the need to establish a central authority to oversee the program.[9] According to the OECD, since collection and processing services involve the greatest program costs, it is the single largest component of the packaging waste management chain on which to ensure competition to foster program efficiency and effectiveness.[10] In addition, continuous government oversight—through planning, monitoring, and enforcement—will reduce free riders and other program deficiencies. Oversight is critical to ensure data transparency, program operational effectiveness, and course correction when needed.

1994 EUROPEAN PACKAGING DIRECTIVE

Sparked by the 1991 German Packaging Ordinance, other EU member nations implemented their own EPR packaging systems. France passed its law in 1992, followed by Austria in 1993, and Belgium and Sweden in 1994.[11] To harmonize these programs and increase efficiency and effectiveness, while also allowing for country variations, the European Parliament and Council issued a directive on December 20, 1994, that covered packaging not only from households but also from the commercial, institutional, and industrial sectors. It included primary packaging, such as a cereal bag (high-density polyethylene plastic), as well as secondary and tertiary packaging, which in this example would be the cereal box (chipboard or fiberboard) and the large cardboard box (corrugated fiberboard) shipped to a store containing multiple cereal boxes, as well as the plastic shrink wrap holding a dozen large cardboard boxes in place for safe transport.

The Packaging Directive was updated in 2004, 2008, and 2018,[12] and a new draft directive was published on November 30, 2022, with approval expected in early 2024. The directive requires all EU member nations to manage packaging waste in accordance with the waste management hierarchy of reduction, reuse, recycling, recovery (including combustion that generates energy), and landfill disposal. To enhance material recovery, the directive requires that uniform markings be placed on packages or labels to indicate that the materials used in the package can be reused, recycled, or

9. Quoden, email communication, February 10, 2021.

10. OECD, "Extended Producer Responsibility: Updated Guidance for Efficient Waste Management" (2016), 232, http://dx.doi.org/10.1787/9789264256385-en.

11. Joachim Quoden, EXPRA, email communication, February 9, 2021.

12. The most recent update to the EC Packaging Directive started in 2020 and is expected to be approved in 2023. Joachim Quoden, EXPRA, email communication, March 31, 2021.

composted.[13] Noncompliant packaging cannot be sold into the EU market. The directive also specifies program measurement standards to detect heavy metals and other dangerous substances in packaging to reduce their use and prevent their release into the environment. Bio-based materials, materials suitable for reuse, and compostable biodegradable packaging are also defined and promoted. In addition, it fosters the use of recycled material in new packaging through recycled content standards.[14] The directive even lays out rules for conducting life-cycle analysis on packaging, which can provide a company with information on the relative environmental benefits of one package type over another.

From the start, the EC recognized EPR as a policy approach that creates a network of multi-stakeholder accountability on which the whole system of waste management relies. The initial 1994 directive identified packaging producers as having the greatest degree of responsibility for financing and sustainably managing packaging waste, with transporters, recyclers, government agencies, and consumers having other specific roles to play, each sharing responsibility for the success of the entire system.[15] Brand owners choose the materials that go into the packages in which they sell their products. They hire manufacturers to make packages to exact specifications and then sell products into the market. Other key players don't have the same degree of control over packaging design. Consumers must buy the products they want in the packaging that brand owners design. Local officials educate residents about which packaging to recycle. They collect whichever materials their residents put in the collection bins, whether those materials are part of the program or not, and they must budget for recycling costs that fluctuate with secondary commodity markets. Recyclers, finally, manage contaminated loads from consumers who may be confused about what is recyclable, and haggle with end markets and municipalities about devalued material.

Each of these entities has an important role to play that is coordinated by a central oversight authority. Without a network of accountability, the post-consumer materials recovery chain will be fragmented, inefficient, and ineffective—much like it is for packaging in the United States today. The 1994 EC Packaging Directive addressed this systemic problem by connecting the

13. As of July 2021, there is no EC-wide requirement to label packaging. However, those member states that do require labels have to use a standardized EU label. It is likely that standardized labeling will be required by the EC in the near future. Joachim Quoden, EXPRA, email communication, March 31, 2021.

14. Recycled content standards became mandatory for polyethylene terephthalate (PET) bottles through a separate EC Single-Use Plastics Directive, not the EC Packaging Directive. Mandatory recycled content for certain other packaging is expected to be part of the new revision of the EC Packaging Directive in 2023. Joachim Quoden, EXPRA, email communication, March 31, 2021.

15. "European Parliament and Council Directive 94/62/EC of 20 December 1994 on packaging and packaging waste," 4.

dots between stakeholders, defining their roles, and tying them together in a comprehensive system. The EC took the basic principle of responsibility and elevated it by knitting together stakeholder groups into a team. The directive laid out the rules by which all team members—each stakeholder—relate to one another. With this bedrock policy, the EU established a new system of accountability for each stakeholder to acknowledge and embrace their unique responsibility to manage packaging waste, prevent pollution, create recycling jobs, and transform a disposal economy into a circular economy.

European government officials did not wait until they knew all the answers and had designed the perfect system before embarking on a sustainable path. Instead, they laid out basic principles of producer responsibility and multi-stakeholder accountability and built on that foundation over time. They sought to create a level playing field that fostered competition among program service providers and prevented unfair practices, then refined policies based on new information, experience, and intent. Their aim was to harmonize EPR packaging laws throughout Europe and prevent adverse environmental impacts on all member states.[16] By outlining key elements of EPR legislation for packaging, the directive provided a road map not only for Europe, but also for the United States, Canada, and other countries.

To meet that vision, the 1994 EC directive, as well as subsequent updates, set short-, medium-, and long-term packaging reduction, reuse, and recycling targets for all EU member nations while also recognizing that countries would proceed at different paces. They relied on continuous improvement in meeting those goals and recognized the need to develop markets for materials collected. With foresight, the directives established a harmonized system for data compilation to evaluate programs and refine packaging policies across the EU, with particular attention to assisting small and medium enterprises. The EC was also required to develop a formal report to the Parliament, along with optional policy recommendations.

A striking advancement in policy noted in the 2004 directive is its insistence that packaging waste exported out of the EU be recycled or recovered "under conditions that are broadly equivalent" to those prescribed by EU member states.[17] Data were sought specifically for materials exported for recycling and recovery outside the EU. The directive also sought to address the economic, environmental, and social impacts of litter from packaging waste, particularly plastics. These requirements exhibit a moral and economic consciousness that counters the "out of sight, out of mind" disposal and recycling perspectives prevalent in all countries. The public does not want its

16. "European Parliament and Council Directive 94/62/EC of 20 December 1994 on packaging and packaging waste," 5.

17. "Directive 2004/12/EC of the European Parliament and of the Council of 11 February 2004 amending Directive 94/62/EC on packaging and packaging waste," 3.

recyclables or household waste to cause harm. But without data that proves safe management, we do not know the ramifications of our good intentions. Requirements like those in the EC Packaging Directive have laid the groundwork for the development of data tracking software that provides greater system transparency. If data is not required by an oversight agency, it won't necessarily be available to that agency to update policies and bring our good intentions in line with reality.

Each EC directive further defined and emphasized the environmental, economic, and social benefits of waste reduction, reuse, and recycling. EPR became a cornerstone policy to reduce the consumption of primary raw materials and the energy needed to extract them. Each successive directive, issued by the central authority of the EC, was like a stone plopped into a lake, with the concentric ripples of policy spreading out from the EC to each EU member country.

2018 EUROPEAN PACKAGING DIRECTIVE

The most recent EC Packaging Directive, passed on May 30, 2018, built on the elements outlined in previous directives and ushered in a host of new or expanded policy concepts. For example, by December 31, 2024, EPR systems must be implemented for all packaging waste in EU member states, solidifying EPR as the centerpiece of the circular economy. It pays homage to the ability of EPR to reduce health and environmental impacts and shines a light on the need to efficiently consume natural resources, promote circular economy principles, use renewable energy, and reduce dependence on imported resources. These goals were promoted not only as socially and environmentally beneficial but as key political and strategic EU imperatives. Achievement of these goals was intended to reduce net costs for businesses, governments, and consumers, thus fostering global competitiveness, all while reducing greenhouse gas emissions.

In one of the most significant aspects of the 2018 directive, all EU member states are mandated to establish "eco-modulated fees," which require producers to apportion recycling system fees to brand companies based on the cost to manage their packaging following consumer use. The main fee criterion for single-use packaging is recyclability. Fees that producers pay on their packaging materials in an EPR system will be lower if the packaging materials have a high value on the commodity market after they are processed in a recycling facility. Conversely, higher fees will be charged for single-use packaging for which there are few or no current markets. For example, a yogurt cup made from polypropylene (which has higher value once recycled) will cost the manufacturer less under an EPR system than one made with polystyrene (which has lower value once recycled). Also, if a yogurt cup is made of light-colored polypropylene, it will cost less than one made of dark opaque polypropylene, since lighter colored plastics can be made into a greater number of resin colors than a darker plastic resin, thus ensuring more markets and higher value.

These fees are not seen on a customer's receipt. Instead, they are paid by the producer as their share of the cost for the recycling system. There are several ways in which these fees can be established. For example, producers may, within their stewardship plan, propose these fees to an oversight agency for approval, often with multi-stakeholder input. Alternatively, the oversight agency may develop detailed regulations for eco-modulated fees that go through stakeholder review and comment prior to being finalized. The main point is that these fees are not paid by a consumer at the point of sale. In fact, a consumer will not know a fee has been charged. The role of eco-modulated fees is to incentivize producers to select materials for their packaging that have the least impact on the environment and cost less for the recycling system. This mechanism is perceived as the strongest opportunity to change the design of products and packaging, since producers will pay less to use materials that have less cost and impact on the recycling system and the environment. Even the cost of litter, which adds system cost through cleanup and has additional unquantified social impacts, can be included in these systems by charging extra fees on certain materials (e.g., plastic bags) because they result in litter.

In the United States, we also use the term *eco-modulated fees* for fees that are set based on material, weight, recycling system costs, and other factors that incentivize the use of sustainable materials, such as those made with recycled content and materials that result in lower environmental impact as measured through life-cycle assessment. France uses the term *disrupter fees* for materials that cause high system cost or are otherwise undesirable; producers using those materials are charged extra fees. In most EPR systems in the EU, companies making reusable packaging do not pay *any* fees, although how reuse systems are verified is currently being debated.[18] The topic of modulated and eco-modulated fees is in a rapid development phase globally and is expected to receive significant attention in the next iteration of the EC Packaging Directive. As greater public and governmental pressure is put on producers to place eco-friendly packaging materials on the market, financial incentives through various fee systems will be tested through the next decade, attempting to connect packaging management fees beyond direct system costs to the wider social and environmental costs and harm.

18. Joachim Quoden, EXPRA, email communication, February 9, 2021.

The 2018 Packaging Directive further emphasized the need to reduce waste generation by offering multiple ways to encourage reuse, including deposit return systems and lower producer fees, and by setting reuse goals. In addition, recycling targets were once again increased to bring them in line with the vision of a circular economy. Minimum packaging recycling targets, to be reached by December 31, 2025, were increased to 65 percent by weight of all packaging waste, with specific targets set also by material (e.g., plastic, wood, ferrous metals, aluminum, glass, and paper and cardboard; see table 5.1). Recycling rates to be reached by the end of 2030 for all packaging, by material type, were also provided. Along with reuse goals, these aggressive recycling goals clearly signaled that all packaging materials put on the market should eventually be reusable, recyclable, and recycled, to achieve a circular economy and reduce greenhouse gas (GHG) emissions. Even the recycling rate definition itself was specified in great detail to more accurately assess the percentage of materials sold into the market that are actually turned into recycled products.[19]

Significantly, the 2018 directive eliminated the recovery goal (the goal for the amount of material collected and processed) that had been included in previous directives, which had promoted combustion with energy value. Previous directives stated that "Member States shall, where appropriate, encourage energy recovery, where it is preferable to material-recycling for

Table 5.1 Packaging Material Recycling Targets from EU 2018 Directive on Packaging and Packaging Waste

Packaging Material	*2018 Targets*	*By 2025*	*By 2030*
All packaging	55%	65%	70%
Plastic	25%	50%	55%
Wood	15%	25%	30%
Ferrous metals	50% (incl. Al)	70%	80%
Aluminum	—	50%	60%
Glass	60%	70%	75%
Paper and cardboard	60%	75%	85%

Source: "Directive (EU) 2018/852 of the European Parliament and of the Council of 30 May 2018 amending Directive 94/62/EC on packaging and packaging waste," 147.

19. As defined in the 2018 directive, the recycling rate definition includes biodegradable packaging if treatment results in a recycled product, material, or substance. Although mandatory material recycling rates were increased in the 2018 directive, the method of determining what counts as recycling has changed. Instead of counting all materials at the entrance gate of the recycling facility as having been recycled, the 2018 directive considers recycled material to be the material entering the remanufacturing process to be turned into a new product or package. The difference between what comes into the recycling plant and what is sent to end markets represents the degree of contamination (e.g., material that must be disposed and not recycled).

environmental and cost-benefit reasons."[20] By eliminating the recovery goal, the 2018 directive highlights reuse and recycling as the metrics of superior interest, moving away from waste-to-energy methods while moving toward a more circular materials economy. The EU recycling definition, therefore, specifically does not include combustion. The directive does, however, allow for ferrous metals recovered following waste incineration to be counted as recycling if it meets quality criteria.

2000 EUROPEAN DIRECTIVE ON END-OF-LIFE VEHICLES

The second of four European Commission directives, issued in 2000 and slightly updated in 2018, governs the management of end-of-life vehicles (ELVs), covering passenger vehicles and small trucks,[21] through an EPR framework. This directive applies to the entire life cycle but focuses mainly on those who dismantle, shred, or recycle. ELVs are unique in that any vehicle is the host for numerous difficult-to-manage products, such as batteries and tires, as well as hazardous materials like fuel, oil, cooling liquids, antifreeze, brake fluids, battery acids, and air-conditioning system fluids. The EPR system seeks to safely manage ELVs by specifying storage tanks for hazardous fluids and requiring vehicle tracking systems and certification for facilities and vehicle destruction. The directive also promotes the reuse, repair, and recycling of vehicle components, as well as the reduction of hazardous materials. As with all EC directives, the responsibility to fund and manage systems to comply with the law falls to vehicle producers. The directive required the reuse and recovery of ELVs to reach a minimum of 95 percent, and reuse and recycling to reach a minimum of 85 percent, by January 1, 2015.[22] By 2018, ELV reuse and recovery reached 92.9 percent and recycling and reuse averaged 87.3 percent.[23]

---

20. "Directive 2004/12/EC of the European Parliament and of the Council of 11 February 2004 amending Directive 94/62/EC on packaging and packaging waste," 3.

21. The European Commission ELV Directive legislation applies to passenger vehicles and small trucks but not big trucks, vintage vehicles, special-use vehicles, and motorcycles. "Directive 2000/53/EC of the European Parliament and of the Council of 18 September 2000 on end-of life vehicles," 4.

22. These rates are measured by an average weight per vehicle and year. "Directive 2000/53/EC of the European Parliament and of the Council of 18 September 2000 on end-of life vehicles," 8.

23. "End-of-life vehicles—reuse, recycling and recovery, totals," Data Browser, Eurostat, last update 15, 2023, https://ec.europa.eu/eurostat/databrowser/view/env_waselvt/default/table?lang=en.

2003 EUROPEAN DIRECTIVES ON ELECTRONICS AND RELATED HAZARDOUS SUBSTANCES[24]

In 2003, nearly a decade after the 1994 Packaging Directive, the European Commission enacted a third directive on waste electrical and electronic equipment (WEEE), which required electronics manufacturers selling in Europe to be responsible for recycling all electronics that depend on electricity (i.e., those with a plug, battery, or switch). The directive split WEEE into 10 categories, including large household appliances, refrigerating appliances, small domestic appliances, consumer electronics, information technology and telecommunications, toys, tools, sports equipment, medical devices, monitoring and control instruments, and automatic dispensers.[25] The WEEE Directive, and its accompanying Restriction on the Use of Certain Hazardous Substances in Electrical and Electronic Equipment (RoHS) Directive (which bans certain hazardous materials from the production of electronics), swept EPR into the United States. Since the Product Stewardship Institute's focus coming out of our inaugural conference in 2000 was electronics, we closely followed the legislative developments on electronics coming from Europe.

The WEEE and RoHS directives significantly affect how electrical and electronic products worldwide are designed and managed at the end of life. Any manufacturer that wishes to sell in the European Union must ensure compliance[26] with the national transpositions of these directives in the EU's 27 member states. The WEEE Directive requires each individual producer (brand, manufacturer, or importer) to meet its obligation by establishing an individual program or joining a collective scheme. Under the directive, producers apportion responsibility for the cost of managing "historic"[27] end-of-life electronic waste from private households[28] arising from prod-

---

24. This section on European Electronics Waste Management was originally published in the chapter: Scott Cassel, S., "Product Stewardship: Shared Responsibility for Managing HHW," in *Handbook on Household Hazardous Waste, 2nd edition*, ed. A. Cabaniss (Lanham, MD: Bernan Press, 2018), 167. The section was written and updated by Scott Cassel with input from Pascal Leroy, Director General of the WEEE Forum.

25. Directive 2002/96/EC of the European Parliament and of the EU Council of 27 January 2003, http://eurlex.europa.eu/LexUriServ/LexUriServ.do?uri=CELEX:32002L0096:EN:HTML.

26. Non-EU manufacturers usually ensure that the importer complies with the law.

27. Under the WEEE Directive, besides "historic" products, producers also share collective responsibility for "ownerless" and "abandoned" products. "Ownerless" products are those manufactured by companies that are no longer in business and have not been purchased by another company. "Abandoned" products are those whose brand name cannot be identified.

28. The costs associated with the management of historical WEEE from sources other than private households are also borne by producers in most cases. "By 13 August 2005, financing is to be covered by producers in the case of waste from holders other than private households and placed on the market after that date. In the case of waste from products placed on the market before 13 August 2005, management costs are to be borne by producers. However, Member States may provide that users be made responsible, partly or totally, for this financing." http://www.weee-forum.org.

ucts sold prior to August 13, 2005. ("Historic" products are those that were sold prior to the date of any newly implemented system.) These costs are assigned according to the current market share (i.e., percentage of sales) of the same or similar equipment types. All member states have implemented national WEEE registers to which each producer must report on the amount of electronic and electrical equipment sold by weight. The register calculates each producer's market share (per type of equipment), which becomes that producer's take-back obligation. The producer is required to report its collection volume of WEEE to assess compliance. In some countries, individual producers conduct the registration and reporting tasks, while in other countries compliance organizations provide these services for the producers.

The WEEE Directive also requires a company to take back its own products sold after August 13, 2005, and provides flexibility in complying with this responsibility either by contracting with an approved compliance organization or through a company-specific scheme. While the national derivations of the directive maintain this flexibility, joining a producer responsibility organization (PRO), also known as a stewardship organization, is the only compliance option in most countries for producers of household WEEE. However, individual compliance is feasible in many member states for those manufacturers of non-household WEEE. Until 2011, producers were allowed to charge their customers (retailers or wholesalers) a visible fee to recover product end-of-life management costs, and retailers were allowed to pass this visible fee on to consumers. However, most countries chose not to implement visible fee schemes.[29]

In many countries, retailers and municipalities must provide collection sites for the return of WEEE at no cost to the consumer. In many member states, municipalities have entered agreements with compliance organizations to structure the return of WEEE. Some of them are structured to incentivize municipalities to increase per capita collection. Another important source of WEEE volumes for compliance organizations are retailers, which are obligated to take back WEEE of the type they sell.

The implementation of the WEEE Directive sparked the setup of multiple take-back compliance schemes. In almost all countries, several schemes compete with each other, which has significantly reduced take-back cost. These schemes are established either by producers or by companies involved in logistics or waste management. A revision of the WEEE Directive ("Recast Directive") was adopted in 2012 and member states had to transpose it into national law by February 2014. The Recast Directive notably revised the collection targets—since 2019, producers must collect 65 percent of the average weight placed on the market in the preceding three years or 85 percent of

29. Pascal Leroy, WEEE Forum, email communication, February 16, 2021.

WEEE generated. It also extended retailers' take-back obligations for small WEEE and merged the 10 WEEE categories into 6 to align them with the 6 collection groups typically found in member states.[30]

The companion RoHS Directive restricts the use of six hazardous substances in electrical and electronic products that are covered by the WEEE Directive. The RoHS Directive was adopted in February 2003 and took effect on July 1, 2006. As of that date, new electrical and electronic products that contain more than the permissible levels of lead, cadmium, mercury, hexavalent chromium (or chromium VI), and polybrominated biphenyl (PBB) and polybrominated diphenyl ether (PBDE) flame retardants were banned from sale in the European Union.[31] Similar regulations have been introduced in other parts of the world. In the United States, the California Electronic Waste Recycling Act (enacted in 2003) references the RoHS Directive, as does China's equivalent electronics waste management law (China RoHS) in 2006.

2006 EU BATTERY DIRECTIVE

Following EC directives on packaging in 1994, ELVs in 2000, and WEEE and RoHS in 2003, the EC issued its fourth directive in 2006 for single-use (primary) batteries and rechargeable (secondary) batteries, which the Europeans call "accumulators" since they store energy for multiple uses. This directive covers nearly all batteries for residential, commercial, and industrial uses regardless of their chemical nature, size, or design, including those used in hearing aids, watches, clocks, flashlights, laptop computers, cordless power tools, small portable equipment, backup power, bicycles, automobiles, solar panels, photovoltaics, and other renewable energy applications. On the heels of the producer responsibility policy that was laid out in previous directives, the Battery Directive also requires producers[32] to finance and manage systems to recover used batteries and lays out a framework of responsibility for other stakeholders as well, including local and national governments, collectors, and recyclers.

This law features many of the same themes that were previously written into the other EC directives, including a desire to harmonize laws across EU member states, reduce trade barriers, and reduce environmental impacts posed by hazardous materials in household, commercial, and industrial

30. Leroy.

31. Directive 2002/95/EC of the European Parliament and of the EU Council of 27 January 2003, http://eurlex.europa.eu/LexUriServ/LexUriServ.do?uri=CELEX:32002L0095:EN:HTML.

32. As is common with all effective US EPR laws, the Battery Directive includes exemptions for small producers to achieve program efficiency.

batteries. The Battery Directive again emphasizes the waste management mantra of reduce, reuse, and recycle and establishes common labeling of batteries to notify users about their toxic contents (e.g., mercury and cadmium) and locations where batteries can be recycled. It also requires that batteries be easily removable from products. In addition, the directive prohibits the sale of certain batteries containing mercury or cadmium and bans the disposal of industrial and automotive batteries in landfills and combustion facilities.

As with any effective EPR policy, the directive requires producers to meet minimum rates of collection[33] and recycling of waste batteries[34] in a system that is convenient[35] and cost-free to consumers,[36] and it sets a common method for calculating the collection rate based on battery consumption. It also specifies the methods of recycling to be used and specifically excludes energy recovery from the definition of recycling. In a nod toward aspects of EPR that encourage acceptance by producers, the directive promotes technological developments that improve the environmental performance of batteries throughout the entire life cycle as determined through detailed audits. There is also a level of protection given to non-EU citizens regarding batteries exported to countries lacking the same legal protections as those in the EU. Under the law, exported waste batteries can be counted as recycled "only if there is sound evidence that the recycling operation took place under conditions equivalent to the requirements of [the Battery] Directive."[37]

The directive contains other elements that have also become standard in effective US EPR programs. For example, EU member countries must report to the European Commission every three years on their progress to meet program requirements, including the issuance of penalties for noncompliance. The European Commission issues its own consolidated report evaluating the law's effectiveness across the European Union, along with recommendations for improvement. This reporting chain has also become standard in all US EPR laws, with oversight agencies (usually the state but sometimes local agencies) reporting to the appropriate legislative body and recommending changes to the law as needed over time.

33. Minimum collection rates were set at 25 percent by September 26, 2012, and 45 percent by September 26, 2016. Collection rate is defined as the percentage of battery material collected compared to the amount of battery material put on the market.

34. Minimum recycling efficiencies were set at 65 percent by average weight of lead-acid batteries, 75 percent by average weight of nickel-cadmium batteries, and 50 percent by average weight of other waste batteries. Recycling efficiency is defined as the percentage of battery material recycled compared to the amount of battery material collected.

35. The directive references convenience as "accessible collection points" in relation to "population density," which has become the basis for setting convenience standards in many US EPR laws.

36. "Cost-free" means no fee for the consumer to recycle; consumer fees at the point of recycling discourage recycling and often result in illegal dumping or trash disposal to avoid the fee.

37. EC Battery Directive, Article 15, Exports.

2020 PROPOSED EUROPEAN BATTERY REGULATION

On December 12, 2020, the European Commission proposed a revision to the 2006 Battery Directive that tackles the greatest environmental challenge of our lifetime and that of future generations: climate change. Through this proposed directive, the EC recognized the importance of battery technology in reducing greenhouse gas (GHG) emissions and stabilizing our global climate, and went further still by declaring EPR as the system needed to return batteries to the circular economy and conserve resources. The 2020 battery proposal is, therefore, a key component of the European Green Deal[38] to reduce greenhouse gas emissions across the continent. "In the EU, transport is responsible for roughly a quarter of GHG emissions and is the main cause of air pollution in cities."[39] To reduce greenhouse gases and achieve e-mobility, the EU plans to increase the use of electric vehicles by developing a sustainable battery value chain. To do this, the World Economic Forum estimates that "there is a need to scale up global battery production by a factor of 19 to accelerate the transition to a low-carbon economy."[40] More batteries manufactured will mean more resources extracted for battery production and a greater need to recover those batteries in the circular economy to reduce further resource extraction and consumption.

The 2020 battery proposal is designed to manage a range of battery types—those that are portable (including household use), those used for light means of transport (e.g., e-bikes, e-scooters; what those in America call "e-mobility devices"), and automotive, industrial, and batteries used in electric vehicles. For each battery type, the European Commission proposes a circular approach that extends both upstream and downstream of the battery production process. There are three main objectives of this revolutionary proposal: (1) achieve sustainability throughout the battery life cycle, (2) ensure strategic security by closing the loop on the EU battery supply chain (e.g., keeping battery materials in Europe once they are manufactured, used, collected, recycled, and remanufactured), and (3) reduce environmental and

38. The proposal describes the European Green Deal as "Europe's growth strategy that aims to transform the Union into a fair and prosperous society, with a modern, resource-efficient and competitive economy where there are no net emissions of greenhouse gases in 2050 and where economic growth is decoupled from resource use." "Proposal for a Regulation of the European Parliament and of the Council concerning batteries and waste batteries, repealing Directive 2006/66/EC and amending Regulation (EU) No 2019/1020," 23.

39. European Commission, "Transport Emissions," Climate Action, https://ec.europa.eu/clima/policies/transport_en#tab-0-0.

40. "Proposal for a Regulation of the European Parliament and of the Council concerning batteries and waste batteries, repealing Directive 2006/66/EC and amending Regulation (EU) No 2019/1020," 2. The original source is the World Economic Forum and Global Batteries Alliance, A vision for a sustainable battery value chain in 2030: Unlocking the potential to power sustainable development and climate change mitigation, 2019.

social impacts throughout all stages of the life cycle of batteries, including activities that take place outside the EU.

The 2020 proposal goes beyond harmonizing European country regulations to avoid trade restrictions and increase efficiency. A comprehensive European strategy is viewed as essential to transition to a circular economy, which is expected to contribute to fostering "innovative and sustainable European business models, products and materials."[41] Innovations in battery storage and distribution technologies are expected to significantly reduce greenhouse gases, and EPR policies are the vehicle for recovering, reusing, and recycling critical battery materials, making the system profitable, sustainable, and politically secure. Through this proposal, the European Commission places EPR at the center of the circular economy. It creates clear rules of engagement so that accountability is spread among all actors in the supply chain, not equally, but proportionate to the degree of responsibility they have for preventing harm.

The 2020 proposal seeks to increase investments in battery technology, increase battery recycling, and reduce environmental and social impacts related to battery production. It highlights the need for a reliable EU-wide database by which to compare and evaluate national battery EPR programs. It also requires that, by January 1, 2026, the EC set up a sortable and searchable data management system for third-party use that contains data and information on rechargeable industrial batteries and electric vehicle batteries covered under the final law. The proposal also seeks to track batteries across their life cycle. It requires that each industrial battery and electric vehicle battery placed on the market after January 1, 2026, and with a capacity higher than 2 kWh, have an electronic record ("battery passport") that contains basic characteristics of each battery type and is stored in a central data management system.

Those implementing US EPR programs can relate well to the need for better, and more accessible, data. As EPR laws pass in Europe and are adopted in Canada, the United States, and other countries, a need has emerged for national and global data systems for program accounting and comparison. In the United States, we have seen a steady increase in the number of companies seeking to provide user-friendly data management systems that help companies transition to a system that integrates vast amounts of data from multiple sources to ensure system compliance, oversight, and efficiency. As governments clarify what data they need to actualize their vision of a circular economy, additional requirements will be included in EPR laws globally, which will rely further on an expanded network of integrated data management systems.

41. "Proposal for a Regulation of the European Parliament and of the Council concerning batteries and waste batteries, repealing Directive 2006/66/EC and amending Regulation (EU) No 2019/1020," 4.

In the United States, for example, state and federal regulators lack adequate and sufficient data to evaluate how EPR programs compare across the country. Data that *does* exist in the United States is often centralized through PROs, not reported in a multistate manner, and not accessible to all stakeholders. In addition, governments have yet to require data on important aspects of a product's life cycle, such as impacts incurred from extraction, manufacturing, and sourcing of raw or recycled materials used in the manufacturing process. Sustainable materials sourcing will need to be defined and the data management systems developed to provide the information that is needed to continually reduce adverse environmental, social, and economic impacts throughout the product supply chain—starting upstream and extending to the manufacturing facility and then downstream to, and past, the consumer to domestic and international processing of recyclable materials and remanufacturing of products and packaging.

One of the most striking aspects of the 2020 battery proposal is that it completes the metamorphosis of prior directives by addressing each aspect of the circular economy. By referencing the 2020 Circular Economy Action Plan, the battery proposal represents a new generation of EPR directives that is more holistic in the management of the targeted product. It addresses all aspects of the battery circular life cycle, including upstream aspects such as performance and durability; consumer challenges that affect the collection rate, such as the ease of battery removal and replacement; access to information to assist in battery repair; and market incentives, such as recycled content requirements.

To actualize the concept of a circular economy, the proposal lays out options for increasing the collection and recovery of lithium,[42] as well as cobalt, nickel, copper, and lead.[43] To incentivize material recovery and stimulate the market for these recovered materials, the proposal includes targets for battery collection[44] and recycling efficiency.[45] A high level of recovery of

---

42. The material recovery target for lithium is 50 percent by the end of 2027, raising to 80 percent by the end of 2031. "Legislative Train Schedule: New Batteries Regulation" (February 20, 2023), European Parliament, accessed March 2, 2023, https://www.europarl.europa.eu/legislative-train/theme-a-european-green-deal/file-revision-of-the-eu-battery-directive-(refit).

43. Material recovery rates for cobalt, nickel, lithium, and copper, respectively: 90 percent, 90 percent, 35 percent, and 90 percent in 2025. "Proposal for a Regulation of the European Parliament and of the Council concerning batteries and waste batteries, repealing Directive 2006/66/EC and amending Regulation (EU) No 2019/1020," 10.

44. Collection targets for waste portable batteries are 63 percent by the end of 2027 and 73 percent by the end of 2030. Collection targets for e-mobility batteries are 51 percent by the end of 2028 and 61 percent by the end of 2031. "Legislative Train Schedule: New Batteries Regulation," February 20, 2023, European Parliament, accessed March 2, 2023, https://www.europarl.europa.eu/legislative-train/theme-a-european-green-deal/file-revision-of-the-eu-battery-directive-(refit).

45. Recycling efficiency targets for waste portable batteries are proposed at 65 percent by December 31, 2025, and 75 percent by December 31, 2030. "Proposal for a Regulation of the European Parliament and of the Council concerning batteries and waste batteries, repealing Directive 2006/66/EC and amending Regulation (EU) No 2019/1020," 10.

these materials is essential for the manufacture of new batteries, since "over 50 percent of the global demand for cobalt and over 60 percent of the world's lithium is used for battery production."[46] The proposal also spotlights the collection and recycling of waste batteries as critical to capturing valuable materials for use in the manufacture of new products and keeping the value chain—materials and manufacturing—inside the European Union. Three and a half years after the regulation becomes effective, portable batteries incorporated in appliances must be designed to be "readily removable and replaceable by the end-user," while e-mobility batteries must be "replaceable by an independent professional." The Commission will be required to assess, by the end of 2027, the feasibility and potential benefits of setting up deposit return systems for batteries, particularly for portable batteries.[47]

The proposal goes further in also requiring producers of electric vehicle batteries and rechargeable industrial batteries with internal storage and a capacity above 2 kWh to report the life-cycle carbon footprint (excluding use) of batteries placed on the market. Since cobalt, lead, lithium, and nickel are not readily available in the EU, and some are considered critical raw materials by the EC, the commission's strategy is to recover those materials from used batteries as security against potential supply disruptions. "Enhancing circularity and resource efficiency with increased recycling and recovery of those raw materials, will contribute to reaching that goal."[48] The proposal also includes recycled content targets for batteries to be met in 2030 and 2035 to stimulate the market for use of recycled cobalt, lead, lithium, and nickel in the manufacture of new batteries.[49]

The proposal acknowledges the potential health and environmental impacts related to mining and refining of cobalt, nickel, lithium, and graphite, which are critical raw materials for battery use in the European Union, "and their sustainable sourcing is required for the EU battery ecosystem to perform adequately."[50] To ensure proper attention is paid to the upstream practices of extraction and processing of materials, the proposal also requires adherence

---

46. "Proposal for a Regulation of the European Parliament and of the Council concerning batteries and waste batteries, repealing Directive 2006/66/EC and amending Regulation (EU) No 2019/1020," 33.

47. "Legislative Train Schedule: New Batteries Regulation," February 20, 2023, European Parliament, accessed March 2, 2023, https://www.europarl.europa.eu/legislative-train/theme-a-european-green-deal/file-revision-of-the-eu-battery-directive-(refit).

48. "Proposal for a Regulation of the European Parliament and of the Council concerning batteries and waste batteries, repealing Directive 2006/66/EC and amending Regulation (EU) No 2019/1020," 27.

49. "Proposal for a Regulation of the European Parliament and of the Council concerning batteries and waste batteries, repealing Directive 2006/66/EC and amending Regulation (EU) No 2019/1020," 27–28.

50. "Proposal for a Regulation of the European Parliament and of the Council concerning batteries and waste batteries, repealing Directive 2006/66/EC and amending Regulation (EU) No 2019/1020," 34.

to internationally recognized due diligence principles outlined in the *Ten Principles of the United Nations Global Compact*,[51] among others. The battery proposal, however, goes even further by seeking to offer protection for biodiversity, human rights, labor rights, children, and gender equality.[52]

While the 2006 EC Battery Directive provided flexibility among countries to achieve important targets, country-specific variations in implementation created an uneven playing field that thwarted investment in battery production, recycling, and sustainable material sourcing. The 2020 EC battery proposal seeks to replace that fragmented approach with a harmonized regulatory framework that the European Commission hopes will result in a unified market for investments. While flexibility might work well in some situations, the uneven results across the European Union have been a major impediment to a circular economy, a consistent climate strategy, and a sustainable growth strategy for the continent as laid out in the European Green Deal.

In December 2022, a provisional agreement was reached on the new European Union Battery Directive, and Member States ambassadors to the European Union endorsed it on January 18, 2023. It still needs formal approval by the Environment Council and the European Parliament. Once approved, the regulation will establish a general regulatory framework, with certain technical aspects requiring secondary legislation.

EUROPEAN EPR DIRECTIVES: IN SUMMARY

The four European directives—covering packaging, vehicles, electronics, and batteries—were groundbreaking. But they are also imperfect. Their implementation is incomplete, uneven, and full of challenges. Taken together, however, they present a vision for preventing pollution and protecting human health by managing the expectations and actions of multiple stakeholders in complex systems. The goal of a fully protected population is one we won't reach in the near future, but we can be satisfied that we are moving swiftly in a direction that is more in harmony with our natural environment. Nature will never be defeated. It only responds to our actions in relation to it. Either we take action to correct our missteps or natural forces will do it for us. The environment is in distress, and the four directives were Europe's response to, in part, correct the course.

---

51. The *Ten Principles of the United Nations Global Compact*, available at https://www.unglobalcompact.org/what-is-gc/mission/principles. These principles cover human rights, labor, environment, and anti-corruption concepts.

52. "Proposal for a Regulation of the European Parliament and of the Council concerning batteries and waste batteries, repealing Directive 2006/66/EC and amending Regulation (EU) No 2019/1020," 35.

Each law passed by European government agencies represents an evolution of thought and practice. These agencies sought to take the vision as far as they could at the moment while recognizing the need for continuous improvement. The focus on data, definitions, and methods of collection and recycling indicates an understanding that clarity in meaning will result in achievement of goals. There are no pleas for voluntary action or vague requirements entrusted to companies to meet. The specificity of the laws reflects a strong intention to safeguard people and the planet, and it contains an understanding about the nature of businesses operating in a range of European economic systems. Businesses are nurtured as the backbone of a sound economy. Government assistance comes with a clear social contract. These laws constrain the production of products and packaging not unlike other laws constrain behavior to temper negative aspects of our common human nature. In the four EC directives, there is an unspoken premise that companies are expected to extract resources for human use and enjoyment and that governments are expected to reasonably constrain these businesses to prevent harm.

These EPR laws, emanating from Europe, have laid the foundation for a worldwide embrace of a balance between controlling nature and protecting it, with an eye toward conserving resources for future needs. Their focus is only on four types of wastes. Even so, they provide strong examples that guide manufacturers of other products to set up integrated systems that conserve resources and energy for sustained and sustainable production. These laws assign primary responsibility to the actor that puts products on the market, but also ties other key stakeholders together in a solid web of accountability. Without producers having a direct financial incentive to design their products and packaging for least harm to the environment and the public, frustration will continue for recyclers, governments, and consumers who experience a lack of viable options.

As these European waste-related EPR directives were debated, passed, implemented, and updated, the waves of policy and experience reached waste management practitioners in Canada, the United States, and other countries. Their basic premise of assuming product responsibility may not have been new, but the application to producers as those most responsible was a novel approach to managing billions of tons of waste generated each year around the world. The European directives were systematic and based on sound principles. Each country could adapt these basic principles and approaches, and the system tied together all those who manage waste.

The global trash crisis did not emerge as a simultaneous tidal wave in all lands. It permeated our collective consciousness at different times with different strengths in different places. In the United States, the garbage crisis has grown and bubbled up for decades, with recycling stagnated, costs rising, and consumer frustration mounting. The global uprising against plastics pollution and the restriction on recyclables by China has recently blown the issue wide open. Similar forces in other countries resulted in greater attention to the need for a paradigm shift in the way resources and waste should be managed. The United States and other countries are seeking new ideas. EPR, as practiced in Europe, captured our attention. These laws represented hope and propelled a movement of care for the environment. They have created a legacy to pass on to future generations. Canada rose to the occasion, and the country did not take long to incorporate these new laws and techniques.

CANADIAN PRODUCT STEWARDSHIP

As European producer responsibility systems were evolving, similar waste management dynamics began to unfold in Canada. Faced with an increase in solid waste generation and financial constraints, municipalities pressured provincial governments to place responsibility on producers to find end-of-life solutions for their products. In 1994, at the time of the first European-wide directive on packaging, the Canadian province of British Columbia (BC) enacted the country's first producer responsibility law, which addressed paint. In 1996, the province established an EPR program for used medications, followed by household hazardous wastes (HHW) such as flammables and pesticides in 1997, aerosols in 1998, and electronics in 2004.[53]

Passing individual product-specific EPR laws was effective but also time consuming. Each time the province wanted to manage a new product under EPR, it would have to go through a lengthy legislative process. Under EPR, however, the legal aspects as they pertain to each product are relatively the same, even if the products require different management methods. For example, residents will always want a product take-back system to be convenient for them, wherever they live. The principle of convenience does not need to be established repeatedly. It is one of the most basic elements of all effective EPR laws. What changes is the way in which each product—whether paint, carpet, mattresses, packaging, or pharmaceuticals—will be conveniently collected and managed.

The Legislative Assembly of British Columbia recognized the pattern of basic EPR elements. In 2004, the assembly passed a landmark law known as the Environmental Management Act, which established basic standards for consumer convenience, performance goals, program financing, education, enforcement, and other requirements in a framework law governing all

53. Recycling Council of British Columbia, *B.C. Product Stewardship Programs–Feb 2018 Update,* accessed March 23, 2023, https://www.rcbc.ca/recycling-programs/epr.

products.[54] At the same time, the act left the details of program implementation to the Ministry of Environment and Climate Change Strategy, the provincial agency that administers laws. Under British Columbia's framework approach, the assembly did not need to introduce and pass a new piece of legislation each time it wanted to manage a new product. Instead, the act vested the Ministry of the Environment with authority to add products under an EPR system. To implement the act, the Ministry developed a recycling regulation in 2004, thus establishing EPR as the way in which all waste would be managed in the province.

Under the act, the Ministry designates new products for inclusion in the regulation to be managed under EPR, such as textiles, and then holds a lengthy consultation process with the textile industry and other stakeholders to discuss ways to implement the new program. Based on these discussions, producers submit a detailed draft stewardship plan for approval to the BC Ministry that explains how it proposes to meet all requirements of the law, including program financing, performance goals, a convenient collection network, and comprehensive consumer education program. This systematic approach is a more efficient way to manage multiple products under a broad EPR umbrella since Ministry officials can propose new products under the framework and approve stewardship plans without repeated approval from the legislature.

This new framework concept helped British Columbia further expand its EPR program to include vehicle tires (2007), thermostats (2009), household batteries and cell phones (2010), lead-acid batteries (2011), large appliances (2011), lamps and fixtures (2012), outdoor power equipment (2012), small appliances and electrical power tools (2012), and packaging and printed paper (2014).[55] Each industry sector is managed by one or more entities created by the producers that develop a draft stewardship plan for submission to the BC ministry for approval. These entities, producer responsibility organizations (PROs), manage the program to achieve compliance on behalf of their members (called stewards across Canada), which are typically brand owners or companies that import products and packaging for the first time into a province. These entities, also known as "responsible parties" in some legislation, can also be a retailer, owner of a franchise, or online seller.

Environment and Climate Change Canada (ECCC), the Canadian department that administers federal environmental laws, provided important information on Canadian and European EPR developments to Canada's provincial

54. Recycling Council of British Columbia, Video: "Evolution of Industry-Led Product Stewardship Model in British Columbia," accessed March 8, 2021, https://www.youtube.com/watch?v=RGWNfMfoSjU/.

55. Recycling Council of British Columbia, *BC Product Stewardship Programs—Feb 2018 Update*, accessed March 23, 2023, https://www.rcbc.ca/recycling-programs/epr.

governments, particularly through a series of national biannual EPR conferences.[56] ECCC, however, has limited authority to influence waste management policies in the provincial governments.

ECCC's effort was supplemented by the Canadian Council of Ministers of the Environment (CCME), which established the Canada-wide Action Plan for Extended Producer Responsibility (CAP-EPR) in October 2009. This plan provided a national framework and common EPR elements and principles. With the issuance of the plan, EPR was officially adopted as a key component of Canada's national waste management strategy.[57] CAP-EPR also included nonbinding provincial commitments to apply an EPR approach to manage designated groups of products within a given time frame in two phases. Phase 1 product categories include packaging, printed paper, mercury products (including lamps and thermostats), electronics, and household hazardous waste, including a range of automotive products (used oil, filters, lead acid batteries, lamps, tires, refrigerants, antifreeze, and other toxic fluids and their containers). All these products were to be brought under EPR management by 2015. Phase 2 product categories include construction and demolition materials, furniture, textiles, carpet, and appliances (including ozone-depleting substances). These were to be brought under EPR management by 2017.[58]

In its 2014 Update Report, CCME noted a significant increase in the number of EPR programs and requirements throughout the country since the CAP-EPR was issued in 2009. By July 2014, there were 94 product categories (e.g., electronics, paint, pharmaceuticals, etc.) covered by legislated EPR programs or requirements, compared to only about 33 product categories at the time of CAP-EPR adoption in October 2009.[59] By 2017, as reported by EPR Canada, a nonprofit organization, Canada had over 120 full EPR programs and a few packaging EPR programs in which local governments shared financial and/or management responsibility.[60]

Over the past 25 years, Canadian provinces have adopted relatively similar EPR programs, with more EPR laws and programs established in British Columbia, Ontario, and Quebec, and fewer programs in the other provinces and territories.[61] Each province has its own brand of product stewardship. Alberta, the last province with voluntary stewardship programs, enacted an EPR framework law on November 30, 2022, starting with packaging, printed

56. Duncan Bury, former EPR lead for Environment and Climate Change Canada, email communication, March 22, 2021.

57. Canadian Council of Ministers of the Environment, "Canada-Wide Action Plan for Extended Producer Responsibility" (2009), 1.

58. Canadian Council of Ministers of the Environment, 45.

59. Canadian Council of Ministers of the Environment, 12–13.

60. EPR Canada, *Extended Producer Responsibility Summary Report* (September 2017), http://www.eprcanada.ca/reports/2016/EPR-Report-Card-2016.pdf.

61. Canada is divided into 10 provinces and 3 territories.

paper, and single-use products, as well as hazardous and special products, all of which are expected to launch in spring 2025.[62] Other provinces with framework EPR laws include British Columbia (established in 2004), Ontario (established in 2016), Quebec (established in 2012), and Manitoba (established in 2009). All of these provinces regulate packaging and printed paper.

In contrast to European governments that quickly embraced EPR as the region's key waste management policy, Canadians took a more gradual path to EPR. Both have ended up at nearly the same place, with a vision of full producer responsibility in which producers fully finance and manage their products and packaging within the context of government regulatory oversight that seeks effective, efficient, and transparent programs. Not surprisingly, the United States is also moving toward that same position of full producer responsibility, although more slowly since it has not traveled the EPR path for as long as either Canada or Europe. In all countries, however, there is acknowledgement that EPR is the most effective way to return materials to the loop that constitutes a circular economy. National EPR leaders also recognize that cultural acceptance must accompany the transition from municipal waste being managed by governments to a system managed and financed by producers. For the transition to take place, businesses and other stakeholders operating under EPR laws need to accept this new form of governance and the new roles expected of them in the product supply chain that extends upstream to mining and downstream to end-of-life management. These changes do not occur overnight but take place along a continuum that will be different for each country.

Canada, like the United States, uses the terms *product stewardship* and *EPR* in similar ways, although there are key differences. In both countries, product stewardship includes initiatives that can be undertaken individually by either government or industry. For example, retailers voluntarily taking back plastic bags, producers collecting batteries, or governments banning plastic bags are considered product stewardship in each country. Once Canadian and US governments regulate industry under a producer responsibility system, the term "EPR" kicks in. In both countries, to be considered EPR, a policy must require producers to take on a degree of financial and/or management responsibility, and the term *EPR* includes both partial and full EPR systems. In the United States, EPR is considered a regulatory *subset* of product stewardship, which is a broad term that includes both regulatory policies and voluntary initiatives. However, in Canada, EPR is considered *distinct* from product stewardship; it is a more advanced form of producer

62. "Extended Producer Responsibility: Information for Albertans," Alberta Environment and Parks (October 5, 2022), accessed March 5, 2023, https://open.alberta.ca/dataset/a8608a9c-4cfb-47c5-a141-415d82349ce9/resource/80dcf617-a550-49f9-ad77-5923d14139c3/download/aep-extended-producer-responsibility-information-for-albertans-2022.pdf.

responsibility that occupies the stage *after* product stewardship along a progression of waste management approaches (see figure 5.1).

The term *EPR*, as used in Canada, is sometimes divided into two phases: partial EPR and full EPR. Partial EPR is when producers partially finance and/or manage the systems needed to safely reuse, recycle, or dispose of their products and packaging. Full EPR is when producers *fully* finance and manage those systems. For example, the consumer-funded paint EPR programs in Canada are known as partial EPR since consumers fund the system and

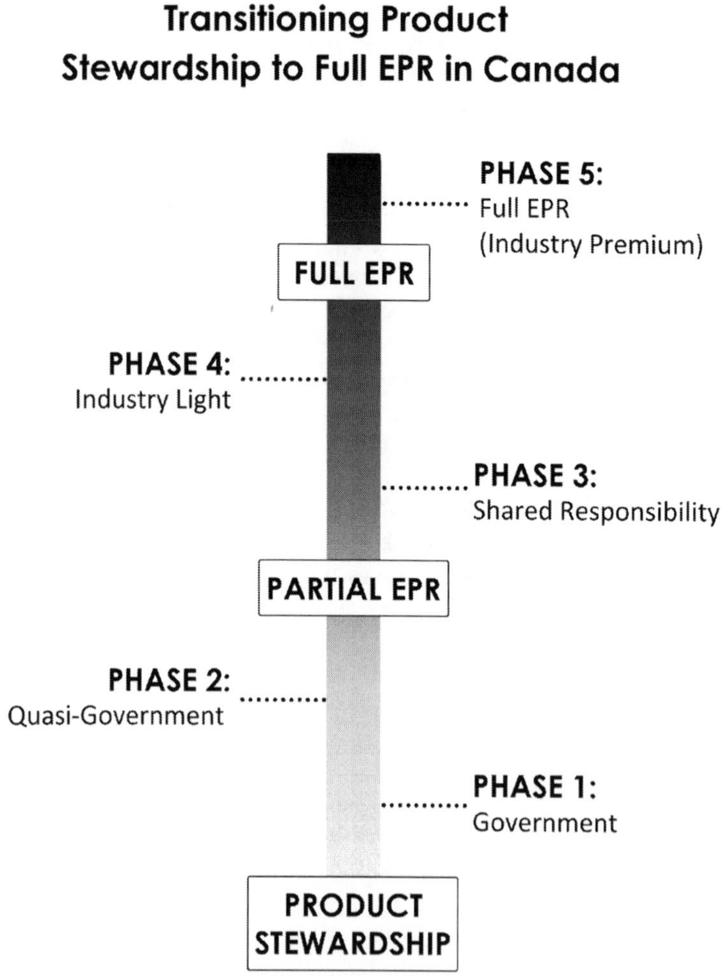

Figure 5.1 Transitioning Product Stewardship to Full EPR in Canada

Source: © 2023 Product Stewardship Institute, Inc. Redesigned from EPR Canada, *Extended Producer Responsibility Summary Report*, September 2017. Available at www.eprcanada.ca/reports/2016/EPR-Report-Card-2016.pdf.

producers manage it. The United States uses "EPR" to encompass both partial and full EPR systems. However, full-producer-financed programs are considered preferable by most US governments because there is a direct financial incentive for producers to change product design and operate efficiently. Requiring producers to internalize full recycling system costs will usually result in producers seeking a significant level of system control. In both countries, the limitations, as well as the benefits, of a consumer-funded but industry-managed program are still being debated, although the momentum is clearly in the direction of full responsibility in both.

The construct used by EPR Canada to describe the evolution of Canadian waste management systems, from product stewardship to EPR, is instructive for how the transition also occurred in the United States. In fact, the transition in both countries, as well as in Europe, continues along this basic path. Much as the development of US EPR has occurred state by state, Canadian product stewardship programs vary by province and have shifted over time. According to Duncan Bury, formerly the EPR lead at Environment Canada, waste management programs have evolved over time to include five phases, starting with product stewardship and ending in full EPR based on the degree of producer responsibility and government control: (1) Government, (2) Quasi-Government—delegated agency, (3) Shared Responsibility—industry and municipalities, (4) Industry "Light," and (5) Industry "Premium." The last stage is also called "Full Producer Responsibility."[63] Using terms from figure 5.1, the first two phases are considered product stewardship, while phases 3 and 4 are partial EPR, and phase 5 is full EPR.

The first stage, **Government model**, involves a consumer fee that is collected at the point of purchase (i.e., store checkout). The fee could eventually go into the general fund and be subject to government budgeting pressures with no guarantee that the funding will be appropriated to the recycling and management of the product on which the fee was levied. The **Quasi-Government model** involves delegating the program to a government agency that manages the program through inclusion of a variety of stakeholders. Industry (including manufacturers, importers, and retailers) will sometimes play an advisory role only and have no direct responsibility for program funding or operation. Under this model, advanced recycling fees are either set by government regulation or by the quasi-government agency and are dedicated to end-of-life management. Commonly, there is no municipal role since many of these programs rely on private or public depot collection systems (drop-off sites) originally set up for beverage containers under

63. Duncan Bury, Environment Canada, Presentation to the Association of Municipal Recycling Coordinators, Extended Producer Responsibility (EPR) Funding and The Future, Hockley Valley (February 14, 2007).

deposit legislation. Municipalities can still offer their collection systems for a fee paid by the agency. There is also no connection between producers and end-of-life management of their products. Therefore, this model is considered to be a weak product stewardship system.

Under Canada's **Shared Responsibility model**, industry and municipalities share operational and funding responsibilities. Industry can be assigned some of the net recycling system cost (after the sale of recyclable materials) or a part of the operational system, although in most cases collection remains a municipal responsibility. For example, municipal property taxpayer dollars might pay for half the total system costs, while industry funds the remaining portion. In other programs, municipalities pay for collection, while producers cover transportation and end-of-life management, which occurs in some Canadian EPR programs for HHW. A classic example of this second model is the original Ontario "Blue Box" multi-material recycling program that was fully operated by municipalities but with 50 percent of its funding provided by Stewardship Ontario, the stewardship organization representing the packaging and printed paper industry.[64] Similar multi-material, shared responsibility recycling programs were launched in Quebec,[65] Manitoba,[66] and Saskatchewan.[67] Ontario and Quebec have both begun to transition their packaging and printed paper (PPP) programs to full EPR, with plans to complete it by the end of 2025. Manitoba, as well, is in the midst of a transition from a shared responsibility system to one that is full producer responsibility.

The Shared Responsibility model is often used to share and offset costs of existing programs and avoids replacing the existing recycling system with one that is producer managed. Since municipalities commonly fund and operate collection facilities under this model, public perception is that municipalities are still 100 percent responsible. In reality, municipalities get reimbursed for their effort. Since larger municipalities have more capacity to run efficient programs due to the greater volume of materials collected, they are often paid more per ton of recyclables than smaller municipalities whose operations are often less cost effective. Since producers are footing the bill, they require verification of legitimate costs to authorize reimbursement, which makes this model burdensome, especially for smaller municipalities. It is important for trust to be established between partners that must divide responsibilities and costs, although this takes time to develop. One of the reasons for the shift

64. Stewardship Ontario, accessed March 7, 2023, www.mmsk.ca/stewards.

65. Éco Entreprises Québec, accessed March 7, 2023, https://www.eeq.ca/en/enactment-of-the-regulation-for-the-modernization-of-curbside-recycling-companies-and-partners-invited-as-of-now-to-engage-with-eco-entreprises-quebec.

66. Multi-Material Stewardship Manitoba, accessed March 7, 2023, https://stewardshipmanitoba.org/mmsm/full-epr-plan-development.

67. Multi-Material Stewardship Western, accessed March 7, 2023, https://www.mmsk.ca/stewards.

to full EPR for PPP in Ontario was the long history of contention between municipalities and Stewardship Ontario as to what were legitimate costs, as well as concerns by municipalities about increasing costs and their inability to control them.[68]

In the **Industry Light model**, industry is given a legal mandate by the government oversight agency to operate a producer responsibility program and has control over the funding mechanism, recycling, and promotion. Industry's ability to negotiate program costs and efficiencies, however, may be limited by existing contracts and government interventions that might delay start-up of the program and compromise producer flexibility. Under this model, the provincial government retains control over some key responsibilities, such as goal setting, planning, and enforcement, and municipalities are usually able to determine whether, and under what conditions, industry will assume collection responsibilities.

Under the **Full Producer Responsibility (or Industry Premium) model**, industry has full responsibility to fund and operate the program from collection through end-of-life product management. Industry sets fees, determines the collection mechanism, promotes the program, and contracts for recycling and other services. Government's role is to provide high-level policy direction, set objectives, and enforce penalties for noncompliance and free riders. Under this model, industry provides municipalities with an option to continue municipal collection services and be reimbursed under negotiated conditions, or to have the industry, through its PRO, take over the collection for them. This model requires performance goals and reporting to ensure that government objectives are being met. Examples of this model are the British Columbia product stewardship programs for paint, pesticides, oil, flammable liquids, solvents, beverage containers, pharmaceuticals, tires, electronics, and packaging.

The key distinction between the Industry Light and Full Producer Responsibility models is that the former model can include a consumer fee, whether visible or invisible to the consumer (e.g., for paint and electronics), as long as there is significant industry responsibility. However, Canadian systems are striving to move toward Full Producer Responsibility systems that include the costs of end-of-life product management in the product price. In this case, there would be no consumer fee at the point of purchase, and costs would be internalized by the producers. Canadian systems are also striving to enhance the Full Producer Responsibility model by supplemental legislative and regulatory approaches that drive design, much like the WEEE and RoHS directives work together to achieve end-of-life management and design

68. Duncan Bury, former EPR lead for Environment and Climate Change Canada, email communication, March 22, 2021.

changes. Canadian approaches seek the use of several system tools to develop a full product stewardship program.[69]

Priority materials, such as packaging, have been managed differently by province, but are trending in the direction of full producer responsibility, with producers taking greater management and financial control over the system with government oversight and enforcement. Notably, many of the Phase 2 products still have not been brought under EPR requirements. The core of BC's approach requires industry to achieve prescribed results but allows producers flexibility as to how to achieve their goals. This approach, which at the time distinguished British Columbia from other provinces, typically gives more control to producers, although they are still required to meet clear transparency and accountability requirements. The province of British Columbia continues to have some of the most progressive EPR systems across Canada.

US PRODUCT STEWARDSHIP EFFORTS

The trajectory that producer responsibility has taken in the United States bears similarities to its evolution in Canada and Europe. In all three regions, municipal officials were first to recognize the burgeoning challenge to manage a waste stream that was growing larger and more complex each year. Public works staff experienced waste problems firsthand—used motor oil poured into storm drains that empty into rivers; huge piles of scrap tires left in abandoned warehouses and fields; dead auto batteries leaking acids from their lead casing; and empty soda cans and bottles strewn about neighborhoods. The existing waste management system was no longer functional. It was time for action to replace inertia. Local and state governments, along with environmental groups, stepped in and rallied support, tapping into universal disgust for litter that marred the natural landscape and cityscapes, the massive amount of wasted resources, and the growing piles of trash.

What stood out most amid the millions of tons of waste were beverage containers, which would become synonymous with litter and represent a cavalier transition from reusable items to those that were single use—one and done—and tossed, without a care for the impact on others. The economic force behind this change was powerful. Beverage companies reduced production costs by using lighter, single-use materials. They also cut off the previous circular system of recovering and refilling reusable containers, which saved companies money by eliminating the need to pick up empties so they could deliver new product.

69. Bury.

Beverage companies could deliver their products to stores, continually, in a linear (one-way) "make, sell, dispose" system. They were not required to calculate the cost shift of this massive change in the delivery of beverages and the disposal of empties. There were no prior government policy papers that considered the costs and impacts of the change. Instead, industry acted alone, in its own best interest, to lower their costs in a way that worked for them. The system allowed it. It was fully legal. And so they passed the external cost of litter, recycling, and disposal to governments and their citizens. It would take public outrage to force manufacturers to manage their products after consumers were done with the packaging.

US BEVERAGE CONTAINER LAWS[70]

Some of the first product stewardship laws in the United States targeted beverage containers. In 1971, Oregon became the first state in the nation to pass a "bottle bill," which required consumers to pay a five-cent deposit on each purchase of packaged soda or beer that they got back when returning the empty container. Today, there are 10 states with a beverage container law,[71] also called a deposit return system (DRS), the last one having passed in Hawaii in 2002. There is a wide variation in program design and effectiveness, with Oregon's bottle bill being considered by many to be the most effective in the country.

Beverage container laws typically require beverage distributors to shoulder operational and/or financial responsibility for the redemption, collection, and processing of brands they sell, and require retailers to take back containers of the brands they sell. In some states, this requirement spawned the creation of private redemption centers to provide the collection service. Government monitors the program, enforces penalties against noncompliant retailers and manufacturers, and combats fraud by those who purchase containers in non–bottle bill states but attempt to receive a deposit in a state with a container deposit law. Thus, the bottle bill became one of the earliest experiments in shared responsibility.

Since their introduction, US DRS laws and related bills to expand these systems have been contentious, partly due to concern that containers with

70. A portion of this section on Beverage Container Laws was originally published as a chapter: Scott Cassel, "Product Stewardship: Shared Responsibility for Managing HHW," in *Handbook on Household Hazardous Waste*, 2nd edition, ed. A. Cabaniss (Lanham, MD: Bernan Press, 2018), 167–222.

71. *Redemption Rates and Other Features of 10 US State Deposit Programs*, Container Recycling Institute, accessed March 5, 2023, https://www.bottlebill.org/images/PDF/BottleBill10states_Summary41321.pdf.

high value will be removed from the recycling system and put into a duplicative system.[72] Even so, the deposit is a powerful incentive that significantly changes consumer behavior; in the United States, Canada, and across Europe, jurisdictions with DRS laws have higher collection rates for beverage containers than those without DRS laws.[73] In states with DRS laws, the recycling rate of containers on which the deposit is paid is about triple the rate for containers that carry no deposit.[74] One study has even posited that jurisdictions with DRS laws have markedly lower rates of coastal debris (litter).[75] To date, however, only one state DRS law has been enacted in the past two decades, with at least one being updated (Connecticut), while two DRS laws have been repealed (in Delaware and in the City of Columbia, Missouri).

DRS systems are recognized as effective in producing a clean stream of consistent materials, but they were never meant to clean up the *entire* waste stream. As EPR systems for packaging have become more viable and widespread in the United States, discussions about beverage container recycling have been viewed in the wider context of all packaging waste. Canadian packaging EPR programs, for example, operate successfully alongside DRS laws. In many locations *outside* the United States, the number of DRS laws and programs has increased. This change has also recently come to the United States. In a change from past positions, the American Beverage Association has come to support well-designed collection policies including EPR and DRS, recognizing that changes in public policy are necessary to improve the recovery of material and promote a circular economy.

Current US packaging EPR laws and bills generally exclude containers covered under a state deposit return system, but include other containers not covered by the deposit. In the 40 states without DRS laws, this is of little consequence. Whether a state has a DRS or not, however, effective EPR packaging laws include performance targets for the collection and recycling of all packaging materials and types. The challenge for those selling products in containers and other packaging is to develop systems that recover and recycle at the same high level as under effective DRS laws. These laws produce clean streams of high-value materials in a system that consumers know. Containers purchased in DRS states have a clear mark on the label to indicate a deposit has been paid. People know they can get that deposit back where they bought the product or at a nearby redemption center. Consumers

72. *Redemption Rates and Other Features of 10 US State Deposit Programs*.

73. *Fact Sheet: System Performance*, Reloop, May 3, 2021, accessed March 5, 2023, https://www.reloopplatform.org/wp-content/uploads/2021/05/Fact-Sheet-Performance-3May2021.pdf.

74. *Redemption Rates and Other Features of 10 US State Deposit Programs*.

75. Qamar Schuyler, Britta Denise Hardesty, T. J. Lawson, Kimberley Opie, and Chris Wilcox, "Economic incentives reduce plastic inputs to the ocean," *Marine Policy* 96 (October 2018): 250–255, accessed March 5, 2023, https://www.sciencedirect.com/science/article/abs/pii/S0308597X17305377.

receive their deposit back in exchange for transporting containers to a location that consolidates containers. Since consumers separate containers from other recyclables, the result is a clean stream of high-quality glass, metal, and plastic that is not hindered by consumer confusion about what to recycle in curbside or drop-off bins.

Material contamination rates are far lower in DRS collections than in curbside or drop-off recycling collections,[76] especially for glass, which is a prevalent contaminant in the curbside recycling stream. But a DRS creates a separate collection network that could add cost by diverting materials of value from community recycling bins. Those bins have a separate financing system often tied to a contract with a private collector and recycler that base their fee on the value of the recyclables they pick up. If that value is changed by altering the value of containers in that bin, the contractor price will need to be adjusted. For example, if more containers made of high value materials (e.g., aluminum) are collected at retail stores or redemption centers and removed from curbside bins or drop-off centers, the value of the recyclables collected in curbside bins and drop-off centers will decrease. That increases the cost for contractors, who will pass it on to municipalities and residents. Even so, strong proponents of DRS systems have demonstrated notable instances in which the avoided costs of collection, treatment, and disposal gained in a DRS system result in net cost savings to the municipalities despite losing "value" from the curbside bins.[77]

These factors must be reconciled to develop good policy. The issues are complex and challenging to understand, although they can be broken down into smaller bits of shared information ("chunked") to enable dialogue that can lead to resolution. All the pieces of the puzzle should be identified, and all stakeholders involved in collecting beverage containers brought together to discuss, understand, and decide how best to achieve jointly stated goals. In the context of beverage containers, these unresolved issues have sometimes caused PSI local government members to be at odds with PSI partners from the environmental community promoting DRS bills and PSI partners in the waste management community opposing them.

Fifty years after the triumphant breakthrough of bottle bills into the environmental protection arena, there seems to be growing support for the inclusion of well-designed DRS systems in a comprehensive waste reduction policy approach, along with EPR and recycled content standards. With proper

76. "Bottle bills produce high-quality recyclable materials," part of the Bottle Bill Resource Guide, Container Recycling Institute, accessed March 5, 2023, https://www.bottlebill.org/index.php/benefits-of-bottle-bills/bottle-bills-produce-high-quality-recyclable-materials.

77. Reloop, fact sheet, "Deposit Return Systems Generate Cost Savings for Municipalities" (February 2021), accessed March 5, 2023, https://www.reloopplatform.org/wp-content/uploads/2021/05/Fact-Sheet-Economic-Savings-for-Munis-8FEB2021.pdf.

attention and dedication of resources, remaining issues can be resolved. It is conceivable, for example, that EPR systems could eventually include a consumer financial incentive like a DRS on containers that are recycled at a low rate, as long as all stakeholders agree that the redefined collection system is efficient and equitably accounts for material value. Or perhaps containers that are collected at low rates will have higher producer fees because lower recycling imposes higher costs for government and taxpayers due to litter and disposal in landfills with limited space. Only time will tell, as solutions await their discovery.

In the United States, DRS laws have a special place. Some consider them EPR that is on par with other US EPR laws. Others, like the Product Stewardship Institute (PSI), consider them more like EPR predecessors that were first to hold producers responsible for their postconsumer products and packaging. Even though most state DRS laws require producers to fund handling fees paid to retailers, significant cost is incurred by consumers through unclaimed deposits, and the system is often overseen by government staff funded by taxpayer dollars. Due to these and other aspects, PSI developed a policy statement that asserts, while "[deposit return] systems are highly effective at recovering beverage containers and providing high quality recovered materials to recycling facilities," they do not fit our model for EPR in which producers take significant responsibility for financing and/or managing their products and packaging. I managed the DRS system in Massachusetts before I started PSI and did not consider it a model for US EPR systems that strive for consistent elements and a network of stakeholder accountability. The bottom line, however, is that the two policies—EPR and DRS—can coexist, and they certainly intertwine.

MAKING TOXICS A PRIORITY

With the idea that those who create pollution should bear the costs of managing it to prevent harm—the polluter-pays principle—originating from Europe, international consumer product goods companies sought ways to reduce chemical use in the production process. Pollution prevention, toxics use reduction, and similar government programs strove to decrease hazardous materials used in product manufacture. These programs encouraged voluntary changes in company practices by showing that up-front investments in new technologies could save money while reducing pollution. 3M Company's employee-based Pollution Prevention Pays (3P) program, initiated in 1975, is widely recognized as one of the earliest successful attempts at preventing pollution while saving the company money.[78]

In the early 1980s, municipal officials around the country grew concerned about other toxic materials that were used in common household products such as cleaning fluids, paints and varnishes, automotive fluids, and pesticides. These products, dubbed household hazardous waste (HHW), captured the attention of governments and advocates due to their toxicity, significant growth in sales, and the inability of agencies to fund programs to divert them from disposal in landfills and combustion facilities. Since most HHW collections were one-day events, funding decisions were often made based on the volume of material and not on the actual content of products brought by residents. If a car was turned away, there was no way of knowing if it contained a jar of elemental mercury, latex paint, or an empty hydrogen peroxide bottle.

Many programs in this new HHW field encouraged residents to collect all products. Over time, as programs became more sophisticated, several states began prioritizing materials based on potential risk and harm. This became a critical distinction as budgets for HHW programs began to shrink. One of the first states to prioritize waste streams was California, which set up "ABOPs," collection sites that focused on antifreeze, batteries (lead acid), oil, and paint. I was another early advocate of HHW product prioritization when I was waste policy director for Massachusetts. In the state's first comprehensive plan for managing hazardous household products in 1996,[79] we set a policy of phasing in the collection of priority materials according to three categories: high volume, universal waste, and low volume. High-volume materials included automotive products (used oil, oil filters, antifreeze, auto batteries, and gasoline), leftover paint products, and household batteries (Ni-Cd and button cell). Universal waste materials included certain mercury batteries, thermostats, and pesticides. Universal wastes[80] are hazardous wastes that US EPA designated to be regulated in a

78. Samuel Perkins, "Pollution Prevention and Profitability: A Primer for Lenders," Northeast Waste Management Officials' Association (1996).

79. MA Executive Office of Environmental Affairs, "Massachusetts Plan for Managing Hazardous Materials from Household and Small Businesses" (July 5, 1996). Massachusetts used the term hazardous household products (HHP) to denote that the goal was to reuse and recycle materials and not have them become a waste.

80. The Federal Universal Waste Rule was developed by US EPA to reduce regulatory barriers to collecting particular waste streams, such as batteries, thermostats, and pesticides, which increased recycling and reduced environmental impact. The Universal Waste Rule was amended in 1999 to include some lamps (i.e., fluorescent, high intensity discharge [HID], mercury vapor). As of July 2021, there are five types of universal waste: batteries, pesticides, mercury-containing equipment (including thermostats), lamps, and aerosol cans. "Universal Waste," US Environmental Protection Agency, accessed March 5, 2023, https://www.epa.gov/hw/universal-waste.

streamlined manner that reduced regulatory barriers to make collection easier, thus better protecting the environment. Low-volume materials included solvent-based glues, metal cleaners, toxic art supplies, chemistry sets, photographic chemicals, and other HHW that pose environmental and health risks in relatively small quantities.

These hazardous household products were presumed safe to use but presented risks if improperly used, stored, and/or disposed. While government pollution prevention experts met with company engineers to reduce toxics during production, municipal solid waste managers scrambled to divert toxic products from being disposed of in household garbage and transported via trucks to transfer stations, landfills, and combustion facilities that were ill-equipped to prevent harm to workers, community members, and the environment.

TIRES, MOTOR OIL, AUTOMOBILE BATTERIES, AND PESTICIDES

In the early 1990s, in response to increased risk from certain products, there was a profound policy shift toward better management of four particular "special wastes": tires, motor oil, auto batteries, and pesticides. Abandoned piles of scrap tires had been burning out of control for days, precipitating public alarm that led to state laws creating funds for the cleanup of existing piles and prevention of new ones. Used motor oil recycling laws sprouted from public concern over oil slicks at marinas, in inland waterways, and in the ocean. Automobile battery recycling laws were enacted to prevent the lead-acid components of these products from polluting vacant lots or wooded land. And pesticide disposal laws created funding and systems to capture outdated and unwanted pesticides from small farms, households, and other locations.

These state laws not only prevented harm but also fostered the creation and expansion of companies to safely manage products that were no longer wanted by the consumer. They also spurred the creation of end markets for some of the collected materials, like scrap tires to construct roads and playgrounds, used motor oil re-refined into new oil, and automobile battery casings turned into recycled plastic products and the lead smelted for reuse. These laws resulted in the recovery and recycling of materials even before the term *circular economy* became a concept.

The impacts from these four product types were vivid and unmistakably tied to the manufacturers that made them, which motivated the companies to develop a solution. Through their associations, the manufacturers of tires, motor oil, and automobile batteries developed, promoted, and encouraged state governments to pass legislation that required consumers to pay visible fees that were collected at retail and deposited into a state-managed fund. Government programs provided grant funding for public and private collection centers, and they covered the cost of handling, transporting, and recycling or disposing of the waste. Masterfully, these industries offloaded the problem onto consumers to pay the cost, onto retailers to collect the fees, and onto governments to manage the programs. As for themselves, aside from taking the initiative to pass legislation (not an insignificant achievement), producers took little to no responsibility for funding or managing the programs needed to keep the public safe from postconsumer product impacts.

The American Petroleum Institute developed a model bill that required consumers to pay an extra recycling fee to retailers upon purchase of motor oil. These "advanced recycling fee" (ARF) laws passed in many states and required state officials to manage the program. Similarly, the US Tire Manufacturers Association was instrumental in getting 35 states to pass scrap tire legislation in which ARFs, collected from consumers upon purchase of a new tire, fund government-managed used tire programs.[81] Battery Council International (BCI) successfully promoted legislation for collecting lead-acid batteries used in cars and trucks. These laws require retailers to take back used batteries. They also require consumers who purchase a new battery without returning a used battery to pay a "core charge," which is a deposit redeemable upon later return of a scrap battery. According to BCI, "in all 50 states, retailers can charge a battery core charge, and in over 30 states it is required by law."[82]

Although all states require pesticide manufacturers to pay a fee to register their products, funds from pesticide registration fees are often used for pesticide use purposes (e.g., training, licensing and registration, and program administration) and not to finance the proper management and disposal of unwanted pesticides. Only 24 states have programs for the collection and disposal of unwanted pesticides that are partially funded by pesticide registration-related fees. Of those, only 14 include funding for household pesticides.[83]

While these four laws reduced some of the immediate environmental risk that emanated from disposal, they relied heavily on government staff, retailers, and consumers. At a time when political pressure heightened to reduce the size

81. "State Scrap Tire Legislation Summary," US Tire Manufacturers Association, accessed March 5, 2023, https://www.ustires.org/sites/default/files/2021-06/US%20Scrap%20Tire%20Management%20Summary.pdf.

82. *State Recycling Laws*, Battery Council International, accessed July 6, 2021, https://batterycouncil.org/page/State_Recycling_Laws (no longer accessible).

83. Product Stewardship Institute, *How-To Guide for Advancing Pesticide Stewardship* (September 2019), accessed March 6, 2023, https://www.productstewardship.us/page/Pesticide-Guide.

of government, these laws put a significant new burden on state and local government officials to educate residents about how used and unwanted products would be collected and the harm posed by improper disposal. Government officials were required to manage grant programs that created the collection infrastructure. They hired contractors to service numerous local collection sites, enforced penalties against noncompliant companies that polluted the environment, and evaluated and publicly reported on program success.

A common feature of these four take-back laws is the dearth of producer responsibility. Manufacturers sell the products and consumers often pay an upfront fee to retailers that pass it to the state government to manage the program. While these laws fit into the broader category of product stewardship in the United States, they are not considered EPR. These laws represent an early generation of policies that addressed problems and lowered harm, but they have not been updated since. The amount of consumer funding that is generated often does not cover the full program management cost, and it is not uncommon for these funds to be funneled into the general fund rather than be used to address product end-of-life management.[84]

Of the four special wastes that required national attention, motor oil and scrap automobile batteries have maintained steady markets, in part due to the relatively higher recovery value of these products. This resulted in less pressure on governments to manage those hazardous products. Tires and pesticides, however, are still considered problem waste streams. Scrap tires are still dumped in piles, although they are much smaller than the massive piles of the past that ignited into blazing fires. Currently, most scrap tires are burned for energy, transforming a waste into a fuel source. However, this form of disposal is more polluting than retreading and recycling, and it contributes to greenhouse gas emissions. Pesticide waste from small farms and households continues to be the second most costly item collected in municipal HHW collection programs (behind paint) and are highly toxic. Funds generated by state laws are used mostly to ensure safe pesticide use, but do not have adequate funding available for safe collection and disposal.

Scant attention, though, has been placed on the potential impacts from all four of these special wastes along the supply chain, both upstream and downstream. It only seems a matter of time until the US public will want to know where used oil is re-refined and to what extent these facilities contribute to air pollution; where the lead in scrap automobile batteries is processed and whether the lead smelters are protecting surrounding neighborhoods; the degree to which tires are burned for energy at the expense of retreading and other, safer alternatives; and the location and safety of facilities in which pesticides are disposed. We are likely to find that polluting disposal facilities

84. Suna Bayrakal, Product Stewardship Institute, email communication, July 7, 2021.

are more often situated in disadvantaged communities, where health and environmental impacts are greatest.

IMPORTING PRODUCER RESPONSIBILITY TO THE UNITED STATES

The 1994 EU Packaging Directive actualized the concept of EPR in Europe, which subsequently spread around the world. In the United States, some academics, federal officials, and environmental organizations sensed the power of this new way of thinking. The Center for Clean Products and Clean Technologies at the University of Tennessee hosted the country's first academic symposium on EPR in November 1994. In cooperation with US EPA, this initiative sought to extract EPR policies from the European context for US implementation.[85] Also in 1994, the Organisation for Economic Co-operation and Development (OECD), started a multiyear research program on EPR that culminated in a 2001 guidance manual for governments that held producers primarily responsible and emphasized responsibilities for other stakeholders such as retailers, distributors, and consumers.[86] This work was also a conduit for expert EPR education in the United States.

As the concept was being understood by US waste policy professionals, the ways in which EPR was defined diverged. Environmental advocates viewed EPR as a means to hold producers and others accountable through regulation for actions to reduce, reuse, and recycle waste. The federal government, however, took a different path. In February 1996, under President Bill Clinton, the President's Council on Sustainable Development (PCSD) issued a report that highlighted the promise of EPR, but called it "extended ***product*** responsibility," redefining it as "a voluntary system that ensures responsibility for the environmental effects throughout a product's life cycle by all those involved in the life cycle."[87] The PCSD report described EPR in terms of "strategic opportunities for resource conservation and pollution prevention,"[88] implying

85. Gary A. Davis, Patricia S. Dillon, Bette K. Fishbein, and Catherine A. Wilt, "Extended Producer Responsibility: A New Principle for Product-Oriented Pollution Prevention" (1997), prepared for the US EPA Office of Solid Waste

86. OECD, *Extended Producer Responsibility: A Guidance Manual for Governments* (Paris: OECD Publishing, 2001), https://doi.org/10.1787/9789264189867-en; OECD, *Extended Producer Responsibility: Updated Guidance for Efficient Waste Management* (Paris: OECD Publishing, 2016), http://dx.doi.org/10.1787/9789264256385-en.

87. Bill Sheehan and Helen Spiegelman, "Extended Producer Responsibility Policies in the United States and Canada," in *Governance of Integrated Product Policy in Search of Sustainable Production and Consumption*, ed. D. Scheer and F. Rubik (Sheffield, UK: Greenleaf Publishing, 2017), 202–223.

88. Bette Fishbein, "Extended Producer Responsibility: A New Concept Spreads Around the World," *Rutgers University Demanufacturing Partnership Program Newsletter* 1, no. 2 (Winter 1996), Grassroots Recycling Network website (archived), accessed April 21, 2021, https://archive.grrn.org/resources/Fishbein.html. This quote is from the PCSD report but reported by Fishbein.

"that responsibility should be shared by all members in the product chain and not be imposed on the producers alone."[89] This redefinition of EPR led environmentalists to argue that "if everyone is made responsible for everything, no one is responsible for anything."[90]

Despite interest in EPR by advocates and academics, there was little fertile ground for the seed of this new concept to take hold in the United States. There were no federal laws then, and still now, that direct the US EPA to regulate the management of solid waste in accordance with EPR principles. Although there was interest in including EPR in the 1992 reauthorization of the Resource Conservation and Recovery Act (RCRA), which would have authorized US EPA to implement EPR systems, that did not happen.[91] Another initiative in 1998 by the Natural Resources Defense Council (NRDC) and others may have been the first federal EPR bill introduced in the United States and was certainly the first on packaging waste.[92] The bill was developed by former policy staff Allen Hershkowitz, US Representative Gerry Sikorski, and US Senator Max Baucus. To promote the bill, Hershkowitz wrote an article explaining the concept of EPR that was published in the *Atlantic*.[93] Without the fertile ground needed for EPR to grow, however, the bill died.

As the concept of EPR was entering the United States in 1994, some states individually latched onto the producer responsibility concept. However, instead of first promoting legislation targeting packaging, as did European governments, US states focused on passing laws to reduce the health threat from toxic products. As governments began to initiate legislation that would place product end-of-life responsibility on producers, industry began to respond, using both voluntary and regulatory strategies to take responsibility for toxic household products, including rechargeable batteries.

PRODUCER RESPONSIBILITY: BATTERIES[94]

In 1994, the rechargeable battery industry established the first national, industry-wide producer responsibility program in the United States. Around this time, US EPA estimated that nickel-cadmium (Ni-Cd) batteries, which

89. Fishbein, "Extended Producer Responsibility." This quote is not from the PCSD report but is directly attributed to Fishbein.
90. Fishbein, "Extended Producer Responsibility."
91. Fishbein, "Extended Producer Responsibility."
92. Allen Hershkowitz, email communication, March 6, 2023.
93. Allen Hershkowitz, "How Garbage Can Meet Its Maker," *The Atlantic Monthly* (June 1993).
94. A portion of this section on batteries was originally published in the chapter: Scott Cassel, "Product Stewardship: Shared Responsibility for Managing HHW," in *Handbook on Household Hazardous Waste,* 2nd edition, ed. A. Cabaniss (Lanham, MD: Bernan Press, 2018), 167–222.

accounted for less than 0.1 percent of municipal solid waste by weight, accounted for 75 percent of the cadmium content in the waste stream.[95] The industry effort was prompted by eight state laws with take-back requirements for rechargeable batteries, growing interest in Europe to ban cadmium from rechargeable batteries, and the passage of comprehensive legislation in Minnesota and New Jersey. "Both states require[d] that rechargeable batteries be easily removable from products, be labeled as to content and proper disposal, and be banned from the municipal waste stream. In addition, they require[d] manufacturers to take rechargeable batteries back at their own expense for recycling or proper disposal."[96]

In response to these laws, the Portable Rechargeable Battery Association established Call2Recycle (known as the Rechargeable Battery Recycling Corporation, or RBRC, prior to 2013) to manage a national program for the recovery and recycling of Ni-Cd batteries. Call2Recycle is a nonprofit public service corporation comprised of more than 300 battery and product manufacturers. The program expanded to include Canada in 1997 and further broadened in scope in 2001 to include all small rechargeable batteries, including Ni-Cd, nickel metal hydride (Ni-MH), lithium ion (Li-ion), and small sealed lead-acid (SSLA). In 2004, Call2Recycle enlarged the collection program to include used cell phones.

Call2Recycle funds the rechargeable battery collection efforts by licensing the right to use the organization's US EPA-certified chasing arrows recycling logo on products and packaging. Manufacturers contribute funds to the organization based on their rechargeable battery sales into the North American marketplace. Company fees are based on volume, weight, and chemistry of batteries sold. These funds are used to conduct an education campaign; establish collection sites at retail outlets, municipal locations, and commercial establishments; and process batteries for recycling. The program is free for consumers, retailers, and municipally operated collection sites, which use special prepaid shipping containers provided by Call2Recycle.

To ensure that the Call2Recycle program became the national model, the battery industry sought federal legislation that facilitated its national rollout. The Mercury-Containing and Rechargeable Battery Management Act (the "Battery Act") became law on May 13, 1996. "This legislation reduce[d] barriers to the battery collection and recycling system and avoid[ed] the need to deal with inconsistent legislation in different states."[97] Although the

95. Bette Fishbein, "Industry Program to Collect Nickel-Cadmium (Ni-Cd) Batteries," Inform Report (1997), https://p2infohouse.org/ref/06/05930.pdf.

96. Fishbein, "Industry Program to Collect Nickel-Cadmium (Ni-Cd) Batteries."

97. Fishbein, "Industry Program to Collect Nickel-Cadmium (Ni-Cd) Batteries," Section 4.1.

federal Universal Waste Rule (UWR) was in effect at this time, each state needed to adopt the UWR, making implementation of Call2Recycle's program cumbersome. The Battery Act established national, uniform labeling requirements for Ni-Cd and certain SSLA rechargeable batteries and mandated that they be "easily removable" from consumer products. The act also made the UWR immediately effective in all states, rather than having to wait for them to adopt it one by one. While Call2Recycle's program is voluntary for retailers, both California and New York State passed laws requiring retailers to collect and recycle rechargeable batteries. The batteries are sorted and recycled through a network of over 15 facilities worldwide.[98] The metals, such as nickel and cadmium, are recovered; the nickel is used to make stainless steel, and the cadmium is used to make new batteries.

The Call2Recycle program was an important step for US industry since battery and product manufacturers voluntarily took full responsibility for their products at end of life. They funded and managed the entire program, developed effective educational materials and outreach programs, and secured participation from thousands of retailers and municipal HHW programs to serve as convenient collection sites around the country. The program has also illustrated serious shortcomings of voluntary programs that have no mandated performance goals or reporting requirements and, therefore, lack accountability.[99] Since the program is not mandatory, some producers choose not to contribute funding to it even though their batteries are collected and recycled along with the brands of those companies that do contribute funding. These free riders increase the cost paid by those who take responsibility for managing their own batteries.

While the number of batteries collected by Call2Recycle increases each year, it still captures only a small amount of batteries sold into the marketplace, varying widely across North America and ranging from a high of over 30 percent collection of batteries in Canadian provinces to single-digit percent collection of residential and commercial batteries sold in most US states, averaging about 15 percent nationwide.[100] In addition, an unknown, but not insignificant, number of rechargeable batteries that remain in electronic equipment (e.g., laptop computers, tablets, and cell phones) are collected and recycled nationally. These additional batteries will undoubtedly increase the collection rate.

Over the past decade, as public outrage over disposable products has heightened once again, governments have sought to implement EPR systems

---

98. "How Call2Recycle Recycles Your Batteries," Call2Recycle, March 5, 2023, https://www.call2recycle.org/flow-chart/.

99. Bette Fishbein, formerly Senior Fellow, INFORM, Inc., telephone communication, August 15, 2007.

100. Call2Recycle, Inc., July 2021.

for all household batteries (single-use varieties as well as rechargeable ones). There have been several notable milestones in the development of US battery EPR. Initial momentum was supplied by the Connecticut Department of Energy and Environmental Protection, which hired PSI in 2014 to develop a battery briefing document[101] and convene the country's first national multi-stakeholder battery stewardship dialogue.

The June 2014 meeting was attended by more than 130 local, state, and federal government officials, recyclers, retailers, and other key stakeholders, including government officials representing 23 states.[102] At this meeting, for the first time, the four associations representing single-use and rechargeable battery manufacturers[103] unveiled a draft EPR bill intended to cover the recycling of all household batteries—both single-use and rechargeable. The industry model bill, along with a model all-battery bill developed by PSI, set the stage for subsequent policy discussions convened by PSI with battery manufacturers and government officials. Eventually, these events led to the first EPR law in the country for single-use batteries in Vermont, which passed in 2014. It would take until 2021 for the nation's first all-battery EPR law to pass in Washington, DC.

Over the past few years, there has been a resurgence of interest in battery EPR bills due to increased incidences of fires caused by lithium-ion batteries coupled with heightened interest in recycling rechargeable batteries used to power lawn mowers, hand tools, bicycles, vehicles, and other products. As increased importance is placed on batteries to store and transmit energy related to solar panels and other products, as well as their central role in reducing greenhouse gas emissions through the products they power, the need for an effective system to recover and recycle battery materials has grown more urgent. As a result, battery EPR legislation is sure to gain further momentum in the coming years.

PRODUCER RESPONSIBILITY: MERCURY PRODUCTS

When I started working for Massachusetts on waste policy in the 1990s, major attention was given to the health effects of mercury in the environment, particularly from products containing mercury. With the release of a seminal document in 1996 by state officials, *Mercury in Massachusetts: An*

101. Product Stewardship Institute, "2014 Regional + National Batteries Stewardship Dialogue Meeting—Product Stewardship Institute" (PSI).

102. Product Stewardship Institute, *Meeting Summary: Battery Stewardship in the United States—Collaboration for Advancing Legislation and Programs, June 11 and 12, 2014, Hartford, CT.*

103. These four associations are the National Electrical Manufacturers Association, Portable Rechargeable Battery Association, Corporation for Battery Recycling, and Call2Recycle.

Evaluation of Sources, Emissions, Impacts, and Controls,[104] mercury products became a top priority for the state. This included thermometers, thermostats, fluorescent light bulbs, and household batteries (which, at the time, were the leading source of mercury from products). In-state mercury releases were estimated to contribute 41 percent of total mercury releases from the air to land and water in the state. Mercury-containing products burned in municipal waste combustors were estimated to contribute the largest share of the in-state mercury releases. These mercury emissions entered already stressed water bodies. With over 70 percent of its solid waste being combusted,[105] and with fish advisories issued on a significant number of state water bodies owing to mercury pollution, Massachusetts set out to remove mercury from the waste stream. As director of waste policy for the state, I met with battery and lamp manufacturers around this time to try to convince them to conduct voluntary initiatives to collect their mercury products. As you know from reading the preface of this book, my offer was refused.

Massachusetts was not alone in its focus on mercury products. States in the Northeast and Midwest (Great Lakes Region), supported by the Mercury Policy Project, Clean Water Action, and other environmental groups, began to lead a national effort to remove mercury products from the waste stream. In 1998, the Conference of New England Governors and Eastern Canadian Premiers (the "Conference") developed a Mercury Action Plan to reduce mercury pollution in the region. To assist the New England states and the Eastern Canadian provinces in implementing the action plan, the Conference asked the Northeast Waste Management Officials' Association (NEWMOA) to develop model legislation that would synthesize approaches and provide a comprehensive and consistent framework for managing mercury-containing wastes.

Starting in 1999, the states in the Northeast and other parts of the country actively began to use NEWMOA's Mercury Education and Reduction Model Legislation to pursue enactment of legislation focused on reducing mercury in products and wastes. The legislation is based on producer responsibility and includes restrictions on the sale of certain mercury-added products, phaseouts and exemptions, labeling, a disposal ban, and a manufacturer collection requirement that reads (in part): "The cost for the collection system must be borne by the manufacturer or manufacturers of mercury-added products. Manufacturers may include the cost of the collection system in the

104. Massachusetts Department of Environmental Protection, *Mercury in Massachusetts: An Evaluation of Sources, Emissions, Impacts, and Controls* (1996).

105. At the time, three solid waste combustion facilities in northeastern Massachusetts contributed to a high concentration of mercury and other pollution, prompting particular concern in that part of the state.

price of the product and may not assess a separate fee for the use of the collection system."[106]

In 1998, Massachusetts went one step further by limiting the amount of mercury emissions from solid waste combustion facilities and imposing a requirement that all such facilities remove mercury-containing products prior to combustion. This regulation,[107] one of the first in the nation and one on which I worked closely, required companies to set up product take-back systems for the collection and recycling of mercury-containing HHW that, for years, went into the household garbage. As a result of this regulation, the state's subsequent mercury products law (modeled after the NEWMOA model bill), and the closing of several solid waste combustors, the state was able to show a significant reduction in mercury pollution in 2007 in its water bodies, particularly in the northeastern part of the state.[108]

THERMOSTATS

In response to rising concerns over mercury and the threat of legislation in multiple states, a voluntary, industry-wide thermostat take-back program was launched. To increase the collection and recycling of retired mercury thermostats, the three largest mercury thermostat manufacturers—Honeywell, General Electric, and White-Rodgers—established a nonprofit entity in 1998 called the Thermostat Recycling Corporation (TRC). A typical mercury thermostat contains three grams of mercury that can be released into the environment if the thermostat is broken or improperly disposed of.[109] The program began in nine states at its inception and became a national program (excluding Alaska and Hawaii) in 2001. The number of thermostats and the amount of mercury collected and recycled annually has increased over time as the program has expanded and taken root.[110]

The TRC program was preceded by a multifaceted take-back program that Honeywell established in Minnesota after the passage of a 1992 state law

106. Northeast Waste Management Officials' Association, "Revised Discussion Document: Mercury Education And Reduction Model Act" (5 July 2007), Section 10 (F), http://newmoa.org /prevention/mercury/final_model_legislation.doc. For information on the NEWMOA Mercury Reduction Program, including updates on laws, see http://www.newmoa.org/prevention/mercury/.

107. Municipal Waste Combustor Regulation (M.G.L. c. 111, Sections 142A through 142M and 150A, M.G.L. c. 21A, Section 18), April 1998.

108. C. Mark Smith, Massachusetts Department of Environmental Protection, PSI Networking Conference Call, "The Fate of Excess Mercury in the United States" (presentation, June 13, 2007).

109. Mercury thermostats contain three grams of mercury per ampoule. Some thermostats contain more than one ampoule. The average thermostat, therefore, contains about four grams of mercury.

110. For up-to-date information, see TRC's website, http://www.nema.org/gov/ehs/trc/.

prohibiting the disposal of mercury thermostats, assigning responsibility for compliance to contractors removing thermostats from households, and requiring manufacturers to provide education and incentives to encourage recycling. Honeywell established a wholesaler-based, reverse distribution system for heating and cooling contractors in 1993; a homeowner mail-back program in 1994; and collections at HHW centers in 1995. The Honeywell mail-back program was never expanded nationally and was subsequently terminated by the company in 1999 based on cost.[111]

TRC used Honeywell's experience, along with market research, to design a program to serve the heating and cooling wholesaler/contractor distribution chain that distributes and installs 75 percent of all thermostats.[112] TRC believed that this supply chain could retrieve the greatest number of thermostats. Homeowners, who replace thermostats with units purchased at retail stores, account for approximately 25 percent of thermostats installed. TRC did not provide options for homeowner collection until later in the program and discontinued all Honeywell-initiated programs except the wholesaler-based reverse distribution system.

In 2004, following concerns expressed by state and local governments about the low level of TRC collection of mercury thermostats and a NEWMOA report critical of TRC's performance,[113] PSI initiated a national dialogue to increase the recycling of mercury thermostats and explore ways to reduce the continued production of mercury thermostats.[114] After conducting extensive interviews with TRC, thermostat manufacturers, the National Electrical Manufacturers Association (NEMA), and other key stakeholders, PSI developed a background summary report that highlighted the problems, key issues, and potential solutions to managing mercury thermostats. PSI convened two national stakeholder meetings in July and October 2004, held in Oregon and Wisconsin, which drew about 30 participants, including state and local government officials from about a dozen states, TRC, NEMA and its member companies, contractors and wholesalers, environmental groups, and others.

As a result of these meetings, the multi-stakeholder group, including Honeywell, NEMA, and TRC, reached agreement on seven priority

---

111. Product Stewardship Institute, "Thermostat Stewardship Initiative: Background Research Summary—Final" (November 18, 2004).

112. After an initial payment for the collection bin, TRC pays for the replacement collection bins, shipping costs, and the cost of recycling the mercury thermostats.

113. Northeast Waste Management Officials' Association, "Review of the Thermostat Recycling Corporation Activities in the Northeast" (November 2001), http://www.newmoa.org/prevention/mercury/TRCreport.pdf.

114. The national thermostat dialogue was initially funded by state and local governments, and the US EPA. Several of the collaborative projects were also funded by US EPA.

initiatives, including expanding the voluntary TRC program to wholesaler chain stores; collecting thermostats at contractor locations; collecting at HHW facilities nationwide (following a PSI pilot project); and conducting a two-state pilot project to test the degree to which a cash bounty would cause contractors to return more thermostats.

While these programs were being implemented, one of PSI's state members, the Maine Department of Environmental Protection (ME DEP), wanted to go beyond voluntary initiatives. Maine had already banned the sale of new mercury thermostats and required wholesalers to recycle those taken out of service, providing leverage for additional requirements. In 2005, the agency introduced a bill that proposed to hold thermostat manufacturers responsible for increasing the collection and safe management of mercury thermostats. Although the industry resisted at first, good rapport was developed with stakeholders through the national dialogue. As a result, I was able to convince the ME DEP, TRC, Honeywell (the manufacturer with the vast majority of thermostat market share), NEMA, the Natural Resources Council of Maine (NRCM), and other key stakeholders to seek consensus on a revised bill. After several months, in one of my first attempts to mediate a solution to reduce product harm through an EPR approach, the group reached consensus on a bill that was signed into law by Maine governor John Baldacci on April 14, 2006.

The new law required each thermostat manufacturer "individually or collectively" to "establish and maintain a collection and recycling program for out-of-service mercury-added thermostats . . . to ensure that . . . a maximum rate of collection of mercury-added thermostats is achieved."[115] The law required manufacturers to pay a financial incentive (i.e., bounty) "with a minimum value of $5" on each thermostat turned in by contractors and homeowners.

In a joint press release celebrating the signing of the law, the thermostat industry touted its achievement. "We feel an obligation to do as much as possible to reduce the potential environmental impacts from our products all across their lifecycle," said Mark Kohorst, TRC executive director. "We are encouraged by this joint effort to capture as many mercury thermostats as possible."[116]

The success in Maine spurred interest from other state governments to move beyond the expanded voluntary initiatives established in the national

115. Maine legislature, "§1665-B. Mercury-added thermostats," Title 38: Waters and Navigation, Chapter 16-B: Mercury-Added Products and Services, 2006, https://legislature.maine.gov/statutes/38/title38sec1665-B.html.

116. Joint press release from the Thermostat Recycling Corporation (TRC), Natural Resources Council of Maine (NRCM), and Product Stewardship Institute (PSI), April 14, 2006.

dialogue, which provided a conduit for communication about Maine's new law. At the request of Honeywell and to avoid a patchwork of thermostat laws around the country, PSI initiated national discussions to develop a best-practices model with our multi-stakeholder group. After eight phone meetings over six months, which included participation by Honeywell, NEMA, and TRC, PSI convened two final calls to wrap up the agreement, which included a menu of policy options from which states could choose.

The model required manufacturers to finance, collect, and recycle mercury thermostats removed from service; pay a $5 bounty per thermostat returned; and submit a manufacturer program plan to the oversight agency for approval that included proposed collection options (e.g., contractors, wholesalers, HHW facilities, and mail-back), performance goals set by the oversight agency, and education and outreach components. It also required a detailed annual manufacturer report, manufacturer funding to cover government oversight expenses, a wholesaler and contractor collection requirement, a thermostat disposal ban, and other key elements. This model bill, finalized March 5, 2007, outlined many of the best-practice elements that were subsequently included, and refined, by PSI for future product and packaging bills.[117]

Honeywell, NEMA, and TRC, however, quietly abstained from the last two conference calls, later claiming that they never agreed to the model legislation because they did not attend these calls. In essence, after initiating discussions and participating actively as a stakeholder, the company walked away from the dialogue to rethink its role in thermostat collections. Honeywell later expressed its concerns on the following issues: performance goals, any type of collection incentive (i.e., bounty), giving the state authority to increase program requirements if goals were not met, lack of mandatory retail collection, and total program costs.[118]

In the year ahead, PSI continued to work with states to introduce legislation aligned with its comprehensive model thermostat bill. Discussions in 2006 in Pennsylvania ended with the state Department of Environmental Protection (PA DEP) wanting a reduced role in setting performance goals, although other model language remained intact. In the 2007 legislative session, however, NEMA convinced the PA DEP to accept a significantly weakened bill, which passed into law. Also in 2007, after lengthy discussions between NEMA and the Illinois Environmental Protection Agency

117. Product Stewardship Institute, *Model State Mercury Thermostat Program Final Text* (March 5, 2007).

118. Product Stewardship Institute, "PSI/Honeywell Relations Timeline" (July 28, 2008), submitted by Product Stewardship Institute to Honeywell in advance of meeting.

(IEPA), on which I played an advisory role to IEPA, NEMA again walked away from its agreement with IEPA, although the legislature passed the bill over NEMA's opposition.

In just one year, the thermostat industry went from a champion of a model EPR law in Maine to opposing other bills based on the Maine law. Despite several attempts by PSI and Honeywell to resolve differences, including a meeting with company executives that I attended at Honeywell's New Jersey headquarters in 2009 with several PSI board members, the company continued to oppose key elements of the model bill. As of 2023, 13 states passed thermostat EPR laws based on the PSI model that require manufacturers to collect and safely manage mercury thermostats from households. On May 19, 2008, Vermont succeeded in passing the second thermostat EPR law with a financial bounty, along with other strong best-practice elements. California also used the PSI model to pass a bill that provided authority to the state oversight agency to require additional initiatives of manufacturers if performance goals were not met.

Today, as the number of mercury thermostats remaining on walls continues to decrease, Maine and Vermont remain national leaders in the rate of recovery on a per capita basis due to the bounty.[119] Since 2014, no new states have passed thermostat EPR laws, although California significantly amended its law in 2021, including the addition of a minimum $30 per thermostat bounty. Of the laws that do exist, TRC continues to challenge the legitimacy of performance goals, which rely on estimates and calculations to determine the number of mercury thermostats available for replacement each year. In addition, laws in five states include sunset clauses inserted by manufacturers that abolish the law by a certain date. In California, state officials were tangled for many years in enforcement negotiations with the industry to pay a penalty for repeatedly missing the state's goals. Most states report sluggish attempts by TRC to maximize the collection of mercury thermostats due to weak outreach and education efforts. Even so, more than 2.7 million thermostats, amounting to more than 26,000 pounds of mercury, were captured from mercury thermostat collections over the past 22 years.[120]

One other result of regulatory efforts also had a prominent impact. The model bills of both PSI and NEWMOA prohibited the sale of mercury

---

119. Thermostat Recycling Corporation, "Summary of State Results," accessed March 7, 2023, https://thermostat-recycle.org/program-info/measuring-our-impact/summary-of-state-results/. PSI per capita thermostat collection calculation using 2020 census data.

120. Thermostat Recycling Corporation, "Program History," accessed March 7, 2023, https://thermostat-recycle.org/about/program-history.

thermostats, and this provision became law in 10 states,[121] leading Honeywell to invent a mechanical, non-mercury thermostat that looked exactly like the company's iconic round unit. This technological advancement was part of the advent of the smart thermostats to which we have now become accustomed. The state mercury thermostat sales bans provided added company incentive to invest in new technology to eliminate mercury from thermostats. Back when PSI and Honeywell were collaborating on the Maine bill, I recall being shown a confidential mechanical "Honeywell round" by a company employee. The new invention had not been publicly announced, and I experienced the best that collaborative discussions based on trust could deliver.

PRODUCER RESPONSIBILITY: CARPET

As states in the Northeast and Midwest targeted mercury products, another Midwest initiative sought to capitalize on existing product stewardship efforts by several carpet manufacturers. In 1999, the Midwestern Workgroup on Carpet Recycling, spearheaded by officials from Minnesota, Iowa, and Wisconsin, with involvement from the US EPA, began a voluntary multi-stakeholder dialogue. The regional effort grew into a national two-year initiative. The resulting National Carpet Recycling Agreement was signed on January 8, 2002, by the Carpet and Rug Institute (CRI), which is the trade association for carpet and fiber manufacturers, state and local government agencies, nongovernmental organizations, and the US EPA.

The agreement set a voluntary nationwide goal of 40 percent diversion of carpet from landfills by 2012, including a 15 percent increase in recycling. The agreement outlined the roles and responsibilities for the Carpet America Recovery Effort (CARE), an industry-led stewardship organization that manages the collection and recycling of scrap carpet and identifies viable markets for postconsumer carpet. CARE publishes an annual report outlining the results of its efforts. By helping to promote markets for secondary carpet fiber, CARE seeks to divert carpet from landfills and combustion facilities into value-added products and to build the collection infrastructure to

---

121. States that restrict the sale of mercury thermostats include California (effective January 2006), Connecticut (effective July 2003, unless the manufacturer submits a plan enabling collection), Louisiana (effective 2008), Maine (effective January 2006), Maryland (effective October 2007), Michigan (effective January 2010), New York (effective 2008), Oregon (effective January 2006; prohibits installation of thermostats containing mercury in commercial and residential buildings), Rhode Island (effective January 2006; labeling requirements, phaseout depending upon mercury content levels, and collection plan requirements), Vermont (effective July 2006), and Washington (unless the manufacturer participates in recycling).

eventually support carpet-to-carpet recycling.[122] Unfortunately, the national carpet agreement goals were not reached and the initiative was disbanded, although CARE still works to increase carpet recycling, including in California, where an industry-sponsored EPR bill passed in 2010.[123]

The carpet industry's efforts in the past decade are an example of an industry supporting a free-market principle that carpet recycling should take place only if markets exist for the collected scrap carpet. The life-cycle impacts of carpet manufacture and postconsumer management have not been part of this calculation. With 80 percent of carpet manufacture taking place in Georgia, the carpet industry is opposed to regulation and only supports voluntary initiatives. Although state and local governments identified carpet as a top waste at PSI's inaugural forum in 2000, other organizations and agencies, such as the Minnesota Pollution Control Agency, the US EPA, and University of Tennessee were engaged with the industry in voluntary discussions a few years before that time.

In 2014, after state bills were introduced in Minnesota, Illinois, and New York, and then defeated by the carpet industry, PSI was asked by its members in those states to facilitate the development of a model carpet EPR bill to be introduced again in those and other states. With funding from the Connecticut Department of Energy and Environmental Protection (CT DEEP), PSI developed a briefing document that included key elements of a carpet EPR bill and invited carpet manufacturers, recyclers, government agencies, and others to a national two-day meeting in Hartford, Connecticut. The document also identified meeting goals,[124] carpet recycling problems and proposed solutions, and other important data.

One day before the meeting, the Carpet and Rug Institute (CRI) sent the CT DEEP Commissioner, the meeting sponsor, a letter stating that neither CRI nor any of its carpet and fiber manufacturer members would attend the meeting. Despite having engaged industry representatives in conversations about the regulatory options to be discussed at the meeting, CRI boycotted the meeting. They did, however, allow the representative of CARE, the industry's

122. Product Stewardship Institute, "Update Report on Negotiations for the 2012 Memorandum of Understanding on Carpet Stewardship" (September 8, 2011), accessed March 7, 2023, https://productstewardship.us/wp-content/uploads/2023/03/2011-09-08_MOU_Carpet_Negotiations.pdf.

123. CalRecycle website, accessed March 6, 2023, https://calrecycle.ca.gov/carpet/law.

124. The five meeting goals were as follows: Goal 1: Maximize the collection and recycling of scrap carpet while minimizing cost. Goal 2: Develop a long-term financing system (e.g., extended producer responsibility) to manage scrap carpet in a manner that alleviates the financial burden faced by governments and supports a sustainable recycling industry. Goal 3: Increase the procurement of recycled products made from scrap carpet. Goal 4: Support local businesses that recover scrap carpet for reuse and recycling. Goal 5: Develop a model carpet bill that can be harmonized in the United States and perhaps throughout North America. *Carpet Stewardship Briefing Document* (May 12, 2014), Product Stewardship Institute.

PRO, to attend the meeting to educate the group about its recycling activities. Over 100 people attended the meeting either in person or via telephone, including government officials from 15 states. Despite multistate interest in introducing legislation in the Northeast and multiple other states, CRI defeated these efforts. CRI also went a step further, hamstringing other EPR initiatives by providing funding to carpet recyclers on the condition that they sign a legally enforceable contract prohibiting them from supporting EPR legislation. Carpet recyclers were vulnerable to this tactic. Since recycling carpet is more expensive than landfilling or burning it, recyclers can only exist if people want to voluntarily pay a subsidy or if there is funding available to cover the differential cost. Since many carpet recyclers were on the verge of bankruptcy, they had little choice but to sign the agreement and take the industry's money. And with that, PSI and its members lost a significant advocate for the passage of carpet EPR laws for the next seven years.

After nearly 25 years of attempts to engage the carpet industry in recycling efforts, the national carpet recycling rate stands at only 9.2 percent.[125] Following California's passage of a weak recycling law in 2010 that was promoted by the carpet industry to preempt stronger EPR legislation, the recycling rate remained stagnant. However, after the passage of two amendments in 2017 and 2019, the carpet recycling rate in California increased to about 28 percent by 2021.[126] Carpet EPR bills based on PSI's state model were introduced in 2021 in Illinois, New York, Oregon, and Minnesota, spurred by the need to meet state recycling and GHG emissions goals, the desire to boost economic activity and jobs, and the advent of new technologies to more efficiently process carpet and return it to the circular economy. In 2022, New York's EPR bill was enacted into law, requiring the carpet industry to internalize the cost of carpet collection and recycling. By contrast, the California law requires consumers to pay a fee on the purchase of new carpet that is intended to incentivize carpet recyclers to collect. The New York law also includes artificial turf, performance goals for recycling and postconsumer content in new carpet, convenience standards, education and outreach, and design changes to reduce toxics.

PRODUCER RESPONSIBILITY: ELECTRONICS

As voluntary and legislated producer responsibility initiatives started bubbling up in the late 1990s and early 2000s in the United States, electronics

125. Carpet America Recovery Effort, "California Carpet Stewardship Program 2021 Annual Report," submitted to CalRecycle on September 1, 2022.

126. "CARE Sustainable Funding Oversight Committee, CARE Update: California Carpet Stewardship Program Q1 2021 Approved Results," 3, updated May 20, 2021.

waste eventually became the country's EPR launching pad. In 2000, PSI identified electronics as the nation's top product in need of an end-of-life management strategy following its survey of state officials and discussions at the national forum. State and local government agencies were concerned about the hazardous components of computers and televisions, the rapid obsolescence of these products, and the significant volume they represent in the waste stream. PSI began to coordinate state and local agency interests in a product stewardship solution we called the National Electronics Product Stewardship Initiative (NEPSI).

The creation of NEPSI caught the attention of electronics manufacturers and others, and in early 2001, the US EPA spearheaded the country's first dialogue on electronics management that sought "to establish a national system to collect, transport and process consumer electronics in a manner that is protective of human health and the environment, and one that is economically sustainable and market driven."[127] The dialogue included 45 representatives, with about a third each from government, industry, and other stakeholder interests. During the three years of meetings, the Electronic Industries Alliance (EIA) and the Consumer Electronics Association (CEA) (now combined as the Consumer Technology Association) represented manufacturers, while PSI coordinated participation and comment from over 20 state agencies and numerous local governments. The Silicon Valley Toxics Coalition played a significant role representing environmental interests. The meetings were facilitated by the Center for Clean Products and Clean Technologies at the University of Tennessee, a contractor to US EPA.

Over three years of countless meetings, the fractured beginnings of the US EPR movement started to emerge. In full-group multiday meetings, as well as government-only and industry-only caucuses, we explored every type of financing solution imaginable—advanced recycling fees (ARFs) only for leaded cathode ray tubes (used in television and computer monitors), ARFs for the full system, ARFs for an initial period of time to be followed by a transition to cost internalization, cost internalization for the full system costs, and shared financial responsibility. We also discussed a panoply of operational issues—performance goals, enforcement, recycling facility certification, collection and processing infrastructure, the scope of electronics to be collected, and other issues.

In the midst of the national discussions, CalRecycle (formerly the California Integrated Waste Management Board), manufacturers supporting an advanced recycling fee (ARF), and others successfully passed the first electronics end-of-life management law in the country in California in 2004, with an ARF financing model. Also at that time, stakeholders negotiated a

127. *NEPSI Compromise Resolution* (February 26, 2004). PSI Archives.

formal resolution facilitated by EIA, PSI, and the University of Tennessee. The resolution outlined a scope of products to be covered by federal legislation and a hybrid financing system that would start with an ARF to cover historic and ownerless products and transition to producer responsibility at a later date based on a series of criteria to be determined. The resolution, which referenced a more detailed memorandum of understanding,[128] put a halt on further discussions until manufacturers could come back with a viable system to which they all agreed. They never came back.

Although the NEPSI dialogue did not result in a national agreement, it was a significant turning point in the United States. Since it was the first large multi-stakeholder product dialogue, it attracted national attention and a large commitment of public and private sector resources. NEPSI also became the first dialogue, along with the national carpet discussions, in which industry consciously committed to engage in product stewardship discussions. This transformation took several years to evolve and was heightened in response to state legislative pressure that increased as the bid for a national negotiated solution waned.

As a result of the increased national attention to the environmental hazards of electronics disposal, companies began to compete on voluntary take-back initiatives. These systems became significant transitional solutions offered by companies to meet the growing public support for producer responsibility. HP, Dell, and a coalition of ARF supporters led by Panasonic, Sony, Sharp, and Philips were the first to offer free periodic collections of scrap electronics. By 2006, a number of electronics companies offered some type of end-of-life collection program for their products, with some providing money-back coupons on the sale of new equipment. In that year, Dell became the first company in the United States to collect its own end-of-life equipment from consumers for free. In August 2007, Sony followed with a collection program in conjunction with Waste Management Inc. to collect Sony products for free at specific locations, while other companies still charged the consumer for take-back service.

Following the passage of the California ARF law and the significant retailer pushback, multi-stakeholder coalitions had begun to form in each state to push for full producer responsibility. HP developed model state legislation and worked with influential environmental groups, including the Silicon Valley Toxics Coalition, Electronics TakeBack Coalition, Washington Citizens for Resource Conservation, and the Natural Resources Council of Maine, as well as charities, retailers, and governments, to pass state electronics laws.

---

128. NEPSI, Draft MOU, February 2004, https://www.productstewardship.us/resource/resmgr/electronics/NEPSI-MOU-FEB-2004-DRAFT-WEB.pdf, PSI archives.

In 2005, Maine became the first state to pass a law based on producer responsibility. Since local governments assumed collection responsibility, it was called "partial producer responsibility" in the United States. Maryland passed the third electronics law, which was partly funded by a manufacturer registration fee. When Washington became the fourth state with an electronics recycling program fully funded by producers, it signaled a clear national trend toward full producer responsibility. By 2018, 26 states passed electronics recycling laws, with only the first one being an ARF, while the rest are some version of producer responsibility. Several, however, only cover computers and not televisions.

The NEPSI process laid the foundation for each of these state laws, as most of the participants learned about the positions and interests of manufacturers, retailers, recyclers, and other stakeholders through the many NEPSI meetings. Every aspect of the electronics end-of-life management issue was discussed during these meetings, including the scope of products to be covered, reuse and recycling infrastructure, financial payments to provide incentives for collection, financing system options, environmentally sound processing standards, performance goals, creation of a stewardship organization, and federal preemption of state laws.

As a result of the numerous NEPSI interactions, strong relationships formed among stakeholders that led to important spin-off projects and initiatives outside the process. For example, NEPSI inspired the development of the Electronic Product Environmental Assessment Tool (EPEAT), which is a global eco-label for the IT sector. As described on its website, "EPEAT helps purchasers, manufacturers, resellers, and others buy and sell environmentally preferable electronic products. EPEAT was developed using a grant from US EPA and is owned and managed by the Global Electronics Council (GEC)."[129] In addition, Northwest Product Stewardship Council member agencies that participated in the NEPSI process launched the Take it Back Network, comprised of small retailers and other businesses that provided collection service and demonstrated the convenience and practicality of retail take-back programs.

To help fill a void by an absence of retailer participation in NEPSI, US EPA initiated a series of key retail pilot projects with Staples, Office Depot, and Good Guys that laid the foundation for retail participation in electronics collections. These projects were part of the EPA's wider Plug-In To eCycling program, which was a partnership between the EPA and consumer electronics manufacturers and retailers to offer consumers more opportunities to donate

129. Electronic Product Environmental Assessment Tool (EPEAT), US Environmental Protection Agency, accessed July 7, 2021, https://www.epa.gov/greenerproducts/electronic-product-environmental-assessment-tool-epeat.

or recycle their used electronics. As mentioned in chapter 3, Staples became the first retailer to provide significant electronic recycling services in 2007 at all of its stores nationwide after working with PSI and the Take it Back Network to pilot the program.[130] Finally, the National Center for Electronics Recycling was formed to help fulfill information needs identified by NEPSI participants. These examples are only a sample of the many activities that built on the knowledge and relationships developed through NEPSI.

This first national product stewardship dialogue has become a reference point for new initiatives on other product categories, as well as a background for continued work on national electronics legislation. While states without electronics legislation continue to push for their own state laws, many groups continue to seek federal legislation that will bring electronics end-of-life management systems to all states in the country, In addition, many who participated in NEPSI have provided key leadership and experience on other emerging product stewardship initiatives.

STATUS OF EPR PROGRAMS IN THE UNITED STATES AND GLOBALLY

The NEPSI dialogue was the center of the US product stewardship movement for three years, and many of PSI's government members were engaged in the dialogue. In fact, the effort was so resource intensive that I could barely get PSI members to join me in any other product stewardship efforts. There were so few government officials skilled in EPR at the time and, with their other work duties, most officials could only work on one product stewardship initiative at a time. I had to draft new government colleagues for each new product on which we worked. Electronics certainly was not the only product on which we worked during PSI's early years. In fact, one of our first projects following electronics was a three-year contract with the US EPA to reduce the environmental impacts from industrial and mobile radioactive devices, which also covered tritium exit signs, smoke detectors, and other radioactive devices.

PSI worked on other products during that time, each one offering new challenges with different industry sectors and different stakeholder groups, all with a desperate need for a solution. During these early years of the product stewardship movement, we sometimes sought to engage an industry in voluntary initiatives. One such project was for the Florida Department of Environmental Protection (FL DEP), which sought a safe recycling solution

---

130. For information on PSI's Staples Pilot Project, see http://www.productstewardship.us/displaycommon.cfm?an=1&subarticlenbr=72.

for pressurized gas cylinders used for barbecue tanks and camping stoves, which often explode in recycling and waste management facilities. The gruesome stories told by recyclers about worker injuries was motivation for all of us to develop a solution. The gas cylinder industry was heavily engaged with our group through many fruitful dialogue meetings, and we came up with viable solutions including prototype equipment that punctured and crushed cylinders to render them visibly safe for recycling.

Unfortunately, with no legislative pressure from government or other stakeholders, the project fizzled. In 2020, though, I got a phone call from one of PSI's board members, Jen Heaton-Jones, Director of the Housatonic Resources Recovery Authority in Connecticut, who wanted to develop an EPR bill on pressurized gas cylinders. It was a pleasure to dig out my files from 2001, which gave us a big jumpstart to develop the first gas cylinders EPR bill in the country, which was introduced in the 2021 session and finally enacted into law in 2022.

While PSI was heavily engaged in electronics, gas cylinders, and radioactive devices, we began to research a product on which I knew a great deal, paint. Not only did I paint houses for several summers, but as Massachusetts waste policy director I also developed a project with Benjamin Moore. After receiving regulatory relief from our agency, the company collected leftover Benjamin Moore brand paint only, which had been dropped off by community members at municipal HHW sites that had received paint storage sheds through a grant with our agency. This postconsumer paint was then tested, filtered, processed, and reblended with the company's own virgin paint at the company's Milford, Massachusetts plant. The paint was sold as virgin paint. This was one of the first product stewardship initiatives on which I worked. The relationship that I built with Carl Minchew and Van Stogner at Benjamin Moore, whose idea this was, later allowed me to turn that small project in Massachusetts into a national initiative. Minchew became a key paint industry supporter in engaging with PSI when I switched from the state job to the nonprofit. The national Paint Product Stewardship Initiative, described in the case studies section, emerged from this small project.

Project by project, the product stewardship movement was built. In the midst of the meetings on paint, PSI worked on mercury thermostats and lamps, and later batteries, which is also covered in the case studies section. Next came phone books (remember them?), junk mail (why do we still get this?), and tires. Over time, the first national EPR wave of interest—on electronics—started to produce dividends with EPR laws, after which the second national EPR wave—on paint—resulted in a national model that led to the passage of paint EPR laws. The paint wave of new laws had barely gotten off the ground when yet another major national product wave was building momentum. Pharmaceutical waste, which began as a water quality issue detected by the US Geological Survey and the US EPA, quickly gathered steam with the tragedy of opioid addiction.

Over the past two decades, PSI has worked on over 20 product categories, developing deep knowledge and expertise about each component of the waste stream that our government members identified as a top priority. Each of the products has been in play in some states around the country at various times. Mattresses, carpet, medical sharps, HHW, and textiles all had their starts and stops throughout the years, with solar panels being the latest product to raise its head in need of a solution. The product that has, after 15 years of PSI effort, ushered in the tsunami of product stewardship, however, is packaging. That product has finally burst open the US door to EPR and, it too, has a spot in chapter 10 in the case studies section.

I have recounted the product work of PSI's staff, board, and members because a movement requires strong and persistent effort. A movement is not a campaign. It is not enough to pulse interest—one year working on one topic and another year working on another topic. The product stewardship movement has been successful because there has been sustained effort by numerous state and local governments, working together with environmental advocates and enlightened industry members to pass EPR legislation and advance voluntary initiatives that lead to real change.

The US EPR movement began with a focus on toxics and bulky products, particularly batteries, electronics, mercury thermostats and lamps, paint, pharmaceuticals, medical sharps, carpet, and mattresses. We have come a long way. In 2000, when PSI was created, six states had passed eight EPR laws, all on one product—household batteries (containing mercury, cadmium, and other toxics) (see figure 5.2). By June 2023, 133 EPR laws on 17 products had been enacted in 33 states, with four on packaging (see figure 5.3). Of those 133 laws, 106 were passed at the state level and another 27 at the local level (mainly as a political strategy to enact state legislation).

The past two decades have seen a steady increase in the passage of EPR laws in the United States (see figure 5.4). While the United States first passed EPR laws on toxics and bulky wastes, Europe started with packaging. With the OECD's broad definition of EPR, many systems not considered EPR in the United States, such as deposit return systems, advanced recycling fees, and taxes, are included in the more than 400 EPR systems operating globally as of 2013, the latest data available (see figure 5.5).[131] Therefore, any comparison with US EPR programs must be done carefully.

131. OECD, "Extended Producer Responsibility: Guidance for Efficient Waste Management" (September 2016).

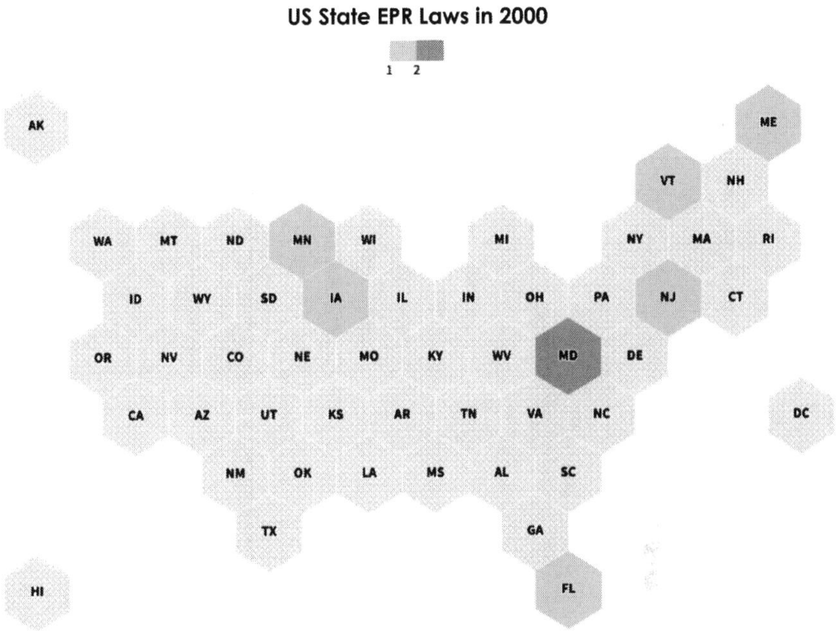

Figure 5.2 US State EPR Laws in 2000
Source: © 2023 Product Stewardship Institute, Inc.

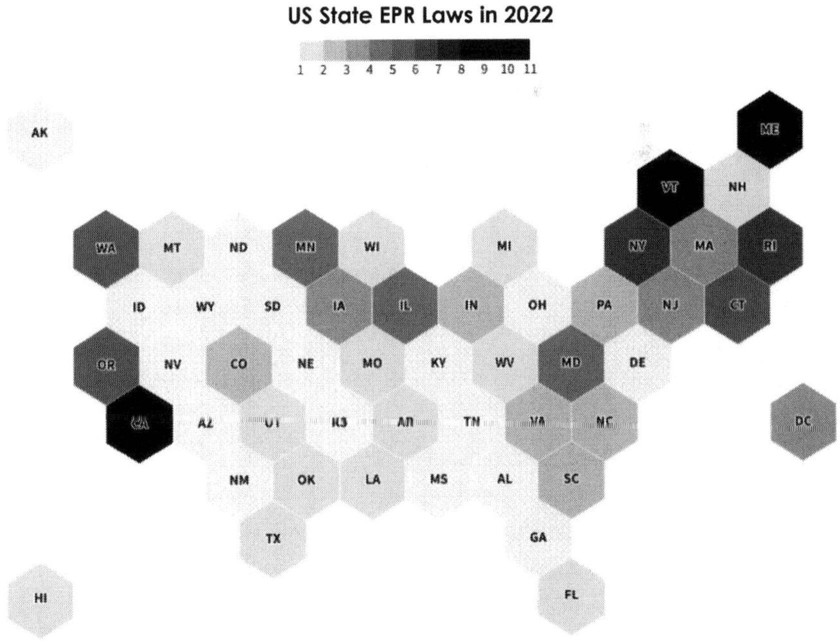

Figure 5.3 US State EPR Laws in 2022
Source: © 2023 Product Stewardship Institute, Inc.

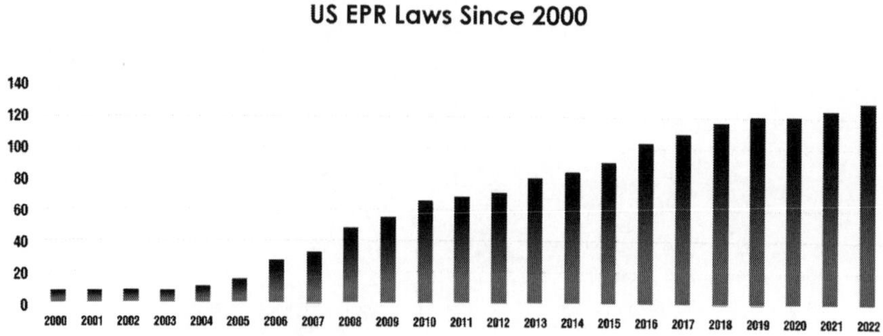

Figure 5.4 US EPR Laws since 2000

Source: © 2023 Product Stewardship Institute, Inc.

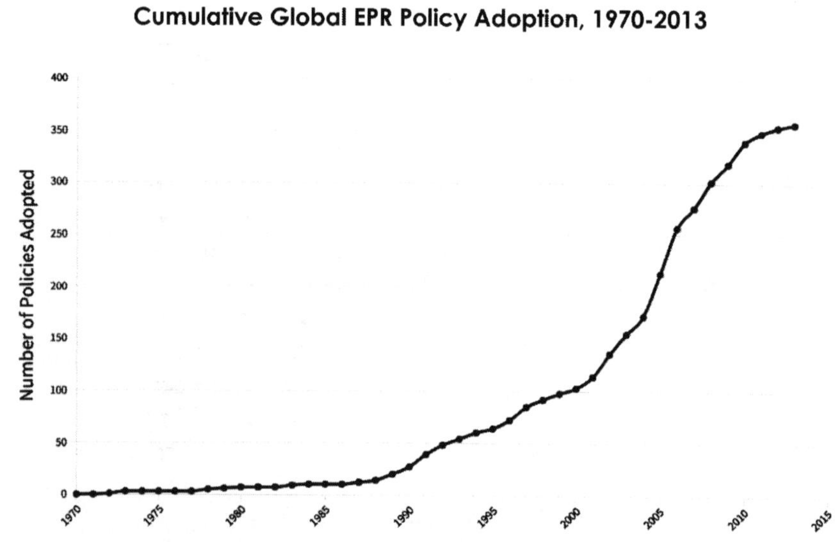

Figure 5.5 Cumulative Global EPR Policy Adoption, 1970–2013

Source: © 2023 Product Stewardship Institute, Inc. Based on data from "What Have We Learned about Extended Producer Responsibility in the Past Decade? A Survey of Recent EPR Economic Literature," Organisation for Economic Cooperation and Development (OECD), 2013, Paris.

Chapter 6

Creating Effective EPR Laws through a Collaborative Stakeholder Process

The will to change a law comes from a desire to solve a problem. Changing laws requires technical expertise with an added measure of passion. Although I was raised to conserve resources, it was a personal habit, not a mission. While I loved the Jersey Shore and the mountains of upstate New York, I did not think of changing laws to protect those areas. It took a trip across the United States and into Canada for me to be transformed into an environmental advocate. After two years of college studies in the humanities, Jack Kerouac beckoned me to a gap year on the road with Jackson, my beloved bearded collie mix. Meeting Niagara Falls, the Grand Canyon, Great Sand Dunes National Monument, Death Valley, and the Pacific Ocean awakened in me an appreciation for the natural environment that John Muir and Rachel Carson knew like their own lifeblood. Spending six months in California opened my mind to antinuclear rallies, recycling, alfalfa sprouts, and a sensitivity to interacting with the environment.

With a newfound passion to protect nature, I went back to college, changed majors to geology and environmental science, joined the recycling club, and embarked on an environmental path. In my first job with the Pennsylvania Environmental Council, a moderate statewide environmental group, I was responsible for convening seminars of about 30 people from businesses, governments, and advocacy groups to discuss challenges and solutions to waste management issues. The seminars were designed to educate participants. I developed the agendas, secured and introduced speakers, brought the whole wheat muffins, and was the first to come and the last to leave. The participants and I learned about waste management laws, problems, and potential solutions. I discovered that each of their varied viewpoints represented different perspectives that made sense to me individually but felt limited as a whole. I could see the pieces clearly, but the full picture was unfocused. I had a limited ability to help solve the problems they identified although I understood the solutions they proposed.

Around this time, just before Microsoft Word software was installed on our one shared office computer, I received a flyer about a national conference on environmental dispute resolution to be held in Washington, DC, and got permission to attend. Of all the luminaries who spoke, MIT professor Lawrence Susskind, the dynamic keynote speaker, caught my attention. Professor Susskind was leading a new field of environmental mediation that brought together industry, government, and environmental advocates to jointly solve seemingly intractable environmental problems. I was hooked and eventually attended MIT and studied with Professor Susskind. About four years after graduating, I was finally able to directly apply my training in consensus-building for environmental issues when serving as waste policy director in Massachusetts. With my colleague John Fischer, we successfully mediated used oil recycling legislation with a multi-stakeholder group. The process used for that mediation became the basis for the stakeholder engagement model that I developed at PSI.

PSI'S ORIENTATION

PSI allowed me the opportunity to solve technical waste management problems by using a process in which I had confidence. I also had the freedom to test approaches with extremely smart and dedicated staff, as well as with colleagues throughout the country who had a deep understanding of the waste problems in their municipalities and states. To build an understanding of PSI's approach to developing policy, we need to investigate the roots of the organization. In 2000, there was no national organization promoting a producer responsibility approach to waste management issues that represented the interests of state and local government agencies. By coordinating highly knowledgeable officials from across the country at hundreds of in-person and online meetings, PSI provided a national context to local issues. Although local differences require unique approaches based on culture, politics, geography, and other factors, there are also many similarities. For example, while recycling cost increases due to China's import restrictions on US recyclables varied by municipality, most increases were significant and represented a nationwide trend. The enormous financial impact on municipal governments called out for a national solution, and PSI was able to funnel widespread discontent into supporting packaging EPR policy as a central part of state and federal legislation.

Since PSI has represented state and local government interests over the past two decades and has a government board, our engagement process starts by first understanding government interests and perspectives. Equally important, mediation training taught me that the most sustainable agreements are ones that include all those with a significant interest in the outcome of an initiative. For example, when addressing the environmental impacts and management costs of leftover paint, it is essential to not only have governments at the table but also virgin and recycled paint manufacturers, retailers, waste management companies, and environmental groups. The one caveat is that *all* participating stakeholders must *want* a solution. In PSI's case, that means a producer responsibility solution aligned with our mission. We do not want to include a disruptive stakeholder for whom the status quo is their goal. The traditional government method of developing policy solutions has been to create a draft policy, put it out for formal public comment, incorporate comments as desired, finalize the policy, and respond to comments in writing to defend the decision. The method we use for developing EPR policy at PSI incorporates what I was taught at MIT—melding the interests of all key stakeholders through a structured dialogue process.

PSI's approach to stakeholder engagement begins by conducting research on the problems associated with a particular product. This step will include online research that involves reading available reports and, more importantly, conducting interviews of key stakeholders, like producers, recyclers, government officials, and other experts. We then design and convene a well-defined multi-stakeholder dialogue. Our goal is to seek consensus by applying the same techniques used by neutral facilitators and mediators. But PSI is not neutral. We are, well, unique. Our organization advocates for producer responsibility, yet we are open to a range of possibilities as supported by our government members. We blend advocacy with a desire to solve solid waste management problems jointly with multiple parties. We design processes that bring different voices together in a professional setting. We meld deep technical subject expertise with process knowledge. We educate stakeholders but let stakeholders also educate each other using their own unique experiences and perspectives. We seek to develop solutions that are progressive while also being possible.

As a nonprofit organization with a state and local government board, PSI walks a fine line between representing our members' interests and also knowing that these officials represent political positions and must speak for themselves. Governments might lean liberal politically in one state and conservative in another. The political orientation of each state can also change, as our democracy allows, being liberal one year and conservative the next. These factors, however, do not influence PSI's orientation when convening a dialogue. Whether convened by PSI or a neutral facilitator, agencies must speak for themselves, as must all stakeholders. PSI's role is to provide a forum for all voices to be heard. If an agency hires PSI to convene a dialogue, the agency will make its interests clear within the context of the forum. PSI's understanding of each stakeholder's interests, particularly those of governments, helps the group to achieve consensus.

However, due to the limitations of a single perspective (in PSI's case, it had a government-only membership base), PSI created a partnership program in 2007 so that other stakeholders aligned with our product stewardship mission—companies, environmental groups, academic institutions, and international governments—could voice their perspectives within the organization's work. One of the biggest values for corporate stakeholders working with a national organization like PSI is that a negotiation can take place in one forum rather than having to negotiate the same issues repeatedly in 50 states.

PSI has coordinated numerous state and local agencies, as well as other stakeholders, throughout the country since 2000, leading to the development and passage of many, if not most, US EPR laws. While we develop the policies that form the basis for many of these laws, we work with thousands of others in a growing chorus of technical experts, advocates, and corporate influencers. A movement like product stewardship and EPR takes an immense number of people, who create their own innovative approaches to connect allies, educate important players, advocate for bill passage, and ensure program effectiveness. While PSI has done a great deal of education and lobbying in conjunction with others, our key role has been the development of EPR policies that incorporate best practices from Europe and Canada into a US context. Here's how we do it.

PSI COLLABORATIVE STAKEHOLDER PROCESS[1]

Professional facilitators and mediators will tell you that no meeting or project is exactly the same. Each situation requires a different approach. However, there are basic techniques that all of us apply in designing a meeting, a series of meetings, or a wider, multifaceted project. I have come to think of this process as akin to ballet choreography, not because I know anything about ballet (other than its beauty and fluidity of motion), but because it includes a structured set of elegant steps to elicit an emotional experience. The choreography that we use at PSI is as structured as ballet but also includes a hefty dose of improvisation. Over the past two decades, we have refined our basic process to account for the growing knowledge of EPR among the stakeholders, as

1. This section borrows heavily from an unpublished PSI publication: Scott Cassel, *The Dynamics of Dialogue: Lessons Learned from the US Product Stewardship Movement* (July 22, 2011).

well as the inevitable limitations of time and funding that do not always allow for a complete process.

Selecting Priority Products

A first step is to broadly figure out which problem(s) you want to solve. If you woke up with a headache in the morning, it is likely that your main focus would be to alleviate that pain. If you are in the solid waste business, the entire waste stream presents a headache, so you have to choose which challenges contribute most to the perpetual throb. Having so many waste problems can be overwhelming. In addition, everyone has limited time, capacity, and resources. PSI works on problems brought to us by our government members and corporate, organizational, academic, and non-US government partners.

When faced with multiple pressing problems, one of the first steps we take with groups is to determine their priority products. Often, these products are considered a priority because of their toxicity, volume in the waste stream, handling difficulty, pollution impacts, and management cost. To an increasing degree, product life-cycle impacts will be a factor to the extent there is agreement among stakeholders about how the calculations are determined and that the results are perceived as unbiased. In addition, funding availability, industry willingness to cooperate, and the degree to which the product represents a problem for multiple stakeholders are all key factors in selecting priority products. There are numerous methods we use to determine a group's priorities, including formal surveys, one-hour facilitated discussions, and all-day educational forums. The purpose of these exercises can range from developing a six-month work plan for a product stewardship council to drafting a policy position for a state solid waste master plan.

Over the past two decades, nearly every product that enters the solid waste stream has reared its head as a problem. As a national organization, we seek to address issues that are problems for a large number of state and local agencies, since multiple state bills create leverage for national change, particularly among recalcitrant industries. Owing to PSI's large government membership (47 state environmental agencies and hundreds of local government entities), an increasing number of manufacturers have worked with PSI in hopes of developing a national model rather than risk state-by-state legislation that could create a patchwork of laws. At other times, PSI will work with one state to pass a strong bill that becomes a precedent, and often a model, for other states.

Since 2000, five product sectors have garnered national attention—electronics, thermostats, paint, pharmaceuticals, and packaging. The first product sector considered a national priority by governments and others was electronics, and 26 state electronics EPR laws were passed from 2003 to 2015, of which 24 are considered EPR by the US definition established by PSI.[2] Significant government interest focused next on mercury thermostats with 13 EPR laws passing from 2006 to 2014, followed by leftover paint with 11 EPR laws passing from 2007 to 2023, then followed by 31 state and local EPR laws on pharmaceuticals signed into law from 2012 to 2023. Over the past few years, multiple state governments have been intent on passing EPR laws on packaging and paper products (PPP), which represent 40 percent of the waste stream. In fact, Maine and Oregon achieved this goal of passing the country's first EPR laws for PPP in the summer of 2021, with Colorado and California joining them in 2022.

Interest in addressing waste problems in the United States has expanded and contracted over time based on domestic and international economic, environmental, and political factors. The same holds for various product sectors. Interest in addressing electronics, thermostats, paint, pharmaceuticals, and packaging has grown over time like swells in the ocean. National interest to address each product has increased like a large wave building in the deep ocean until it is ready to come to shore. PSI watches for opportunities to enhance these swells of interest. For example, we started working on pharmaceutical waste in 2005 after I heard Charlotte Smith, founder of PharmEcology Services, WMSS, speak about the dangers of the ubiquitous presence of pharmaceutical compounds detected in our nation's waterways by the US Geological Survey. Our interest in this product category grew with published US EPA studies of aquatic impacts[3] from pharmaceutical compounds and turned into an explosion of concern when pharmaceuticals stored in medicine cabinets were linked to the opioid crisis.[4]

These five products have received national attention because they represent profound problems that touch individuals in the greatest number of states. PSI's response to this interest has been to build a foundation for problem-solving based on thorough research, deliberate dialogue, and persistent advocacy. We have worked on these five products all throughout our history, with interest in some products, like electronics and thermostats, tapering off nationally after a prolonged surge, while others, like packaging, linger at a lower level for up to 15 years until an incident triggers widespread action. China's restriction on importing recyclables changed the packaging scenario,

2. California's program, passed in 2004, is considered a product stewardship law, not an EPR law, because it is funded by consumers through an advanced recycling fee and managed by the state agency. Utah's program also does not fit the PSI definition for an EPR program.

3. Water Science School, USGS, US Department of the Interior, "Pharmaceuticals in Water" (June 6, 2018), accessed March 8, 2023, https://www.usgs.gov/special-topics/water-science-school/science/pharmaceuticals-water#overview.

4. "Prescription Drug Take-Back Day: The Importance of Clearing Out Your Medicine Cabinets," Addiction Prevention Coalition (April 25, 2022), accessed March 8, 2023, https://apcbham.org/2022/04/25/prescription-drug-take-back-day-the-importance-of-clearing-out-your-medicine-cabinets.

and now packaging brands and waste management companies are scrambling to develop a plan for the inevitable tidal wave of EPR laws that is upon us. Interest in other products, like mattresses, carpet, medical sharps, solar panels, tires, and gas cylinders has bubbled up repeatedly in a smaller number of states, with EPR laws being passed at various times based on opportunity. The importance of prioritizing stakeholder interests is that, when the opportunity does come for change, stakeholders will be ready to move forward, as PSI was when a model packaging bill was needed.

The Project Team

Every initiative needs direction before others are asked to join. Although PSI might be the engine to guide, coordinate, and propel a group to address a priority product, those with whom we work provide specific on-the-ground knowledge and political direction. Together, we form the project team, which is the core group—often involving a public or private sector client—that will work with us to conceptualize the project design. The project team functions like an executive committee that meets regularly during the course of the project to develop, assess, and adjust strategy, and keep the project on track.

To help a client identify and define the direction for the project, PSI starts by asking the project team about the *problem* they are trying to address, the *goals* they want to achieve, the specific products that are the *focus* of the project, the *barriers* to achieving the goals, and viable voluntary and/or regulatory *solutions* they see for overcoming the barriers and achieving the goals. These are the five key foundational pillars on which the dialogue will be built—the problem, goals, focus, barriers, and potential solutions. PSI staff draft preliminary statements about each of the foundational items for the project team to respond to, and edit, in preparation for discussion at a kickoff meeting. I have found that it is much easier for people to react to draft statements rather than provide information from scratch. Using this simple technique has been more efficient than seeking input on key aspects without providing draft text in advance. At the kickoff meeting, I usually ask the project team to envision themselves at the end of the process, pleased with the result—what would that result be? That is the outcome we want to actualize. This is also the point at which the client determines whether they seek a voluntary or a regulatory solution. Since we work with governments in progressive and conservative states, this is an important question.

A project needs a stable foundation on which to build, and the project team creates that foundation for a dialogue. All team members need to be firmly aligned at the start and remain united throughout the dialogue. Otherwise, tensions and differences among core project team members can confuse other stakeholders, and sometimes be exploited. Alignment does not mean that team members must be in absolute lockstep. They can have different opinions, as long as the range of opinions is articulated in a way that is logical to other stakeholders.

The core project team will meet repeatedly at different stages of the dialogue process to articulate and refine their positions and range of opinions clearly. During PSI's national paint dialogue, I often caucused with our core group of government officials before meetings to make sure the presentation of their interests would be perceived as cohesive by paint industry participants. Prior to one multi-stakeholder dialogue meeting, the all-government project team identified three state and local officials who would outline the problems they had with leftover paint. The clarity of their unified presentation helped gain the paint industry's support, leading to an ultimate solution.

After PSI helps the project team articulate its project vision regarding the problem, goals, focus, barriers, and solutions, we will understand the additional information we need to obtain to engage other stakeholders in the dialogue. PSI staff will seek information such as the number of producers, their relative market share, and the amount of material sold; the postconsumer management landscape (e.g., existing reuse and recycling facilities and consumer opportunities); existing regulations that govern the targeted product and any barriers these present; existing outreach materials; and other basic information that all stakeholders need to know to begin a dialogue. This new information, along with the five foundational items, is compiled into a short project summary.

Stakeholder Engagement

Attracting the right stakeholders to the dialogue at the right time is necessary for achieving a successful outcome. At the kickoff meeting, the project team identifies the stakeholders who are essential to include in discussions, and at which stage to include them. At the outset of the project, PSI creates a project contact database that is expanded over time as more stakeholders are brought into the dialogue. PSI uses this database to communicate with the growing stakeholder network throughout the process. Together with the project team, PSI assesses the stakeholder landscape and designs a stakeholder engagement process, which constitutes much of the project design and strategy.

Adding stakeholders, along with their unique perspectives on the problem, goals, focus, barriers, and potential solutions, is like adding ingredients to a food recipe. If done well, it can add texture, depth, and subtle complexity to a dialogue. But if stakeholders are not integrated deliberately and strategically into a policy dialogue, they can create instability. To maintain dialogue stability, PSI interviews each new stakeholder individually or in small groups of like-minded stakeholders. We start with allies, which usually means other state and local government officials who are most aligned with the project team's approach to addressing the targeted priority product. We will then interview others with important perspectives, including producers, collectors, recyclers, and environmental groups.

In the early dialogue phase, we usually interview about 15 people from multiple stakeholder groups, or as many as are needed to further clarify each of the five key foundational aspects. For PSI's first national dialogue, on leftover paint starting in 2002, we conducted 37 interviews with representatives from paint manufacturers, retailers, government agencies, trade associations, recyclers, container manufacturers, and others. For our dialogue with the Flexible Packaging Association in 2019, we interviewed more than 25 stakeholders, speaking with company representatives separately but with government officials as a group. By contrast, in 2020, the second time we worked with gas cylinder manufacturers (following an extensive multi-year effort starting in 2002) we received input from a two-person government project team followed by one call with a dozen local agency officials. We also interviewed two (of only three) gas cylinder companies that manufacture cylinders of the size covered by our dialogue, along with their association representative.

Prior to conducting the first interview after the project team kickoff meeting, PSI sends the stakeholder the draft project summary, which includes the five foundational items along with additional background research. The stakeholder interviews will test the degree to which the framing of the project team's foundational pillars is acceptable to others, or whether an adjustment is needed. Providing the draft project summary in advance also allows dialogue participants an opportunity to consider whether they agree with the technical information and to be prepared to discuss suggested changes.

These interviews set an early tone of collaboration. The process of seeking each person's unique perspective on foundational aspects of the dialogue along with their technical expertise, and confirming with them that we have incorporated it accurately, helps build individual and group trust. The process allows questions to be raised, concerns to be aired, and joint problem-solving to take place. These interviews are also important to assess the degree to which stakeholders are willing to engage in the type of collaborative dialogue proposed and whether they will seriously consider a producer responsibility solution. In the early days of the product stewardship movement, to start a dialogue we would have included anyone genuinely seeking a product stewardship solution. Now it typically means those supportive of an EPR approach.

By interviewing each stakeholder, we will be able to assess the degree to which consensus is possible with each stakeholder and who will be needed to develop that consensus. In developing a solution for pharmaceutical take-back, for example, it was a clear sign that the industry's trade association,

Pharmaceutical Research and Manufacturers of America, never responded to repeated requests to talk, although several of their members did speak with us. We knew early on that any solution would have to be developed without the support of the industry association. This stakeholder assessment phase of a dialogue is extremely important for the project team.

Project Summary, Briefing Document, and Action Plan

A good discussion relies on participants having baseline knowledge about the topics to be discussed. Any discussion will include those who know a lot about the topic and those who know little. Our goal for a dialogue is to educate all participants, giving them the same basic knowledge so they can actively participate, which increases the group's ability to achieve consensus. In PSI's early days, when stakeholders were less aware of the impacts of products, and product stewardship as a solution, PSI did extensive research, including literature reviews and numerous in-depth interviews. Our background briefing documents, or "Action Plans" as we called them, could be 60 pages or more. We would also open our meetings to many stakeholders to educate them about the initiative and build support for a solution. Many meetings included well over 100 people, with about a third or more of the participants piped in through a star-shaped conference phone strategically placed in the middle of the room (long before video technologies were mainstream).

Now, with greater understanding among stakeholders about product problems and EPR, we do not have to conduct the same level of research before engaging stakeholders in dialogue. In fact, in many cases, PSI might develop only a 5- to 15-page briefing paper before engaging stakeholders and, in other cases, we might move right to developing EPR legislation. The degree of speed with which a dialogue takes place depends on the complexity of the product, the degree of prior stakeholder outreach and knowledge, available funding, political momentum, and a host of other factors. The best approach is one that has clear expectations, strong political support, and moves steadily and deliberately toward the goals set forth by the project team and other allied stakeholders.

This steady and deliberate approach uses the project summary as a "single text" document that is continually revised and expanded after each subsequent stakeholder interview. The project summary prepares stakeholders for the dialogue by giving them an opportunity to contribute to the articulation of the five foundational pillars of the dialogue as mentioned above: (1) *problems*, (2) *goals*, (3) *focus*, (4) *barriers*, and (5) potential *solutions* that will overcome the barriers and achieve the goals.

Writing this document is a process that unfolds over time. It is akin to how a crystal grows as atoms combine. PSI starts with our own understanding of the five foundational dialogue pillars, then integrates perspectives of the project team, and then, one by one, integrates input from each stakeholder. Initial information is revised continually and builds on itself as new information is acquired through interviews and research, gradually over time. As the project summary is refined, the project details come into sharper focus. As each stakeholder adds their input to the summary, they become invested in the project, particularly after seeing document changes that reflect their own unique viewpoint. The process is not simply additive but *integrative.* If a new stakeholder offers comments that are not aligned with the existing text, PSI resolves these by speaking with the relevant stakeholders. The revised text must make sense and be consistent, even as the text expands and becomes more nuanced.

By taking this approach, PSI gains trust with stakeholders while also ensuring the accuracy of information. It also builds a desire within the group for a successful outcome, since each stakeholder now gains a sense of ownership of the dialogue. During PSI's first 15 years, we typically expanded the project summary into a more extensive product stewardship action plan or briefing document. Over time, with information more readily available and a greater understanding of product stewardship, this step has become less necessary.

The incremental engagement process described above allows a facilitator to have greater assurance about the degree of stakeholder alignment on the most basic foundational aspects of the dialogue. It is important, before bringing the dialogue group together for its first meeting, to gain a high degree of confidence that the group will agree on the basic problem, goals, focus, barriers, and potential solutions. If at all possible, it is best to interview all key stakeholders before meeting together as a group. Doing so will decrease the chance that a stakeholder will throw a curve ball that can take the meeting off course. Bringing people together to discuss contentious issues is always unpredictable. There will usually be circumstances that arise for which facilitators will need to "think on their feet," or "call an audible" as former PSI board member Andrew Radin always said. Decreasing the chances of distraction from an unexpected stakeholder comment will serve the group well. Talking to every major player before the first dialogue meeting is one way to have greater certainty that the dialogue will stay on course.

Starting a first meeting with agreement on five foundational items that can fit on five presentation slides creates significant group momentum. It is always best to achieve a general agreement on these items *before* getting the group together and then *confirm* this agreement during the meeting. It provides a feeling of progress and confidence in the process and encourages the group to engage further on actual solutions. If additional discussion is needed at the first meeting to reach consensus on these five items, PSI takes some time to do so, but we won't tie up the group.

This approach conforms with PSI's 80/20 rule: 80 percent of a dialogue effort should take place outside the meetings and the remaining 20 percent will include meeting for key decisions that require stakeholder interaction. A facilitator has more opportunity to manage a process by strategically building a level of consensus outside a meeting than by relying solely on the meeting itself to develop that consensus. Otherwise, there is greater potential for dissonance to occur among group members during a meeting, which could result in uncertainty and a lack of confidence in the process. Repeated uncertainty can undermine an entire initiative. This is another reason that PSI seeks common ground on the first four foundational aspects (i.e., the problem, goals, focus, barriers) before discussing the fifth aspect.

To achieve a successful outcome, stakeholders should have common expectations about each step in the dialogue process. It is good practice for a facilitator, at the outset of a meeting, to clearly lay out the steps the group will take during that meeting. It is also important to lay out the work a dialogue group is expected to do together during a multi-meeting dialogue, both during the meetings and between meetings. Figure 6.1 shows the dialogue process map I developed for presentation at our first paint dialogue meeting in 2003 (see Chapter 8, "Paint Stewardship: A US Case Study"). Since we conducted extensive interviews and research, I was able to lay out the actual workgroups for each stage in the process, which were confirmed or modified at each subsequent meeting. By providing a group with clear expectations, and then

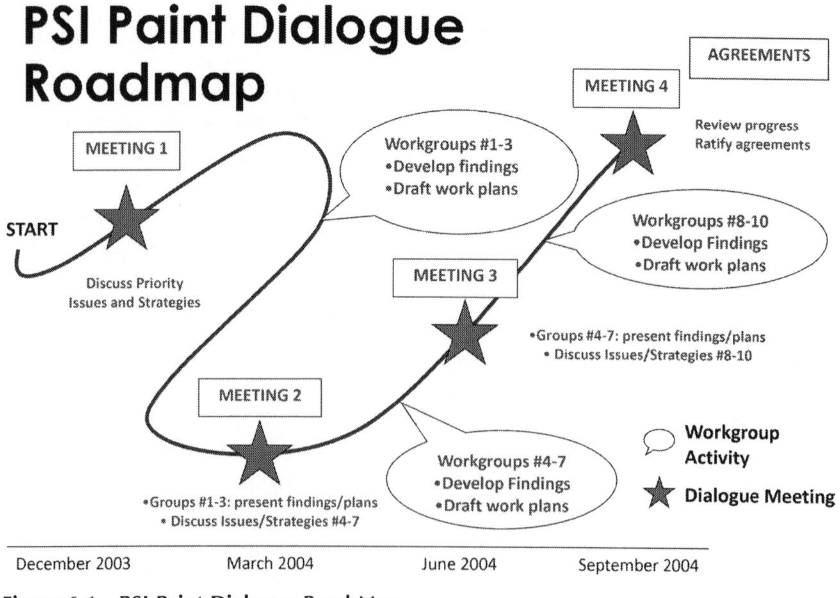

Figure 6.1 PSI Paint Dialogue Road Map
Source: © 2023 Product Stewardship Institute, Inc.

meeting them, facilitators will gain the group's support, which is critical to reaching agreement on difficult topics.

The four full stakeholder meetings were supplemented by multiple workgroup calls. The entire process was completed in nine months before stakeholders signed the first paint memorandum of understanding (MOU), which was signed in 2004 by 29 stakeholders, including US EPA, American Coatings Association, and many state and local governments. As described in chapter 8, the paint case study, that first agreement committed stakeholders to jointly develop 11 projects, including source reduction, reuse, recycling infrastructure, recycling market development, life-cycle assessment comparing latex paint disposal to recycling, and development of a recycled latex paint quality standard. Those projects, eight of which were eventually completed, took two years to complete and led to the creation of a model paint EPR bill and a second MOU in 2008. That second agreement was the basis for the paint EPR laws currently in 10 states and the District of Columbia.

Problem Statement

The foundation for any dialogue is having a common understanding of the problem and associated opportunities related to a priority product. Without a group agreeing on the shared problem they want to solve, their efforts to move to a solution will rest upon a poor foundation. It is important for all stakeholders to agree on why a particular product needs a product stewardship solution. It is necessary for each key stakeholder to not only articulate their own perception of the problem, but also to agree that someone else's perception of the problem is legitimate. The dialogue group needs to agree that all aspects of the problem should be solved, even ones brought up by another party. The individual problems must be melded into one singular problem statement to which all dialogue participants agree.

A problem statement is a short paragraph or set of individual statements that succinctly capture the reasons why the problems related to a product category are a priority. Standard aspects of nearly all product stewardship problems include a lack of consumer awareness about the problem and how materials are collected, lack of collection opportunities and recycling infrastructure, and significant cost burden on government. For example, the opening sentence for PSI's problem statement on leftover paint developed in 2003 stated, "Paint is a top concern based on its high volume in the waste stream, subsequent costs to manage, and high potential for increased recovery, reuse, and recycling."[5] The problem statement agreed to as part of PSI's mediated

5. Product Stewardship Institute, "Product Stewardship Action Plan for Leftover Paint" (Problem Statement, March 18, 2004), 2.

agreement with the Flexible Packaging Association in 2019 was more extensive, including 13 statements (counting sub-aspects) that covered items such as the following three statements: (1) "Flexible packaging is prevalent and visible in the waste stream and as litter, leading to aesthetic impacts, municipal costs, and ocean debris; (2) [Flexible packaging is] sold in countries lacking management infrastructure (global concern); and (3) Producers and consumers do not bear the true life-cycle costs of the goods they buy (true of all products)."[6]

Focus

Another key aspect to developing EPR bills through stakeholder engagement is to define the boundary of the materials the group wants to manage. We call this a "product focus," which includes what are called "covered materials" in EPR legislation. These are the consumer products on which producers (or sometimes consumers) pay a stewardship fee and which are collected, reused, recycled, or safely disposed of as part of the EPR system. The product focus ideally also includes the entities from which covered products must be collected at no cost to the waste generator, known as "covered entities" in EPR laws.

There are always limited resources in terms of staff time and finances, and these limits should be considered at every step of the process. At the start of the product stewardship movement, PSI urged stakeholders to narrow a product problem to the issues they believed were most urgent to address in the near term. For example, the sentence that helped focus PSI's 2011 national dialogue on mattresses stated, "This project is focused on mattresses and box springs from residential sources (including both single and multi-dwelling units), as well as large-scale generators such as hospitals, hotels, universities, military, and other institutions."[7] By defining the type and source of discarded mattresses to be managed, all stakeholders could size up the amount and source of waste products to be managed under an EPR law. Doing so made it more possible to find a negotiated solution.

Now, due to the growing global understanding of the concept of a circular economy and the critical need to recover all materials for reuse and recycling, the focus for each priority product category has widened considerably since the movement started in 2000. This growth in the sustainability movement has already begun to challenge the ability of producers to respond. Just as the concept of EPR was, at one point, new to US producers, the expansion of

6. *Shared Elements of EPR Legislation for Packaging and Paper Products (PPP)*, Product Stewardship Institute and Flexible Packaging Association (November 2020).

7. Product Stewardship Institute, "Mattress Stewardship: Briefing Document" (Scope of Project, July 25, 2011), 4.

EPR concepts and government interest in improving programs has put a strain on even the more progressive producers in the United States. The tendency among producers has been, thus far, to view a first EPR law as the way in which their products should be managed in perpetuity or until they are forced by governments and other stakeholders to enhance the program. Some even claim that a successful program relies on it staying just the way it was from the beginning.

The paint industry has taken a highly collaborative approach with governments and other stakeholders in managing leftover paint. To a lesser degree, so has the mattress industry. Most other industries have been much less cooperative. Both the paint and mattress industries, however, are challenged by changes in stakeholder expectations for their programs to evolve with the times. For example, initial paint EPR laws, which passed starting in 2009, focused on collecting latex and oil-based household paint, regardless of whether the paint could be recycled back into paint or needed to be disposed of because it was hardened and spoiled. Although our research showed that aerosols, marine paints, artists' paints, and other specialty coatings were also part of the problem with consumer paints, the national dialogue excluded those from the focus only because they added another layer of complexity in reaching a basic agreement. We also decided not to include the paint manufacturing process or the ethical and sustainable sourcing of paint ingredients, like titanium dioxide. These exclusions did not mean that PSI or government agencies wanted to forget about those products or the upstream environmental impacts of paint production forever. Instead, it was a recognition that we could not do everything at once, and it was more important to start a good program than to attempt to put every issue into our first bill.

The national paint dialogue group discussed all these issues briefly in 2003 before narrowing the focus to leftover postconsumer architectural coatings. Adding items over time is a growing interest among states with paint EPR laws, as well as those states seeking to pass a first-time law. Both want the latest model, like choosing a new electric vehicle and not the original model that was less efficient. Regarding mattresses, the initial three EPR mattress laws passed in 2013 did not require all mattresses to be managed by the mattress industry, rather only those that were of high quality and could be recycled. This narrowed focus resulted in municipalities having to pay for the disposal of approximately 17 percent of mattresses collected that were wet, soiled, and otherwise not recyclable in California in 2020.[8] Even though all paint collected is covered under EPR laws regardless of recyclability, the

8. In 2020, California disposed of approximately 17 percent of mattresses collected (more than 381,000 mattresses and box springs). *Mattress Recycling Council, July 1, 2020, California Annual Report 2020,* accessed July 12, 2021, https://mattressrecyclingcouncil.org/wp-content/uploads/2021/07/MRC-annual-report-2020-web.pdf.

mattress industry cut a deal with legislators to omit wet and soiled mattresses that were not recyclable.

One of the most important concepts to understand is that our knowledge and ability to tackle product stewardship problems evolves over time. All stakeholders need to acknowledge and plan for it together. In fact, it is hard for a group to wrap its collective mind around every aspect of potential change, since no one really knows what the future holds. Do the mattress or paint industries really know about all the technological developments taking place by each company in their industries? Of course not. But we do know that it is likely that the materials used in these products will change over time, perhaps making them more difficult and expensive to recycle.

We can design EPR laws so that manufacturing changes and other program changes can be reviewed by the government oversight agency and other stakeholders to determine if a change in the law, regulations, or stewardship plan is warranted. Although EPR laws have not yet fully incorporated this transitional concept, best practice is to include periodic review by a multi-stakeholder advisory council, an annual report submitted by producers, and a revised stewardship plan submitted by producers periodically (or at the request of the oversight agency). These statutory and program elements can capture and manage materials omitted in initial laws, like nonrecyclable mattresses and aerosol paints.

Stakeholders will arrive at agreements on EPR bills quicker if they enter a dialogue having joint expectations that the solution to the problem will evolve over time. Currently, producers typically want to start with a more narrowed product focus, since this is more manageable, but resist change to add more sustainable practices over time. By contrast, governments and other stakeholders anticipate this resistance and attempt to include a broad scope at program startup, which may be overwhelming to producers. Best practice for EPR policy is to expect laws to evolve in a time frame articulated in the initial bill. For example, an EPR program could start with a more limited scope, with the law building in the expectation, perhaps with trigger dates, that new stages of the program will kick in over time. A bill must be flexible and forward looking to match the continual evolution of our ideas and technological innovation. This expectation should be reflected in how we develop EPR policies. We can apply a lesson from the evolution of European EPR directives for packaging, batteries, and electronics, which have been continually updated over time to incorporate the concepts of the circular economy, responsible end markets, social and economic justice, equity, and other factors that seek to include the full life-cycle costs of consumer products.

Goals

No business, organization, or other entity should take another step if it doesn't know the direction in which it is headed. Although it might appear simple for each stakeholder to articulate the goals it wishes to accomplish, it is important that all stakeholders commit to moving toward the same destination. This seemingly simple step is important throughout a dialogue. Goals are integrally tied to problems. If one aspect of a problem is that consumers do not know that some batteries contain lithium and should not be put in the trash, a program manager's goal will include increased consumer awareness about the toxicity of batteries, their health impacts, and how they can be collected and safely managed. Each problem has a corresponding goal. A lack of collection and processing infrastructure will require a goal of providing convenient consumer collection opportunities and ensuring recycling locations. A lack of sustainable funding for safely managing solar panels, for example, will require a sustainable financing system, such as producers paying for the collection and recycling of decommissioned solar panels.

As with defining a product problem in a stakeholder group, defining the group's goals is often overlooked in dialogues because it requires facilitators to get all stakeholders to acknowledge that others' viewpoints about the goals of the dialogue are also legitimate. Think of it in terms of taking a vacation with friends or your family. You all need to agree on the destination. There are many other decisions to make, like how to get there, the cost, who pays, side trips, and what clothes to bring. But you all need a joint destination. You all need to be excited to get to that place.

In an EPR dialogue, since stakeholders often differ on the best way to reach their goals, gaining agreement on where the group wants to end up will keep conversations on track. Goals are the place to which any solution must lead and should be continually acknowledged to be sure that stakeholders stay on the same page. Throughout a dialogue I will remind the group of its goals, and I will repeatedly put these goals within the first few slides of a meeting as a 15-second reminder. It is too easy for conversations to wander during brainstorming sessions. One of the goals articulated in the national dialogue on fluorescent lamps, for example, was "maximizing the safe collection and recycling of spent lamps from households and businesses through the development of a nationally-coordinated system that is financially sustainable."[9]

Although a stakeholder group will develop its own unique goals for each product dialogue, there are common goals that the producer responsibility movement has sought to achieve. For example, EPR laws have created

9. Product Stewardship Institute, "Product Stewardship Action Plan for Fluorescent Lighting: Briefing Document" (Proposed Project Goal, June 30, 2008), 2–3.

sustainable financing, saved municipalities and taxpayers hundreds of millions of dollars, increased the collection and recycling of materials, created jobs, and increased consumer awareness. However, EPR laws have not yet achieved the important goals of reducing waste, increasing reuse, or substantially influencing producers to design more sustainable products and packaging. Neither do they incorporate social, economic, and equity considerations. These are all goals that more advanced EPR programs in other countries, and new legislation that PSI is working on in the United States, are only now beginning to address.

In a nutshell, EPR systems have certainly been successful, but they can be improved. In addition, while EPR is considered the centerpiece of a circular economy, it is not the only strategy needed to reduce waste, recover materials for reuse and recycling, and reduce toxicity. Bans on products, materials, and chemicals all have their place, as do recycled content standards and other strategies, including voluntary initiatives. Thomas Lindhqvist, who coined the EPR term, has emphasized that EPR is a policy *principle* to be achieved through multiple *strategies*, and it should not be considered a single policy or strategy.[10] We cannot expect one EPR bill or law to achieve every upstream and downstream goal. EPR, as a concept, requires us to implement multiple strategies over time, in a phased and coordinated manner that allows for new ideas, innovations, and technologies to continually guide our progression toward sustainability.

Not meeting some of the ambitious goals set by the EPR movement does not mean that EPR has failed, and not including every environmental issue in an EPR bill does not mean that the bill, if enacted, would be worse for the environment. These arguments have been made repeatedly by some from the environmental community and industry. Unfortunately, these assertions have thwarted progress toward more sustainable practices, by opposing strong EPR bills that did not meet some people's perceptions of stringency or completeness. We should keep in mind that each "EPR bill" is but one initiative that is not static and should be strengthened over time, just as the four EPR directives have been across Europe. EPR bills and other strategies should all be considered as part of the implementation of the same principle of extending a producer's responsibility to achieve lower environmental impacts across a product's life cycle. Conceptually, it is important for all stakeholders to understand that common interests need to be satisfied by implementing a comprehensive set of strategies that align with the EPR principle. It will be nearly impossible for everyone's interests to be satisfied through one EPR bill.

---

10. Thomas Lindhqvist, email communication, April 4, 2023. Lindhqvist credited Professor Gary Davis, University of Tennessee, with this framing of EPR.

An evolution in stakeholder perspectives takes place in every dialogue and all around the world. We need to be ready to bring stakeholders into dialogue as they are ready, but we should not wait for them if a problem cries out for a solution. The 35-year global track record of EPR has proven that this policy approach is the right path to take. As with all systems, however, programs need to be thoroughly evaluated and refined over time so that all established goals can be achieved. Goals also need to be reassessed over time, and programs must evolve to achieve new goals we cannot even conceive of today. In the meantime, we can be secure in knowing that EPR systems have begun to rebuild the pillars of our economy by establishing a network of multi-stakeholder accountability with producer responsibility at its core.

Key Barriers

Once we know the product stewardship goals we want to achieve, we want to assess why we have not reached them yet. These barriers are usually evident to each stakeholder whose efforts to reach established goals have been frustrated by obsolete regulations, lack of education about the problem or solutions, the lack of infrastructure to collect and process or safely dispose of a product, or the lack of incentives to change consumer behavior so that stakeholders can take advantage of existing solutions. By identifying the barriers to achieving joint goals, a facilitator can help a group understand exactly why they are working on a particular problem. This clarity can help the group summon the resources and strength to break down the steps needed to overcome those barriers.

One of the key barriers identified in PSI's national dialogue on thermostats was that heating and cooling contractors lacked convenient locations to drop off thermostats for recycling. This understanding led to the solution to establish safe collections at contractor locations nationwide. As another example, a major barrier to the take-back of unwanted medications was that the existing law and regulations required the presence of a law enforcement officer at every take-back location. This was not an issue for police stations, which set up early drug take-back programs to supplement their role in confiscating illicit drugs. But it was impractical for retail pharmacies, which are the most convenient drug take-back location for most residents. PSI focused its national dialogue on this key barrier, ultimately gaining consensus among all key stakeholders that the federal law and regulations needed to be changed to make collections of unwanted medications more convenient for residents. It took six years of advocacy by PSI and our public and private sector allies to remove this federal legal barrier. But once we changed the Controlled Substances Act through the enactment of the Secure and Responsible

Drug Disposal Act of 2010,[11] and then upgraded the corresponding Drug Enforcement Administration regulations in 2014,[12] EPR bills could finally achieve our main goal of increasing consumer access to drug take-back sites.

Barriers are often, but not always, related to laws, regulations, and government policies that need to be changed to clear the path for goals to be achieved. For example, although California is the only state with a law prohibiting the disposal of medical sharps in the household trash, most states find sharps take-back programs to be best practice. Since trash disposal of sharps is still allowed in most states, it is important to inform residents that take-back programs are best practice and that trash disposal can be done only as a last resort using specified containers and methods. In PSI's national dialogue on medical sharps stewardship starting in 2008 and subsequent state dialogues in Illinois, Missouri, Oklahoma, and Texas, participants discussed explaining to residents that, while trash disposal is legally allowable, they seek to develop sharps take-back options as best practice. In the meantime, if trash disposal is required because a person has no available take-back option, sharps should be disposed in a manner that limits the chance of needle sticks.[13]

Potential Solutions

There may be 101 ways to educate consumers about how to recycle packaging, but what is the best way to do it for each particular material? And is education more effective than changing regulations or laws to require producers to put consistent symbols on products that match the reality of a consumer's ability to recycle? Or is the best solution to give financial incentives to producers to make products that are more recyclable? Discussing solutions in a structured dialogue makes it possible for the best solutions to rise to the top, while bad ideas sink to the bottom. In PSI's 2019 national dialogue on flexible packaging, for example, participants discussed the barrier that there is no convenient collection infrastructure for this increasingly popular type of packaging, which comprises 19 percent of the municipal waste stream. A main solution to increasing the recycling of this packaging type was to ensure that producer funding from an EPR system would fund new recycling infrastructure to be developed during program implementation.

Asking each stakeholder what he or she thinks are the best solutions to overcome key barriers gives a voice to that stakeholder's unique interests and expertise. By hearing other viewpoints, stakeholders often expand their own

---

11. "Secure and Responsible Drug Disposal Act of 2010," S.3397, January 5, 2010.

12. Federal Register, 79, no.174 (September 9, 2014), Department of Justice, Drug Enforcement Administration, 21 CFR Parts 1300, 1301, 1304 et al. Disposal of Controlled Substances, Final Rule.

13. "Dispose of Household Sharps the Safe Way!," Product Stewardship Institute, accessed March 15, 2023, https://productstewardship.us/wp-content/uploads/2023/03/sharps_disposal_info_sheet.pdf.

concept of what solutions might solve the problems identified. This technique also allows PSI to assess how close, or far apart, stakeholders are to developing an agreement. But while a laundry list of solutions might indicate that the group is energized, a facilitator must tactfully make stakeholders decide which solutions they believe are top priorities and critical to achieving the goals developed by the group. To help the group prioritize, other criteria (such as implementation cost, available resources, and time frame) may need to be developed and agreed to.

In the early days of the product stewardship movement, most governments did not know about product stewardship or EPR, since it was a new concept. We needed to educate them extensively about product impacts and opportunities, while also orienting them to our dialogue process. Back then, PSI developed extensive product stewardship action plans for each product category, in part to educate stakeholders. Prior to engaging the paint industry in dialogue, PSI developed two seminal documents. The first, a background technical document,[14] is packed with original research that included the environmental and human health hazards of leftover paint, paint production data and major industry players, recycled paint markets, regulatory barriers to paint reuse and recycling, and municipal costs to manage leftover paint. A second document, "Product Stewardship Action Plan for Leftover Paint,"[15] describes the PSI dialogue process and includes the problem, goals, potential solutions, and other information to prepare stakeholders for dialogue. Most importantly, we needed to build political support for a solution. This required our organization to lay out both voluntary and legislative solutions. In most cases, governments wanted to first explore voluntary solutions with producers to demonstrate to legislators that voluntary solutions had been considered but did not meet the dialogue goals. These officials felt that they needed to have reasons that voluntary solutions wouldn't work to address the problems.

One of the key changes in PSI's approach over the past 20 years is that we do not have to work as hard to get the attention of industry to take responsibility for their products. During PSI's first decade, staff conducted a great deal of research to convince industry to engage with us in structured dialogue meetings. Back then, governments were also more likely to develop a solution, put it out for public comment, change what they wanted or had to politically, and issue a final rule. PSI's dialogue approach was fairly unique and helped develop more interest in product stewardship dialogues. Now, more agencies seek to have multi-stakeholder input using facilitators that seek consensus.

14. Product Stewardship Institute, "A Background Report for the National Dialogue on Paint Product Stewardship—Final" (March 2004).

15. Product Stewardship Institute, "Product Stewardship Action Plan for Leftover Paint" (March 18, 2004).

These dialogues have also changed over time as stakeholders have become more educated about EPR policy and have an expectation of collaboration. PSI dialogues now are more likely to move straight to legislation in states with experience in EPR but also take a collaborative voluntary approach in states that have little chance of legislative success. Over time, we have truncated the research phase of product initiatives, although we still believe it is essential to start any voluntary or regulatory/EPR dialogue based on the five foundational pillars—problem, goals, focus, barriers, and solutions. When the goal is to develop an EPR solution, PSI has a special process for this as well. Instead of developing a project summary or action plan, we develop what we call "Elements of an Effective EPR Model Bill."

ELEMENTS OF AN EFFECTIVE EPR MODEL BILL

Research allows me to understand a subject in a deep way. Organizing information and reports is a way to present ideas to a reader so they can also understand what I learned. Grouping this information into categories allows me to better communicate it to others. Over time, I used this organizational technique to design stakeholder dialogues. When people talk at the same time, it sounds like chaos, and when information is presented all at once it can be disconcerting. Facilitators have techniques for presenting information and managing groups to ensure respectful dialogue. Chaos calls for order, and those who design meetings take complicated issues and resolve them through a process that creates certainty. In her book, *Civic Fusion* (2012), my wife, Susan Podziba, describes the beginning of a public policy mediation as a jumble of concepts much like viewing a Jackson Pollock painting. I tend to think of the issues that comprise an EPR policy as a hodgepodge of terms.

Newcomers in EPR policy discussions often feel confused. It is overwhelming to be confronted by multiple elements of a bill that, together, seek to solve significant problems. As PSI's experience grew in convening dialogues and developing model EPR bills, we began to organize the basic elements of bills into categories. Early best-practice EPR legislation established on mercury thermostats (2006) and leftover paint (2007) had already begun to use terms that structured bills into elements. The project focus identified as a basic steppingstone for dialogue engagement became two key elements: covered materials and covered entities. Each bill included text that specified which entity was responsible for complying with the EPR law. We came to call those stakeholders the "responsible party." Each bill included a way to measure progress and to evaluate the program's success by including performance goals (often a collection rate and recycling rate), as well as a standard for consumer collection convenience. As experience grew among an expanded government group working on EPR policy, I sought to organize the various terms used by each state to develop legislative language. Together, we began to develop a common language to discuss bills on different product categories and in different states.

Over time, PSI staff refined the elements into a set of 16 standard elements of an effective EPR bill and provided standard definitions for each element (see figure 6.2 and textbox 6.1).

These 16 elements form the basis for a document that we call the *Elements of an Effective EPR Model Bill* (hereafter referred to as the "Elements document"). The Elements document is a four-column matrix that incorporates global best practices along with the knowledge, expertise, and preferences from PSI government members and corporate and organizational partners. It serves as a method to educate stakeholders about all aspects of an effective bill, but also is a tool by which we can incorporate their preferences and a means by which we seek consensus. In essence, the Elements document is a way to systematically negotiate key bill concepts, as well as actual bill language, without having to dissect a bill and get lost in unnecessary details.

An Elements document starts with a definition of each element. Since a great deal of interest has been paid to packaging and paper products (PPP), I will use our Elements document for PPP as an example. The first column in the matrix defines the first element, "Covered Materials/Products," and the definition we give it is "Materials/products that are subject to the EPR program." The second column (called the "base model") provides a basic explanation of that section of the model bill based on best practices, experience, and basic preferences. For example, in our EPR for PPP example, it describes

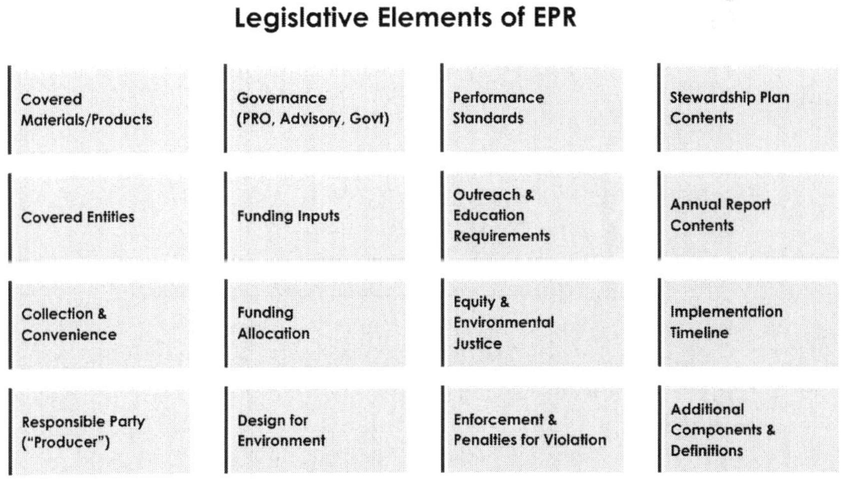

Figure 6.2 Legislative Elements of EPR
Source: © 2023 Product Stewardship Institute, Inc.

TEXTBOX 6.1 ELEMENTS OF AN EFFECTIVE EPR MODEL BILL

1. **Covered materials/products:** materials/products that are subject to the EPR program.
2. **Covered entities:** stakeholders that may use the EPR program free of charge.
3. **Collection and convenience:** the minimum level of collection convenience that a stewardship plan must provide to covered entities.
4. **Responsible party ("producer"):** defines who is responsible for funding and managing the EPR program.
5. **Governance:** defines roles for program operations, administration, multi-stakeholder input, oversight, and enforcement. Includes PRO structure, multiple PROs, advisory council, and government oversight.
6. **Funding inputs:** how funding enters the EPR system. Includes cost internalization vs. eco-fees and material fees (including eco-modulated fees).
7. **Funding allocation:** how EPR program funds are spent.
8. **Design for environment:** provisions beyond eco-modulated fees that minimize environmental and health impacts of covered materials.
9. **Performance standards:** requirements and metrics to gauge the success and progress of the EPR program, including reduction, reuse, collection, recycling, and postconsumer recycled content.
10. **Outreach and education requirements:** provisions to ensure that consumers, retailers, and other key stakeholders are informed about the EPR program.
11. **Equity and environmental justice:** components that encourage equitable and just practices.
12. **Enforcement and penalties for violation:** measures to ensure compliance with the EPR law. Includes disposal bans and retail/sales prohibitions for noncompliant covered materials.
13. **Stewardship plan contents:** minimum components of a stewardship plan describing how responsible parties will implement the EPR program.
14. **Annual report contents:** minimum components of an annual report that responsible parties will submit to the state.

15. **Implementation timeline:** schedule for the submission, review, and approval of stewardship plans.
16. **Additional components (including definitions):** legislative provisions to ensure the EPR law is compatible with existing laws. These components include protection against anti-trust violations to allow responsible parties to collaborate; whether any preemption provisions are allowed to ensure program consistency; authority to promulgate regulations; protection of financial data and other proprietary information; procurement; and definitions for essential terms in the EPR law.

Source: © 2023 Product Stewardship Institute, Inc.

the "Definition of Covered Materials" as "Covered Materials include all packaging and paper products regardless of recyclability." That section also includes definitions of both "packaging" and "paper products."

The third column in the Elements document (called the "recommended bill language") provides actual legislative text we recommend based on our research, often referencing the state bill from which the language is taken. A fourth column includes "options and additional considerations" to acknowledge that, while there are basic elements all states will want to include, some may want to go beyond the base level and incorporate additional policies.

PSI has developed this tool for many reasons. For starters, it makes it possible for all stakeholders to understand the various elements included in any effective EPR bill. By breaking down a policy into 16 distinct elements, it is easier to comprehend each isolated element. This reductionistic approach is like looking at a landscape through a camera lens; it allows us to stay focused on what is most important to be discussed at that time. Over time, we have come to realize that the same issues come up for each product, even if the manner in which the issues are addressed differs by product and by state. The Elements document also allows PSI to develop EPR policies on new products. By starting with the basic outline of a bill, we know the content that needs to be covered. In most cases, we now have standard language for each element that can become a starting place.

Having a standard set of elements also makes it possible to bring together multiple stakeholders and develop a model agreement on EPR legislation, either a federal bill or a harmonized set of state bills. For example, using consensus-based techniques in conjunction with the Elements document, PSI was able to reach a mediated agreement with the Flexible Packaging Association (FPA) on eight elements of a packaging EPR bill relevant to FPA.

This agreement was possible because PSI and our government members in multiple states spent a year and a half developing the Elements document for PPP well before any industry group was willing to have a conversation with us about a packaging EPR bill.

By the time that FPA members were willing to engage in discussions, PSI already had a draft outline of a bill for consideration. PSI then designed a dialogue specifically for FPA and its members, starting by gaining agreement on the benefits of flexible packaging, the problems that flexible packaging presents, goals the group wants to achieve ("end state"), and attributes of an effective system for managing that specific type of packaging. Once agreement was obtained on these modified foundational aspects and trust built, PSI and FPA used the Elements document to determine the extent to which there could be agreement on an EPR packaging bill.

The Elements document was the main tool through which PSI/FPA dialogue participants became educated about each aspect of a packaging EPR bill, including best practices and options for consideration. PSI structured the dialogue so that stakeholders learned about best practices for each component of a model bill and options for consideration, allowing participants to build the policy they wanted. PSI used the Elements document as a focal point of discussions about the range of stakeholder interests so that each point could be resolved. FPA has used the agreement in legislative discussions to indicate the elements of an EPR for PPP bill that need to be included to gain the association's support (see Chapter 10, "Packaging EPR in the United States").

As with all well-run dialogues, stakeholders who participated in the PSI/FPA dialogue were influenced by the respectful discussions that took place. The magic of a deliberate, structured dialogue that is based on information contributed by all key stakeholders is that people become more informed. Those who enter the discussions with an openness to being educated will come away with a deeper understanding of the problem and the solutions. They often change their perspectives. This does *not* mean that one side sets out to convince the other side, and the stakeholder with the most convincing argument wins. Dialogues seek discussion based on solid facts, actual positions, and real possibilities. When starting a dialogue, if all stakeholders truly seek to solve the identified problems, reach the goals, clarify the focus, identify barriers, and discuss potential solutions together, agreements can take place. I have seen this magic happen time and again.

PSI also uses the Elements document as a best-practices model against which to compare EPR bills introduced across the country. Doing so allows us to quickly develop bill testimony by determining how closely a bill matches our model. We might take a position of strongly support, support, support with modifications, or oppose based on how a bill matches up to our Elements document, which we have now developed for each of about 20 product categories.

Over the past 22 years, PSI staff and I have refined the process and methods described in this chapter. This approach has helped develop the policies behind many of the current EPR laws in the country. A variation of these techniques can be used to run single meetings or a series of meetings, whether in person or virtually. They can be used to implement product stewardship projects or achieve consensus on voluntary producer responsibility initiatives. The approaches described undoubtedly will be refined and improved by others. We have not followed a specific road map but instead have adapted basic techniques used by all facilitators and mediators and applied them in a different way.

KEY PROCESS ELEMENTS

Training someone how to facilitate a complex, multi-stakeholder dialogue is a monumental task, and not one I can fully take on with this book. Even so, there are several topics that are useful to cover.

Invest in a Good Process

Whether you are a government official convening stakeholders to develop an EPR policy or a recycler seeking a policy to help increase the collection of scrap materials to make your business viable, you should invest resources in a good process. Companies that invest tens of millions of dollars in recycling facilities or data management software are just now starting to realize that investing in the process by which that policy is developed is necessary for business success. Whether you use in-house expert facilitators or contract out, it will be helpful if the steps outlined here are covered.

I have facilitated countless meetings, and participated in many others, that convened stakeholders in multiple configurations. I am a firm believer that government agencies should be the convener of most product stewardship policy dialogues because they are charged with developing a policy solution for their jurisdiction. These agencies have the option of using in-house staff or contracting out. Whoever is the facilitator, though, must include all key stakeholders and ensure that all important topics, like EPR, are addressed in a meaningful way.

Dialogue meetings convened by other stakeholders can have great value. As You Sow, an environmental nonprofit, has used shareholder resolutions to drive companies to negotiate meaningful change through dialogue. The US Plastics Pact, convened by the Ellen MacArthur Foundation and The Recycling Partnership (TRP), was instrumental in securing the support from over 100 consumer brands for EPR, deposit return systems, and recycled content standards as basic policy solutions to address the plastic pollution crisis. AMERIPEN, a trade association that "represents the entire packaging value chain including raw material suppliers, packaging converters, brand owners, and recovery experts," convened their members to develop their own EPR principles, as did the Sustainable Food Policy Alliance, which includes Danone North America, Mars Incorporated, Nestlé USA, and Unilever United States. The American Beverage Association and other packaging brands and associations have also developed positions on EPR. And, finally, the Ocean Plastics Leadership Network (formerly Soul Buffalo), convened a dialogue in 2020 of consumer brands and environmental groups which, with PSI assistance, educated participants on EPR as a means to address plastic pollution.

While each of these efforts has been important in moving to a more sustainable recycling and waste management system, dialogues convened by producers, environmental groups, and other nongovernment entities are rarely the best forum for solving complex public policy disputes with multiple parties. There is a tendency for these entities, particularly industry, to seek to control the process and achieve a particular outcome. Adding the right stakeholders from state government, local government, environmental groups, collectors, recyclers, retailers, and others can be a daunting proposition. That, however, is exactly what is needed to develop good public policy.

Government agencies, if they function as they should, will often seek to solve product stewardship problems from a wider perspective because they represent all stakeholders, including businesses, municipalities, and environmental advocates. This is not to say that governments always follow the best process in developing these policies. However, governments have the legal authority to develop policies and pass laws, and these are our chosen institutions for developing EPR laws and related product policies. This authority makes government the right forum to convene dialogues. Overlaying that authority with a structured process, like the one described in this chapter, can often lead to successful policy outcomes.

Dialogue Design

Much of what I discussed already falls under the category of design. A key concept to keep in mind is that meeting participants always want to know that their valuable time is being used wisely. They want to have clear expectations of what is about to take place so that they are fully engaged in discussions. If a facilitator is not clear on the goals of a meeting or dialogue, the agenda constantly changes as a meeting unfolds, or the process does not follow a clear road map, stakeholders will lose faith and check out. For this reason, every meeting needs a tight agenda that can be altered—but only as needed and with the consent of the group. Participants should be able to expect that meetings will start on time, the agenda will be followed most of the time, and the meeting will end on time.

Another important aspect of keeping groups engaged, as mentioned earlier, is to conduct 80 percent of the work outside of meetings, and spend the remaining 20 percent in meetings interacting with one another in person or online to make decisions. Repeating agenda items or information from one meeting to the next or introducing information during an in-person meeting that could have been provided through a short document or email will sap group energy. Particularly during online meetings, where it is harder to maintain group focus, facilitation should be fast and crisp, and stay on the path expected. Conducting as much work as possible outside of meetings will also be the most efficient way to run the dialogue since all participants will be most prepared coming into the next in-person or full online group meeting.

Involving Stakeholders

Deciding who to invite to a dialogue, and when, is an important decision. When PSI started out more than 20 years ago, we invited many entities involved in the manufacture, distribution, sale, use, and end-of-life management of a product. PSI did not limit the number of participants in its dialogue efforts since each stakeholder with an interest in an issue has a role to play. In addition, more stakeholders meant that we were educating more people about product stewardship. In PSI's early days, governments did not know about product stewardship or EPR, since it was a new concept. We needed to educate them about the problem caused by a product and discuss voluntary and legislative solutions. For leftover paint, which was the first national dialogue we facilitated, we routinely had about 35 to 50 people in the room or on the speakerphone (well before web-based meetings and Zoom calls became prominent). During those meetings, the American Coatings Association participated on behalf of all paint manufacturers. Market leaders, such as Sherwin Williams, Benjamin Moore, PPG Industries, and other brand names, also took part. Additional participants included state and local government officials from multiple states with experience managing leftover paint, recycled paint manufacturers from Canada and the United States, and a few other groups. The success of the paint dialogue was due, in part, to having the right stakeholders participating.

The paint dialogue became a model for other PSI dialogues regarding which stakeholders are needed for a successful outcome. Participants for the "inner circle" usually include producers, recyclers, government officials, environmental groups, retailers, and other key stakeholders, although these will vary from product to product. I have found that it is best to include the association(s) representing producers, along with a handful of key market players. Industry associations representing producers are critical because they represent all their members' interests and have a broad view of industry-wide interests, but it is also important for influential member companies to participate actively in the dialogue.

Although an association representative can provide a broad industry view, individual company executives are often best at articulating important perspectives. Also, company leaders who have participated actively in a dialogue can support the association staff person by speaking on behalf of the dialogue in association board meetings. In this way, the burden of communicating about the negotiations does not fall entirely on either a few company representatives who might have their own interests, or the association representative, who will need to sell an agreement to the entire membership base. When it comes to gaining approval on a draft agreement, having both the key association staff person and several influential members present when a vote is taken is often a critical strategic opportunity.

The most effective association leaders are those who have authority to enter into negotiations and are also respected enough to have influence over their members. There is unique skill needed by those leading associations so that they can both negotiate for their members' interests while also satisfying interests of other stakeholders. There are also times when an association leader will need to bring an important member into the fold to obtain their support so that an agreement can be achieved.

Alison Keane, who led the American Coatings Association (ACA) throughout PSI's multi-year national paint dialogue, is one such person. Later, as CEO of the Flexible Packaging Association (FPA), she led the flexible packaging industry into another PSI-mediated agreement. In both dialogues, Keane engaged with PSI due to our ability to convene the right state and local government policy officials, most of whom we knew personally. PSI could also develop national policy models that did not require ACA and FPA starting at the beginning with each state. She trusted the process used in both dialogues because she had the confidence and support of her members to represent them in discussions into policy scenarios not yet explored.

In both instances, it was the first time the industry's members engaged with government officials and other stakeholders in a policy dialogue that gave all stakeholders the opportunity to discuss their view of the problems, goals, focus, barriers, and solutions. Any stakeholder, including Alison Keane, could have walked away at any time. They were at the table voluntarily. What kept everyone in discussions was the continual sense that they were learning from each other, that progress was being made, and that the group was moving toward meaningful agreements. These were not discussions with a goal simply to calm animosity among typical adversaries. The goal in both cases was to develop meaningful EPR policy, and the ingredients to succeed included involving the right stakeholders at the right time in the right process.

Retailers are also a powerful stakeholder group since they make decisions about which companies' products they will sell. However, since retailers sell an abundance of products, it is more difficult to attract them to the table for a negotiation on a single product category. Typically, retailers will enter later in the negotiations and often through their lobbyist, particularly as a bill is being developed. It is critical to keep state retail association leaders informed of developments and seek input often, and definitely before a final agreement is reached.

Messengers Matter

I have learned through hard-knock experience that there are times when I am the right person to introduce a topic or provide information, such as when I kick off a dialogue meeting, discuss the meeting agenda, or explain a concept. Other times it is best for other stakeholders to speak. Their voices are often more legitimate than mine. For example, I had difficulty conveying to the paint industry the full array of problems with leftover paint, until it occurred to me to ask certain state and local government officials to speak: Sego Jackson, City of Seattle, Washington (who worked for Snohomish County, Washington, during the paint dialogue), Theresa Stiner, Iowa Department of Natural Resources, and Jen Holliday, Chittenden County, Vermont. I chose them because they represented state and local governments, were from the West Coast, East Coast, and Midwest, and also were respected by the group.

Jackson, Stiner, and Holliday provided firsthand experience about the cost to their agencies of managing leftover paint, the numerous resident calls they had to field about how to manage the paint, and the lack of solutions available. As they spoke, I could sense that their words were having a noticeable positive impact on industry participants. That was the moment when the paint dialogue cleared a big initial hurdle. Leftover paint was finally perceived as a real problem and the paint industry was going to help. I learned throughout that dialogue and others that PSI's strongest role is to create the space for respectful discussion among stakeholders, provide important technical information, clarify and summarize issues, orchestrate the meetings and dialogue flow, and drive the group toward decisions. But positions held by government officials, producers, recyclers, haulers, and environmental groups should come straight from individuals representing those perspectives.

Meeting Access

Just as a meeting agenda and how it is facilitated will make dialogue participants feel comfortable or not, a meeting convener also has a role in making people feel comfortable and included. Designing the meeting experience so that everyone can "see" and "hear" one another facilitates conversation. When establishing a physical room layout, all participants must have a clear line of sight to the screen for projected images and presentations. I strongly prefer a hollowed square or rectangular shape of tables that are bowed like the bow of a ship. Rooms with pillars and ones that are too tight or dim, have low ceilings, and are not situated with easy access to doors make people grumpy. If the meeting is small enough to use a flip chart, this will often keep people focused enough on what is being written for them to follow the discussion. If a meeting is large or online, we will take notes in real time and project them on screen to the group. Visuals (seen or described for blind or visually impaired stakeholders) help hold peoples' attention and keep group energy unified and engaged.

Providing remote access is something that PSI felt was important from our very first dialogues. We used a microphone on a small table-top stand pointing down at the once iconic Star Phone (conference room speaker phone) or, if lucky, our meeting room had a wired system that picked up room conversation through ceiling microphones. I used at least three hand-held microphones to manage the conversation, requiring people to raise their hand or turn their name placard on its edge to get in a queue to speak, and at times we had to rent large speakers. Microphones were often the only way people on the phone could hear conversation in the room. Planning meeting details with audiovisual (AV) staff at an on-site location is one of the most important, and often overlooked, aspects of a successful meeting. PSI staff arrive the night before as often as possible and set up, making sure the AV is tested and working, with extension cords taped to the floor and available for computer and cell phone power around the room. Agendas, participant lists, and other information are laid out for participants in advance. Most of the time, despite sending a setup diagram in advance, we have to fix the room setup. Doing all this in advance reduces the stress of facilitating the meeting.

Coordinators also must ensure that the meeting venue is accessible to participants with disabilities, including ensuring the physical location is accessible to wheelchair users, providing closed captioning or ASL interpretation for deaf and hard-of-hearing participants, as well as ensuring that slides and documents are designed with people who have color-blindness or other visual impairments in mind.

Lastly, don't forget food! At the least, coffee, tea, water, and light healthy food for breakfast, a solid lunch, and refreshments for breaks are the fuel needed to keep good conversation going. PSI typically asks registrants for dietary restrictions during the registration process. But increasingly, PSI simply plans to provide a menu that by default accommodates those who must, or choose to, maintain vegan, vegetarian, nut-free, gluten-free, and dairy-free diets. We also ensure that foods are labeled so that those with allergies have the information they need to make safe food selections. Post-meeting receptions are my favorite way to debrief, unwind, and build the personal relationships needed for participants to begin to open their minds and shift positions so that an agreement can become possible.

Following a meeting, it is standard form to develop concise notes with decisions highlighted, along with next steps, key comments, and a general summary by meeting section. Developing a meeting summary is a highly skilled task whose importance is often overlooked. Doing this task efficiently is an acquired expertise. If you can synthesize comments on the spot during the meeting and resist the tendency to dictate the conversation word for word, you will save a great deal of time. Meeting summaries capture important comments, decisions, and other highly essential aspects that will not only describe what took place but propel the group forward toward the next meeting. These summaries will be a useful reminder for attendees and the next best thing for those unable to attend, which will maintain group cohesion. Good notes will keep all stakeholders engaged and ready to continue the discussions. They also visibly record the group's steady progress, which is an important psychological aspect when a dialogue stretches over many months.

Meeting Size

The size of meetings has never mattered much to me. Although 35 is typically considered manageable, for our dialogue on pharmaceuticals we had 120 people in the room, including a half dozen federal agencies regulating pharmaceuticals, and another 20 participants on the phones. Our other dialogues also routinely involved large groups of stakeholders, most in the room but also many on the phones. These meetings provided national momentum and created a sense of urgency to act, and we did not need to continually add new stakeholders. Through experience, I found that it is better to allow people to attend if they commit to the project goals and honestly seek a solution rather than omit them from discussions. Those omitted will almost always imagine the worst. It was sometimes intense to manage, but very efficient and the educational value was immense.

Today, in our post-COVID world, people are used to both large and small virtual meetings, with multiple options to communicate with each other and the meeting hosts. No longer will meeting facilitators need to continually check in with phone attendees to make sure they are still on our conference call or ask whether they have a question or comment. The multiple functions of virtual meeting platforms, like hand raising, polling, chat, and question boxes have enhanced our opportunities to communicate. With these options also comes the need to plan to use the type of functions that will best serve your meeting goals.

Sealing the Deal

An agreement reflects the knowledge of those who participate and the skill of the facilitator or mediator. If a dialogue does not include all key stakeholders, it might have value, but it will most likely be insufficient to solve the majority of problems identified at the outset. Agreements must also be recognized publicly either through a formal memorandum of understanding signed by each participant, a press release from key stakeholder groups, or another mechanism that publicly acknowledges the agreement. To avoid a participant backing out of an agreement, it is important for the facilitator to continually check in with all key stakeholders to make sure they are not waning in their support.

One way to ensure that stakeholders stay tied into the outcome is to make sure they have authority to speak on behalf of their agency, company, or organization. Alternatively, the dialogue participant needs to have direct access to a decision maker in their company or organization. Otherwise, their comments, while of value, might not represent an important stakeholder needed for the agreement. Any substitutions during the dialogue must be onboarded by briefing them on past meetings, important decisions, and other dialogue aspects.

PASSING EPR LEGISLATION

Developing an agreement on an EPR bill is a big accomplishment, but getting it enacted into law is another important hurdle. Choosing a bill sponsor who has the political skill and power to guide the bill through the legislature is critical, as is developing the coalition in support of the bill. Assembling a multi-stakeholder coalition that can speak with a coordinated multi-faceted voice presents a strong case for bill passage. Addressing stakeholder interests through group meetings and individual conversations will reduce the dissonance that often accompanies public policy deliberations. Disagreements on policy require legislators to either resolve the differences or side with one stakeholder versus another. The fewer the disagreements, the more likely it will be that a bill will be passed.

Even so, I have experienced the frustrations along with my colleagues regarding numerous instances where EPR bills have extensive support and still take years to pass into law. Paint EPR bills are prime examples, despite being based on the same model bill mediated by PSI in 2007 with full stakeholder support. Our first two attempts at passing the consensus paint EPR bill involved near unanimous passage through the Minnesota legislature *two years in a row*, only to be vetoed *twice* by Governor Tim Pawlenty (2008 and 2009), who was seeking to become the Republican vice presidential running mate during those years. Governor Pawlenty was running on a "no new taxes platform," and he viewed the paint eco fee, which is not a tax but a consumer fee, as politically problematic. The paint bill was finally signed into law in 2013 by Governor Mark Dayton. Even with industry support from the American Coatings Association, paint EPR bills languished for seven years before passing in New York in 2019, and seven years before passing in Washington State in 2019. Paint bills in New Jersey, however, perhaps fared the worst; even after having sailed through the legislature four years in a row—from 2018 to 2022—the bills were vetoed all four times, twice each by Republican and Democratic governors. By contrast, Colorado enacted paint and packaging laws both times on the first attempt.

Addressing each issue raised about every product category discussed over the past two decades would require volumes of reports and data. It is sufficient to say that, while the concept of producers taking responsibility for reducing, reusing, and recycling the postconsumer products they put on the market has taken hold, there is a continual need to educate legislators and other stakeholders. Social change takes time and EPR acceptance has taken more than two decades to produce a massive cultural shift, and one that the public seems to be fully embracing.

PRODUCT STEWARDSHIP COUNCILS

One of the key engines behind the EPR movement in the United States has been the success of state product stewardship councils (PSCs) that are comprised of local governments at their core, but sometimes include business and environmental representatives. PSCs play a key role in educating and gaining support for EPR bills and voluntary product stewardship programs from those most impacted by postconsumer product waste—local governments. Often meeting monthly, PSCs form the nucleus around which important discussions regarding product policy take place. Working by consensus, they decide their state product priorities and develop and execute strategy for the passage of priority bills. Some of the local government PSC members are registered lobbyists, but all provide the technical support and direction needed to pass bills or develop voluntary initiatives. PSI and the Product Policy Institute (now Upstream) created most of the PSCs together or separately. Since then, PSI has facilitated and/or provided technical expertise and coordination support to multiple PSCs.

The PSC concept was created in 1998 by the Northwest Product Stewardship Council (NWPSC), a group of solid waste professionals in the Seattle area, which has grown to include local governments in Washington State and Oregon. NWPSC was the original model for other state and regional councils around the United States. The Product Policy Institute (PPI) and members of PSI's board of directors helped local governments in California to form the California Product Stewardship Council (CPSC) in 2006. CPSC galvanized local government support for producer responsibility in California and lobbied for bill passage using many PSI policy models and developing a few of their own. California has 11 state EPR laws and 16 local EPR ordinances.

By 2007, other regions of the country began to mobilize support for their own product stewardship councils. The Vermont PSC (established by Jen Holliday of Chittenden Solid Waste District, PSI, and PPI in 2008) has driven the passage of most of the nine EPR laws in Vermont, including some of the most effective programs in the country. PSI has facilitated and advised this group since 2016. The Vermont PSC is comprised of nearly all government solid waste districts in the state, with the state Department of Environmental Conservation playing a non-member technical advisory role. The New York PSC, started by Resa Dimino and the New York Department of Environmental Conservation (DEC) under the auspices of the New York State Association for Solid Waste Management in 2009, includes a board comprised of local governments and businesses, with the state DEC playing an advisory role. The New York PSC has been instrumental in passing New York's six state EPR laws and three local EPR ordinances. PSI has provided technical advice, coordination, and fiscal sponsor services for the New York PSC for many years.

The Connecticut PSC, started by Tom Metzner of the Connecticut Department of Energy and Environmental Protection, local governments, and PSI, helped the state become the first in the nation to enact EPR laws on mattresses and gas cylinders, in addition to enacting laws on paint, electronics, and mercury thermostats. The Illinois PSC was started by the Illinois Counties Solid Waste Management Association and Illinois-Indiana Sea Grant, along with Walter Willis of the Solid Waste Agency of Lake County, other local governments, and PSI in 2010. The Illinois PSC has been instrumental in enacting several laws, including an amendment to the state electronics EPR law and a state pharmaceuticals EPR law. In Massachusetts, the PSC, which was established by PSI and MassRecycle, is currently working on a number of active EPR bills such as paint, mattresses, packaging, and electronics.

PSCs are a key vehicle for local governments to discuss EPR product priorities and map out strategies for the passage of legislation. PSCs in non-regulatory states are also critical in working together to plan and implement voluntary stewardship initiatives, as PSI has done with state and local agencies in Missouri, Oklahoma, Nebraska, and other states. In 2018, PSI and several Missouri solid waste management districts (SWMDs) jointly developed the Missouri Product Stewardship Council (MO PSC) with grant support from St. Louis-Jefferson SWMD, the Mid-America Regional Council SWMD, Ozarks Headwater Recycling and Materials Management District, Mid-Missouri SWMD, and other solid waste districts. Together, we have conducted voluntary pharmaceutical take-back programs throughout the state, as well as laid the groundwork for programs on paint, mattresses, carpet, and other products. In 2022, together with the paint industry, we introduced the state's first EPR bill, on paint. PSI also partnered with the Oklahoma Department of Environmental Quality and local governments to form an ongoing PSC in 2009 to work on voluntary take-back programs for pharmaceuticals, medical sharps, mercury-containing thermostats and lamps, electronics, and paint.

PSCs often form the nucleus of state-based EPR coalitions that also include environmental groups and allied recyclers, waste management companies, and other stakeholders. While PSI has played the role of technical policy advisor to many PSCs, we have also supported PSCs and others in advocating for bill passage, often partnering with statewide environmental groups that have extensive lobbying experience in a state, as well as with our own local government members who are registered lobbyists. A few notable environmental groups with whom PSI has partnered to seek passage of EPR bills include Citizens Campaign for the Environment, New York League of Conservation Voters, Sierra Club, Illinois Environmental Council, Natural Resources Council of Maine, Massachusetts Public Interest Research Group, Trash Free Maryland, and the Missouri Recycling Association, among many others.

PSI's role on EPR legislation varies by state. In most states, we play a facilitation role to develop a bill based on our technical experience and with the input of key stakeholders. Once a bill has been developed, we often switch to playing a support role to those with political experience and deep contacts in the state. PSI will conduct research, facilitate meetings, develop advocacy materials, educate legislators and their staffs about EPR bills, testify in hearings as a national expert, and seek to resolve differences among stakeholders to strengthen the coalition and increase the potential for bill passage. Having strong personal relationships with state and local agency officials, and many nongovernment stakeholders provides PSI with leverage to unify a coalition around a bill. While we provide advice from a national perspective, we will usually defer to those who work daily in their states as to which political strategy is best for the passage of a bill.

IN SUMMARY

When PSI started out in 2000, we sought to develop product stewardship agreements with all stakeholders. Corporations have greater political influence in the United States than they do in Europe and Canada, which required a unique approach to implementing product stewardship policies. PSI brought a collaborative perspective and approach to discussions with those interested in translating producer responsibility to the US context. As described above, PSI's approach involves a deliberative process that seeks to engage producers, retailers, and other businesses in understanding how their products and practices cause environmental harm and fiscal pressures on other parties, and what they can do to alleviate those impacts. The process is based on objective research, but also involves face-to-face meetings, conference calls, and other forms of communication that seek to share information in a way that has the power to change individual and institutional positions.

When we first began our work, however, it became apparent that most producers in the United States opposed taking responsibility for preventing postconsumer impacts from their products; a greater number opposed EPR legislation. Although the arc of acceptance of EPR policies by producers has bent toward accepting the need for an EPR solution, the main reason for acceptance has been the real and present threat of legislation passing. As a result, an increasing number of companies have now taken a proactive approach. They have begun working with groups like PSI, industry associations, government agencies, and others to negotiate solutions that solve joint problems, meet joint goals, and reduce harm to people, animals, and the environment. EPR laws are also increasingly viewed by companies as a means to meet their own voluntary goals regarding recycled content, recyclability, and zero waste, as well as to help companies control valuable materials needed to make new products.

I have met many individuals working for industry who want to reduce the impacts from their company's products and support EPR but were not in a position themselves to change company policy. Some of these people, once retired, volunteer their time to work for EPR advocates like PSI, Sierra Club, and others that value their corporate expertise, insights, and environmental commitment. Others have touted the company line and thwarted well-intentioned efforts to engage in the type of dialogue process outlined in this chapter. Until about 2020, only two industries had agreed from the outset, and persisted, in meeting with government officials to develop a joint EPR solution—the paint industry and the Flexible Packaging Association (FPA). Today, many producer associations are engaged in supporting EPR legislation that meets their members' interests, exemplifying the significant transition toward producer responsibility that is taking place in corporate America.

One notable example of corporate change involves producers of small, one-pound gas cylinders used with camping stoves. In 2002, producers of these single-use products engaged with PSI in a national dialogue to develop a voluntary producer responsibility solution but did not take action. Then, in 2021, Worthington Industries, the company making the vast majority of single-use cylinders, opposed an EPR bill introduced in Connecticut that was developed by PSI and Connecticut state and local government officials. But by the next year, Worthington took a proactive approach and negotiated a bill that was enacted into law. Simultaneously, the Propane Gas Association of New England, representing producers of 20-pound propane gas cylinders used to fuel gas grills, proactively enhanced their longtime reuse/refill program to capture the small percentage of tanks that escape the system.

Other industries have been motivated by the strong global current of producer responsibility. For example, the battery industry, long a supporter of some battery EPR laws, has negotiated increasingly expansive laws in Vermont, the District of Columbia, California, and Washington. The mattress industry was forced to negotiate a deal on a bill in 2013 only when they realized they would lose during the legislative process. Since then, they have been willing to negotiate but still resist changes to the 10-year-old model bill that PSI and Connecticut state and local officials negotiated with them, forgoing changes based on experience and best practices. Still others, such as the electronics industry, are in a wait-and-see mode. The industry engaged readily in a national dialogue in 2001 and helped to enact 24 electronics EPR laws in the country before opposing the expansion of EPR laws to other states after about 2008. Other industry sectors still oppose EPR legislation, such as those making carpet, pharmaceuticals, tires, and hazardous household products. And finally, there are those industries, like solar panels and wind turbines, that will be swept up in the momentum if they are not paying attention to the policy changes around them.

For our country to pass good EPR policies that will return materials to the circular economy and truly protect health and the environment, we need industry leaders who have the confidence and skills to, on behalf of their members, enter discussions with other stakeholders and seek a reasonable solution that has the ability to evolve over time. We also need government leaders who are willing to both flex political muscle and also engage in a process that allows meaningful stakeholder input throughout the development of a joint EPR policy. All parties need to trust a process that has the potential to deliver a fair, negotiated agreement.

PSI set out in 2000 to bring the new concept of EPR to the United States, seeking to harmonize programs nationally through the cooperation of the very group—producers—from which it was seeking significant changes. Essentially, one of PSI's core roles is to establish cooperative agreements among various stakeholders to reduce human health and environmental impacts from consumer products. The organization's ultimate goal is to ensure that all parties involved with a product's life cycle, particularly producers, share in eliminating those impacts.

The process described in this chapter somewhat idealizes the overall approach that PSI takes to develop EPR policies for each of the approximately 25 product categories on which we work. In some cases, the organization has followed this process closely. In other times, owing to various factors, including initial alignment on goals or funding and time constraints, PSI has altered the process to fit the circumstances. Conducting a well-designed dialogue is important, but it is also resource intensive. PSI is continually working with governments that have scarce resources. It is easy for them to go through the motions of seeking stakeholder input but limiting interaction and dialogue. Or they fall back on traditional approaches that rely on pure political might.

Limited funding, staff resources, knowledge, and time, as well as competing priorities will always be challenges to an effective stakeholder dialogue. Ultimately, we need to know there are ways to develop EPR policy differently and include all stakeholders in a deliberate manner. Doing so will not guarantee an agreement that all stakeholders will accept. Pure political forces will always be part of the game. We need to keep in mind that our efforts do not only seek to develop good laws that reduce harm from products upstream and downstream. We are also seeking cultural change, so that laws that are external to our actions are not the only motivating influence. Internally, in our thinking and habits, we must truly understand why we are making those changes.

Chapter 7

Implementing EPR Laws and Measuring Progress

INTRODUCTION: THE CHALLENGE

Implementation of a law is only as good as the law itself. Bad laws will lead to bad results and good laws provide the opportunity to achieve good results. Laws provide structure, just like an architectural blueprint combines vision with the meticulous slicing of space to create a future home. My father-in-law used to marvel at well-designed apartment buildings in New York City as having "good bones," which meant they had a good structure. Contractors implement an architect's vision through attention to detail and by resolving unexpected conflicts in the architect's plans. Under EPR, legislators set the structure in statute, the oversight agency might provide structural details through rulemaking (i.e., regulation), and producers propose an implementation pathway (i.e., stewardship plan) with active engagement from other stakeholders.

US waste management laws, policies, and programs have been continually updated over the past 150 years to protect human health and the environment. There were periods of innovation, then stagnation, followed by renewed innovation and change. We know that our basic waste management structures (our laws) need to change again in significant ways—and their implementation must also change. We need innovation in implementation as much as we do in the laws themselves.

Any law is only as effective as its real-world results. So much effort goes into identifying problems, debating solutions, and enacting legislation that we often turn our attention to other pressing issues once a law is passed. There are so many other legislative priorities waiting in line that it is tempting to move on. But successful implementation of laws, policies, and programs is as important as passing a law. Waste reduction and recycling still falls squarely on the tens of thousands of local and state government officials in our country. Poor implementation gives these officials the lampooned label of bureaucrat. Successful implementation, often overlooked, is still the heart of good governance and public service.

Let's recall why EPR laws have become so prominent. In a government-managed program, agency officials are primarily responsible for meeting the requirements of a law or policy. Agencies, however, are dependent on funding for staff resources to conduct the work, which is often hampered by budget allocations, hiring freezes, and staff cuts. This dynamic has often led US governments to vehemently object to having to assume more responsibility without added funding. The term *unfunded mandate* became a rallying cry long ago for municipal leaders seeking to protect their local governments from the unreasonable expectations that they could take on more work without more resources. The inability of local *and* state governments to meet the increased demands placed on them by an increasingly complex waste stream has led to the surge in interest to pass EPR laws to alleviate government's implementation burden and place it primarily on the producers.

Although many governments, environmental groups, collectors, recyclers, and producers share the main goal of returning materials to the circular economy, these stakeholders have often struggled with the transition to their new roles. It has taken time in the United States for the waste management paradigm to shift to EPR—from governments paying for, and managing, residential recycling to producers paying for, and taking, a stronger management role. After all, prior to EPR, producers did not have to pay attention to, or spend money on managing, their postconsumer products.

Even though many global and US producers have set ambitious company-specific goals, there is a strong perception among those in government and the environmental sector that companies, industry-wide, seek weak EPR legislation, which translates into having to spend fewer resources managing their postconsumer products. Corporate opposition to US EPR laws should be no surprise. Having society pick up a chunk of a company's externalized costs is ingrained in our waste management laws. It has been hard for our culture to break this shackle.

As PSI and others developed best-practice legislation, industry lobbyists often whittled down bills during the legislative process to be much less effective once implemented, if they passed at all. In fact, some carpet producers have used their political influence to get a bad bill enacted, then claimed EPR didn't work because that law was ineffective. Ironically, until recently, it was all too common for global companies that operate in the United States to oppose EPR legislation in our country while taking pride in their performance under a similar EPR law in another country. Most industry sectors have resisted engaging with governments unless absolutely necessary. Some producers, including those making pharmaceuticals, carpet, and tires, still steadfastly oppose taking responsibility under an EPR system, as had packaging manufacturers and brand owners until recently. Other producers, such as those making batteries, mattresses, and gas cylinders, have engaged to varying degrees with PSI and other stakeholders.

The reasons why some industry sectors engage and others do not are based on multiple factors that, at the most basic level, relate to their approach to reducing business risk, which allows the company to focus on their core business—production. Some businesses seek to best position themselves by trying to control the legislative process through lobbying and political influence. Others engage half-heartedly, hoping to run out the legislative clock through slow-walk negotiations, pushing the issue into the next legislative session. Others tire of this game and realize they can actually reduce their risk by engaging with legislators, government agencies, and other stakeholders to craft legislation that provides clarity and certainty.

Two industry sectors sought a new path from the outset but eventually took different routes. The US paint industry readily agreed to enter a national PSI-facilitated dialogue in 2003 with governments and recyclers, eventually signing a mediated agreement that put in place a model paint EPR reuse and recycling program that the industry continues to support today. There are now paint EPR laws in 10 states and the District of Columbia based on that model. Evolving that initial agreement, however, is a new challenge for the paint industry, even as they continue to sponsor EPR bills and take pride in their decade-long PaintCare program. The electronics industry also engaged immediately with the US EPA, PSI, and other stakeholders in 2001, leading to 24 electronics EPR laws and two additional electronics recycling laws, although since about 2008 they have resisted assisting in passing new laws.

While the challenge for most corporations has been accepting their responsibility under EPR laws, the major challenge for some governments, particularly under packaging EPR systems, has been the potential loss of managing their recycling program. Many take great pride in their ability to provide these services, which has brought them accolades from residents even though many know it is insufficient. As producers have begun to accept responsibility for funding packaging EPR programs in the United States, they also want more control over managing those systems. They believe that, if they are required to pay for collection, recycling, education and outreach, government oversight, and other costs, they should have the flexibility to ensure an efficient system that they manage.

Packaging EPR programs have also raised concerns among companies and local governments that own and/or operate equipment and facilities that collect and process recyclable materials. These public and private entities have invested large sums of money and want assurance that those investments will not be abandoned in the move to new recycling infrastructure. Many waste management companies have developed long-term relationships with municipalities that provide recycling services for residents, and directly with residents in many areas of the country where the municipality does not provide services. These companies are concerned about losing customers, as well as facility investments, if producers manage the recycling system. Municipal and some state officials are also concerned about handing over program management to packaging brand owners that have opposed EPR systems for years.

This issue of "program governance" (i.e., who has the authority to make program decisions), is the most important aspect to be resolved in the United States regarding EPR. It is not a one-time problem to be figured out, but instead it is a transition of roles to be continually acknowledged and nurtured. There will always be a tension in what governments want and what industry is willing to do, particularly if the industry is paying for the work that governments are requiring them to do. The most important aspect is that companies acknowledge the importance of taking responsibility for their product impacts. As mentioned in chapter 5, the transition will not be the same for each industry sector. EPR systems for diverting and safely managing paint, mattresses, carpet, pharmaceuticals, and other special wastes for which there was little previous collection and management infrastructure offered new business opportunities for waste management companies.

EPR systems for packaging and paper products, however, will replace long-established waste and recycling systems. Discussions around this transition have been more extensive because more is at stake. Producers often don't trust governments to run an efficient recycling system and don't want to pay full price without greater system control. Governments don't trust producers to take over a system they believe that they have run well, even if the program has been underfunded. And waste management companies are concerned that producers will pull the rug out from under them and upend their businesses.

As governments transfer the burden of financing and managing waste to producers under EPR systems, they must assume a greater oversight and enforcement role. Their role thus becomes more managerial than operational. Whereas governments want producers to finance and implement waste management programs, producers want a greater degree of control over how their money is spent. This change is akin to a person who, having too much work and too few helpers, now has an entire team. The person is elated to finally have much of the burden off their shoulders but finds that the team members do things their own way. The person receiving help must establish guidelines and protocols and a new relationship must be forged. This analogy is not unlike the inevitable tensions between a government oversight agency and producers that take over many of the implementation tasks under an EPR system. This relationship is further complicated by waste management collectors and recyclers who must also establish new relationships and adjust to new systems.

Whether the interaction among state and local government agencies, producers, and waste management companies is productive depends a great deal on the culture of those entities, as well as the personalities of those in leadership positions. Some governments are more willing to relinquish control to producers because they have already abdicated recycling and trash service to companies that contract directly with individual households. Other governments, however, may not trust those in industry to carry out program implementation, particularly when these same companies have vehemently resisted assuming this same responsibility for years. Residential recycling programs have been implemented for so long by state and local governments that many are not ready to hand over the implementation of these systems without assurance that resident access, equity, and public education will be maintained. Local governments deal with angry community members all the time. They do not want residents to become angry over a change in waste management and recycling.

Changing to an EPR system requires a commitment to communicate—not one time, but over time—so that the vision of more sustainable resource use can be actualized. Visions are important. Goals are critical. But neither means anything if it doesn't connect to reality. Successful EPR programs require multiple stakeholders to commit to taking concrete steps outlined in the legislation and stewardship plan. Each needs to hold the others accountable. All stakeholders need to believe in the importance of the vision and in achieving the goals to take the difficult steps toward implementation. Each stakeholder is important to the whole system.

Experience from 133 US EPR programs over the past two decades has shown that this major transformation of how programs are implemented has many benefits. Collection and recycling infrastructure has been established where once there was none, product recycling rates have increased, toxic materials have increasingly been diverted from the waste stream and safely managed, thousands of jobs have been created, and hundreds of millions of dollars have been invested into product management by producers rather than taxpayers and governments. US states and localities have passed EPR laws on 17 product categories (i.e., industry sectors). By consistently engaging with more than 25 industry sectors in the United States for more than two decades, PSI and others have laid the foundation for future EPR programs throughout the country. Passing laws demonstrates what is politically possible for addressing the problems identified. Implementing these laws provides experience that allows new possibilities to unfold and improvement to take place.

When bills are implemented, we have an opportunity to continually learn so that we can refine the steps needed to reach our goals. US companies may finally be easing up on their resistance to EPR. Packaging brand owners and manufacturers have begun to engage seriously in discussions, and an increasing number have even supported packaging EPR bills. We only need to look at the transition to EPR in Europe and Canada for packaging, electronics, and batteries to understand how changes will likely take place in the United States over time. European packaging EPR systems have been operating for up to 35 years, and those in Canada for over 15 years. The United States can learn from these experiences and avoid making the same mistakes. But we must also recognize that we won't get to where these countries are in an instant, and our way of implementing these programs will reflect our own governments, cultures, history, and experiences.

Our best chance at having effective EPR laws is to pass laws that contain best-practice policy elements. Successful implementation hinges on companies genuinely taking responsibility, and government political will to ensure that they do. It takes a team approach that does not end when the law is passed but continues to evolve through program implementation, evaluation, and refinement. Our collective goal must be that all stakeholders eventually take pride in the implementation of EPR programs. These programs protect people's health and the environment. So how do we get there?

ACHIEVING PROGRAM GOALS

As covered in the previous chapter, collective stakeholder agreement on the problems that need to be addressed provides the foundation upon which change takes place. The second most important element is agreement on program goals, which are the results sought through bill development and implementation. The cycle, from identifying a problem to gaining support to address it through legislation, is then followed by program implementation, evaluation, course correction, and monitoring. It is a process of continuous improvement. Ideally, throughout this time, stakeholders will collaborate to stay on track and achieve their collective goals.

A host of challenges always threatens program implementation, including staff turnover and shifts in company and agency priorities. Stakeholders can overcome these delays through continual education, hiring new staff, mentoring, and written agreements that establish institutional continuity of intent. The more difficult challenge to overcome is company or industry lack of commitment to meeting the goals of a law. This failure to commit will lead to lax interpretation of bill requirements and responsibilities, discord with the

TEXTBOX 7.1 STANDARD SET OF EPR PROGRAM GOALS

- Reduce harm to people and the environment
- Reduce material use
- Increase material reuse and recycling
- Ensure safe disposal
- Reduce greenhouse gas emissions and other environmental impacts
- Achieve social equity
- Achieve efficiency
- Internalize costs into product price

government oversight agency and other stakeholders, halfhearted attempts at implementation and, at times, agency enforcement action.

An example of a general set of program goals, which ideally have been developed jointly through a multi-stakeholder group, is provided in textbox 7.1. These general goals are then specified through the development, enactment, and implementation of an EPR law. The important thing is that each EPR law will have its own goals, and every aspect of EPR program implementation will be directed toward achieving them.

STEWARDSHIP PLANS

The central component of every EPR law is a stewardship plan that producers, either individually or collectively, submit to the government oversight agency for review and approval. This plan represents the blueprint for how a program will be implemented and how the program goals will be met. It includes a detailed description of how producers plan to meet the requirements in the law. All stewardship plans must be reviewed and approved by the oversight agency within a given time period as prescribed in the law. If it is not approved, there is a process and a timeline for the agency to provide comments to the producer(s) and for the producer to resubmit the plan.

Let's use paint EPR laws as an example. PaintCare, a nonprofit organization that represents paint manufacturers (i.e. the industry's producer responsibility organization, or PRO), will submit a paint stewardship plan that specifies existing and planned locations from which a contractor to PaintCare will collect leftover paint. The plan will show how the convenience standard will be met; paint laws typically require 90 percent of residents in a state to be within 15 miles of a permanent collection location (e.g., a paint retailer or household hazardous waste collection facility). The plan will also include the locations of facilities where collected paint will be reused, recycled, or (in the case of unusable paint) disposed of. PaintCare will propose a minimum number of annual paint collection events for rural areas to ensure convenient program access. The plan will also specify the target audiences for the type of consumer education and outreach that will be implemented, as well as languages into which materials will be translated. If the law requires that the program be evaluated to show an increase in consumer awareness and proper disposal behavior, the methodology for this study will be proposed in the plan.

The stewardship plan is also an effective way to provide flexibility to producers around program factors that cannot be determined when the bill is written. For example, paint laws require that consumers pay an eco-fee to fund the program but do not specify the exact amount those fees should be. After bill passage, PaintCare will estimate the full program cost that is based on a range of factors including number of collection sites needed, volume of paint expected to be collected, transportation costs, and education and outreach needs. Once the total cost is determined, PaintCare will propose a fee scale based on paint sales volume (i.e., quart, one-gallon, five-gallon). All this, and a lot more, will appear in a stewardship plan.

In general, a good EPR law will set out the requirements that producers must meet, providing the program's basic structure. A well-written stewardship plan, submitted by the producer(s) to the oversight agency, will put "meat on the bones" and specify exactly how the requirements will be met. The general concept to understand throughout the development of an EPR bill and its implementation is that achieving results is not a static process. Goals are defined in legislation to the best of a stakeholder group's ability. What cannot be fully articulated during the legislative process will need to be defined during program implementation. The key is that the legislation must be written so that it *requires* the development or refinement of goals during implementation. In legislation, establishing an institutional mechanism, such as an advisory council, will help to assure multi-stakeholder input throughout the process.

For example, if it is clear that an agreed-upon recycling goal is 75 percent by 2028, it should be put in the legislation. If it is not clear, the law can require that a recycling rate be determined through the stewardship plan process. For another example, consider that there is general consensus among most producers that postconsumer recycled (PCR) content requirements can stimulate the market for specific recycled materials. What might not be readily known is the viable percentage of PCR content for a given material (e.g., polypropylene) in a given product (e.g., yogurt cup). Stakeholders will need expert analysis and group consultation to derive the percentage of PCR that is achievable for that polypropylene yogurt cup, so that they can specify a content requirement.

Thus, a law might require producers to outline, in the stewardship plan, their methodology for proposing a PCR content goal by a certain date and include a timeline for implementation. The plan might also include multiple dates for the rate to be increased so that there is an expectation of continuous improvement. Although uncertainty may exist when a law is written about the specific PCR rate, producers can commit to meeting a goal that will continue to be stretched. The stewardship plan, along with an ongoing advisory council, are the vehicles for realizing this commitment.

The plan represents an opportunity for an agency to clarify legal ambiguities or fill in details that are purposefully left flexible in the legislation. An important aspect of effective EPR legislation is that a bill gives an oversight agency the clear authority to require producers to establish additional collection sites, additional educational materials, more outreach, and other initiatives to meet the goals and intent of the law. Unless this authority is clearly articulated in the law, however, producers will usually contest being required to do additional work. Best-practice EPR laws require that stewardship plans be periodically resubmitted by producers and reapproved by oversight agencies, usually every five years. If a major program change takes place, producers are often required to resubmit a plan at that time, even if it is earlier than five years. Some laws have allowed oversight agencies to waive plan resubmittal if a program is operating smoothly, since the process is time consuming and costly for producers. This stage is important, however, since it is a built-in mechanism to reevaluate a plan. It is a recognition that with the passage of time comes change—and with experience, new perspectives will emerge.

PRODUCER COMPLIANCE: INDIVIDUAL VS. COLLECTIVE

Once an EPR bill is signed by the governor and becomes law, one of the first questions people ask is, "When does the program start?" Producers have a prescribed time, set in statute, to submit a draft stewardship plan to the government oversight agency that describes in great detail how they plan to comply with the law and meet pertinent requirements. To begin the program, producers must have a plan approved by the oversight agency. Since most US EPR laws allow producers to comply either individually or collectively through a producer responsibility organization (PRO), most producers choose to collectively hire a PRO since it is more cost effective and less complex. Instead of each producer providing their own stewardship plan and hiring their own contractors to collect and process their share of products, producers can collaborate on hiring a PRO to perform these tasks on their behalf. This is a particularly attractive option for industry sectors in the United States that have extensive experience working with their trade association, which will often set up another organization, usually a nonprofit, to serve as the PRO.

Similarly, most government agencies would prefer to review one stewardship plan from a single PRO rather than multiple plans submitted by multiple PROs. For this reason, Connecticut requires producers to join single PROs to manage paint and mattresses. In addition, three of the four states enacting packaging EPR laws—Maine, Colorado, and California—require producers to form a single PRO, with Colorado allowing additional PROs after the fifth year of operation, while California allows additional PROs after the seventh year. Even so, most states prefer not to create a monopoly, or a system that appears as one, even though they might prefer interacting with one PRO. They will instead allow multiple PROs to operate, each performing compliance tasks for multiple producers. But these states might also make the multiple PRO option less attractive by requiring each PRO (or individual producers) to meet the same statewide convenience standards, performance goals, and other requirements. Although requiring each PRO or individual company to provide collection coverage throughout the entire state is duplicative, it serves as a not-so-subtle way to encourage a collective arrangement with other producers. At the same time, allowing multiple PROs is also a hedge against producers being stuck with a PRO that does not serve their interests and allows flexibility for industry sectors with effective existing take-back programs (e.g., agricultural containers and barbecue tank cylinders).

For this and other reasons, most current US EPR laws have been implemented by one PRO per industry. For example, the American Coatings Association (ACA) set up PaintCare, a nonprofit organization that has the same companies on its board of directors as ACA, to act as the PRO for companies obligated to comply under paint EPR laws. Thermostat manufacturers had already set up a PRO, the Thermostat Recycling Corporation (TRC), for its voluntary stewardship program, and continued to use this organization when take-back programs became mandatory under EPR laws. The International Sleep Products Association (ISPA), which represents mattress manufacturers, set up the nonprofit Mattress Recycling Council (MRC) as its PRO; the Carpet and Rug Institute (CRI) set up the Carpet America Recovery Effort (CARE); and the automotive industry set up the End of Life Vehicle Solutions Corporation (ELVS) to collect mercury vehicle switches and other substances of concern. The onset of US packaging EPR bills has made the issue of program governance more prominent. We have, therefore, seen a plethora of proposed options, including a single government-authorized PRO under the Maine packaging EPR law, a single PRO during a program's initial years before allowing multiple PROs (five years in Colorado and seven years in California), and multiple PROs allowed in Oregon.

MANAGING MULTIPLE PROS

Although most US EPR laws allow producers to form multiple PROs, US governments are only now beginning to contend with the complexity of having a second PRO enter the product take-back market. MED-Project was the PRO established for the pharmaceutical industry (Pharmaceutical Product Stewardship Work Group) after EPR laws that first passed in 2012[1] required the industry to finance and manage drug take-back programs. Years later, Inmar, a company that collects and disposes of outdated pharmaceuticals from retail pharmacies, succeeded in establishing itself as a second PRO in the pharmaceutical take-back industry.

Inmar forced the question for US governments of how they would manage multiple PROs. One of the early lessons learned by European EPR programs was that, once there are two PROs, there is a need for a coordinating entity. Once MED-Project and Inmar became competitive PROs in the pharmaceuticals EPR take-back market, PSI received several requests from our government agency members for advice on how best to coordinate two PROs. A main interest of governments is to have a seamless program, which means that the educational materials provided to consumers are consistent and the sites at which pharmaceuticals are collected are available to all consumers. The most important program aspects to be coordinated are the collection kiosks, mail-back envelopes, program signage, a single website (owned by the government but managed and funded by industry), a call center, and educational materials.

One option for government agencies is to coordinate the multiple take-back programs themselves and ensure that the convenience standard is met. They will then need to become the arbiter between two or more companies to assure they all pay an equitable (and defensible) cost share. Although some governments might choose this option and accept this responsibility, others may prefer to leave the coordination to the PROs themselves. Although this was not a natural option, it became the preferred approach by Inmar and MED-Project. They reasoned that working together as competitors was still better than another layer of government intervention. As long as a

1. The first US pharmaceutical EPR law was passed by Alameda County, California, in 2012, but implementation was held up until 2015 by legal challenges from the pharmaceutical industry that ultimately failed.

government agency *requires* PROs to work together and submit a consistent plan, they will have to do so.

The electronics industry chose another option—an industry "clearinghouse," which has only been implemented once to date in the United States, in Illinois. This strategy became part of a negotiated agreement between the Consumer Technology Association (CTA) and Illinois state and local governments to shift the management of their electronics EPR law from one requiring producers to meet a per-pound performance goal to one guided by a consumer convenience standard. A key element of the amended law is that producers, coalesced under six service areas, are assigned responsibility by a clearinghouse to collect and recycle scrap electronics from consumers at levels equal to their market share obligations. The clearinghouse function is performed by a nonprofit organization, the National Center for Electronics Recycling, which is under contract to a private entity set up by CTA.

The management of scrap electronics under EPR laws, first passed in 2004, presents an interesting case. These laws were some of the first of a new generation of EPR laws. Back then, electronics manufacturers did not assemble themselves under either one or multiple PROs. At that time, many brand companies already had established unique business relationships with retailers and recyclers and were reluctant to join with other companies in a collective PRO. Several companies, however, decided to form their own PRO in 2007; Panasonic, Toshiba, Sharp, Sanyo, and others formed the Electronic Manufacturers Recycling Management Company (MRM), which has now grown to 40 companies. This PRO still only represents a fraction of the market.

States with electronics EPR laws that put a premium on providing convenient and ongoing collection service to residents, rather than meeting only tonnage goals, recognized the need to have a system that coordinated producers. These states, most notably Maine, Vermont, Connecticut, Washington, and Oregon, gave producers a choice: either they meet a convenience standard individually or through a PRO, or they would have to pay a government-contracted recycler to collect and recycle their market share of scrap electronics. These government-run programs have since become some of the best performers nationally due in part to the control they maintain over the program. Being government run, however, is perceived by CTA as being more expensive than other programs. As a result, CTA introduced legislation in 2021 in Vermont and Maine that sought to replace the current systems with ones governed by a private clearinghouse. Neither bill advanced in the state legislatures. Although governments typically do not want to play a prescriptive role in implementation, they also need assurance that the overall goals set out for EPR programs (e.g., waste reduction, maximum reuse and recycling, sustainable financing) are being met. In 2022, Oregon sought to modernize its 2007 E-Cycles program by amending its law to, among other things, phase out its state contractor program while allowing for multiple PROs to operate, with required coordination on certain aspects of program implementation. The future of US electronics EPR programs will be determined by how well the stakeholders navigate this landscape that continues to evolve at a fast pace.

EUROPEAN AND CANADIAN EXPERIENCE

With limited experience in the United States in managing multiple PROs, we have looked to guidance from European and Canadian EPR programs. For example, among EU countries in 2018, 26 nations had packaging EPR programs. Of these, 10 countries had one PRO managing the program, while 16 countries had two or more PROs.[2] The need for a clearinghouse (i.e., "clearing center") or an independent entity is recognized as essential to operating an efficient EPR program with multiple competing PROs. One study from the Centre for Economic and Market Analyses concluded that "a clearing center with very strong supervisory responsibilities must be set up to establish binding conditions" for an EPR system.[3]

European systems, which have evolved over the past 35 years, still have a significant amount of variation in whether a country's EPR system is managed by one PRO or multiple PROs, whether the PRO is nonprofit or for-profit, and whether the for-profit PRO is governed solely by producers, a waste management company, or a combination of entities. In Germany, which has 11 PROs managing its packaging EPR programs, a clearinghouse owned by the PROs divides costs among PROs by market share. In addition, the government established a "central agency" (private foundation vested with sovereign rights) in collaboration with industry to oversee and enforce competition and compliance from companies and PROs.[4] Austria has six PROs managing residential packaging, although each is responsible for different materials so there is no competition. The country also has seven PROs for industrial packaging that are coordinated in yet another way. Slovakia has 12 PROs that must be authorized by the government and adhere to strict rules.

2. Jiří Schwarz, Aleš Rod, and Pavel Peterka, *One, or more? How to set up the optimal system at packaging waste management industry,* The Centre for Economic and Market Analyses (CETA Czech Republic, 2018), 17.
3. Schwarz, Rod, and Peterka, *One, or more?,* 20.
4. Central Agency Packaging Register—ZSVR, accessed March 21, 2023, https://www.verpackungsregister.org/en.

Canada, for its part, has experience mostly with one PRO, although this is beginning to change. For example, the Health Products Stewardship Association manages EPR programs for unwanted pharmaceuticals and used medical sharps in four Canadian provinces. The Electronic Products Recycling Association manages electronics for producers in eight provinces. And Product Care manages paint EPR programs across Canada except in Québec, where another PRO, Éco-Peinture, operates the program. Likewise, Éco Entreprises Québec has represented producers of containers, packaging, and printed paper regarding their financial responsibilities for curbside recycling in Québec since 2005 and will continue as the sole PRO as the province transitions to full producer responsibility over the coming years. Similarly, most other provinces with packaging EPR laws have programs managed by a single PRO. The province of Ontario, however, has started to transition all its EPR programs away from a single PRO system to one managed by producers in a competitive system with one or more PROs, administered and enforced by an independent regulator. Ontario's packaging EPR program, which was partially financed since 2002 through a single PRO, Stewardship Ontario, began its multiyear transition to full producer responsibility (including system management) in 2021.

One of the most fundamental issues remains whether it is more efficient for producers to coalesce under one PRO or multiple PROs. This question is still in debate in Europe. Again, citing the Centre for Economic and Market Analyses from data taken from select EU countries, "a competitive system not only does not generate better results than a single operator system but also brings a number of disadvantages such as higher transaction costs, low system transparency, higher administration requirements, higher motivation to fraud, more room for obligation avoidance by producers and system operators, and hence higher demands for regulation and control by the state."[5] Regarding a clearinghouse, the Centre says: "Nevertheless, it is crucial to bear in mind that this center de facto only fulfills the obligations that the PRO does naturally in the single operator system (without the need for additional administrative costs)."[6]

Another study, conducted in 2021 by Adelphi, a consultancy,[7] concluded that competitive systems can create leverage for innovation and efficiency, higher customer satisfaction, and cost containment, although measures that will improve the system through higher costs are often thwarted. By contrast, the study, which focused on European packaging, electronics, and battery

5. Schwarz, Rod, and Peterka, *One, or more?*, 19.
6. Schwarz, Rod, and Peterka, *One, or more?*, 20.
7. Julian Ahlers, Morton Hemkhaus, Sophia Hibler, and Jürgen Hannak, *Analysis of Extended Producer Responsibility Schemes: Assessing the performance of selected schemes in European and EU countries with a focus on WEEE, waste packaging and waste batteries* (Adelphi, June 2021), 86.

EPR programs, found that monopolistic systems run more efficient education and awareness campaigns since increased costs can be passed onto producers without fear of losing customers to a PRO competitor. Monopolies, however, do show evidence of power abuse that can be lessened somewhat through regulation and transparency. The Adelphi study outlined the key factors that influence the performance of EPR schemes, which include the number of PROs operating in a market (i.e., competitive or monopolistic schemes), the legal character of PROs (for-profit or nonprofit) and the different types of ownership structures (producer-owned, producer-controlled, independent, and/or the degree of vertical integration).[8]

PROGRAM EVALUATION AND IMPROVEMENT

There is an adage that you can't manage what you don't measure. And to put measurement in context, you need to evaluate a program. Program implementation requires collaboration among producers, the government oversight agency, and other stakeholders. Small squabbles on data can turn into larger challenges if not managed. Setting joint expectations on data requirements is a key technique to alleviate larger program challenges. To avoid unnecessary tensions, it is best for producers, government officials, and other stakeholders to negotiate this information prior to bill passage. Data metrics would then appear in legislation to clarify expectations.

Alternatively, metrics can be set prior to program implementation through regulation and/or in the stewardship plan, with data being presented by producers in an annual report submitted to the government oversight agency. Annual reports are a standard component of best-practice EPR laws that agencies need to determine whether the program is on track or needs adjustment. Since each data requirement is a cost to producers, its value must be evident and justified. When agencies are unsure of the data they need, they can sometimes seek more than they really need, inadvertently causing pushback and delay from producers.

Data needed in annual reports can also be harmonized across states that pass EPR laws on the same products. For example, in 2014, following the passage of paint EPR laws in multiple US states, PaintCare sought PSI's assistance to standardize data that is submitted in its annual reports to multiple states. The outcome of the all-day facilitated meeting was a set of data metrics that PaintCare staff and state government oversight officials agreed were necessary to determine how the program was operating. Annual reports typically

8. Ahlers, Hemkhaus, Hibler, and Hannak, *Analysis of Extended Producer Responsibility Schemes*, 1.

include information related to program material use, reuse, recycling, and disposal; consumer collection convenience; cost; public awareness; greenhouse gas emissions reduction; progress toward using lower-impact materials; and, finally, whether program changes are recommended. (See table 7.1 for an example of annual report data on consumer collection convenience.)

Standardizing data reporting will reduce costs for all stakeholders. PROs like PaintCare created a model template that staff use to develop standardized annual reports for multiple states, significantly reducing reporting time. State agencies can use that same model and, by knowing their colleagues in other states have also approved the model, will have more confidence in the format and data provided. This type of producer coordination is a basic service of PROs, which provide efficiency in complying with EPR laws, even in a single state. Greater efficiency arises when that same PRO operates across multiple states, which all do in the United States, providing an umbrella administrative function regarding data collection, analysis, reporting, and compliance. Similarly, Circular Materials (previously the Resource Recovery Alliance and, before that, the Canadian Stewardship Services Alliance) harmonizes

Table 7.1 Summary of PaintCare Drop-off Sites and Services

Site Type or Service	*Year 8, FY2020*	*Year 9, FY2021*	*Year 10, FY2022*
Year-Round Sites			
Paint Retailers	650	580	596
HHW Facilities	120	124	122
Transfer Stations	47	49	53
Landfills	6	7	7
Recycling Centers	4	4	4
Other Sites	6	6	6
Paint Recyclers	5	5	5
Reuse Stores	9	9	7
HHW Event Site	0	1	0
Total	793	820	847
Supplemental Sites and Services			
HHW Events	250	274	289
Large Volume Pickups	469	599	646
Recurring Large Volume Pickup Sites	71	70	73
Door-to-Door Programs	16	17	18
Seasonal HHW Facilities	7	7	7
PaintCare Paint-Only Events	11	9	20
Paint Retailers (partial-year only)	8	9	6
Other Sites (partial-year only)	0	2	5

Source: PaintCare, California Paint Stewardship Program Official Documents, accessed March 21, 2023, https://www.paintcare.org/california-official-docs. Annual report data from 2020, 2021, and 2022 PaintCare annual reports, illustrating consumer convenience through detailing drop-off sites and services.

administrative requirements for packaging producers belonging to PROs in multiple Canadian provinces, as does the Extended Producer Responsibility Alliance (EXPRA), which coordinates European nonprofit packaging PROs, the WEEE Forum that coordinates electronics PROs across Europe and other regions, and Electronic Products Recycling Association (EPRA), which coordinates electronics PROs across Canada.

Another useful approach to evaluate an EPR program in relation to its initial goals is to study annual reports from a PRO over time. Analyzing data across years provides trends that can give valuable insights to EPR program managers. Recycling data from multiple years plotted on a bar chart, for example, can quickly reveal whether a program is steadily improving, stagnant, or erratic. This data snapshot gives assurance or provides a reason to further determine challenges blocking steady progress. These analytical studies can be performed by the PRO, by a contractor it hires, or by others using publicly available data.

PSI has conducted several such independent evaluations for various product types. For example, in 2016, PSI analyzed collection rates for all state thermostat EPR programs. These state laws varied from best-practice elements used in Maine and Vermont to whittled-down (weak) laws in Montana and Pennsylvania. Maine and Vermont's successful per capita collection rates are attributable to the fact that they are the only two states with EPR laws that required producers to cover the cost of a $5 bounty given to consumers and contractors upon return of a mercury thermostat.[9] A second PSI analysis that also included voluntary thermostat recycling programs showed that some EPR program laws were so weakened by industry lobbying that they had lower per capita collection rates than some voluntary programs.[10]

Knowing the story behind the data was important since thermostat manufacturers had touted the efficacy of voluntary programs to negate the need for EPR legislation. To test the industry's hypothesis, PSI obtained information about the purported voluntary programs, which revealed that some of the results were due to non-EPR regulations or other components. For example, Massachusetts, which has a weak thermostat EPR law, also passed a regulation on which I worked in the 1990s requiring the state's waste-to-energy plants to remove mercury products prior to combustion. Covanta, owner of multiple waste-to-energy plants in Massachusetts and throughout the country,

9. Product Stewardship Institute, *Comparison of Mercury Thermostat EPR Laws in the US—Program Performance* (March 2014), https://cdn.ymaws.com/www.productstewardship.us/resource/resmgr/1/2014.03.04_Thermostat-EPR-La.pdf. This analysis is the latest comparison of state thermostat EPR laws data available.

10. Suna Bayrakal and Sydney Hausman-Cohen, *Lessons Learned: Voluntary Mercury Thermostat Take-Back Programs,* Product Stewardship Institute (May 1, 2016), 3, accessed July 22, 2021, https://www.productstewardship.us/global_engine/download.aspx?fileid=8A4713F5-F071-4FE7-8640-B4907B37AB36&ext=pdf.

was required to remove mercury thermostats from waste in Massachusetts, but also voluntarily implemented the thermostat bounty concept in Maryland, boosting that state's voluntary collection rate.

PSI's research started with available data, but we sought the reasons behind the varied state program results, which resulted in deriving best practices that could be clearly articulated and backed by facts. This research indicated that any program's performance, whether through voluntary or legislated programs, could be improved with the following actions: (1) provide financial incentives (i.e., a bounty) for thermostat recycling as is done in Vermont, Maine, Maryland, and, most recently, California; (2) implement supporting policies, such as in Massachusetts; (3) take steps to increase convenience for residents and contractors; (4) reach out to wholesalers, contractors, residents, and other stakeholders about recycling opportunities; and (5) educate key stakeholders about the importance of mercury thermostat recovery.

Obtaining program feedback from specific stakeholders is also extremely valuable. This information can be gathered in numerous ways, most notably through qualitative surveys and interviews. For example, state retail associations had an early concern about whether paint retailers would participate voluntarily in the PaintCare recycling program. It was critical to provide data about retailers' experience with the program since they are an important part of the paint collection infrastructure. Retailers operate the tens of thousands of neighborhood hardware stores, paint stores, and lumberyards where we go on a regular basis to maintain our homes. They are usually the most convenient location to return leftover paint. Knowing the degree of retailer satisfaction with the program was critical.

To gain insight, PSI conducted evaluations of the PaintCare programs in Oregon (2012), California (2016), Connecticut (2016), and Colorado (2019). Through these program evaluations, PSI found that the overwhelming majority of paint retailers participating in the program loved it because they could provide a valuable community service to customers at little cost beyond staff time and store space. Small retailers compete against large low-cost retailers, like Lowe's, Home Depot, and Walmart, by providing better customer service. Our data showed that small businesses, particularly independent stores, franchises like Ace and True Value, and smaller chains, appreciated the extra foot traffic that the PaintCare program provided. By emphasizing convenience to residents, who almost always have multiple cans of paint stored at home, these retailers were able to enhance their presence in the community while solving a problem for households.

By March 2023, of the more than 2,400 collection sites across 11 PaintCare programs, 76 percent were small retailers while the others were municipal household hazardous waste centers. Currently, no large retailers have agreed to become voluntary collection sites, stemming from concern that collecting paint would set expectations for them to take back other products they sell. The qualitative surveys conducted for PSI's program evaluations provided feedback showing that the program was successful, including direct quotes from store managers and owners about the value of the program to their business. The few negative comments revealed valuable insights into shortcomings that enabled PaintCare staff to identify issues early and address them. For example, some retailers expressed concern about a lack of storage space, which PaintCare now typically addresses by encouraging retailers to schedule bin pick-up before they reach their storage capacity and by scheduling automatic pick-ups for high-volume locations. Other retailers expressed misunderstandings about regulatory requirements, which led them to believe that state laws did not permit them to legally collect paint. PaintCare addressed these comments by developing improved outreach materials, including fact sheets, brochures, posters, and several short videos, to directly address the most common concerns and ensure that retailers are adequately educated about the paint stewardship program.

PSI has also found it helpful to conduct evaluations of all EPR programs in a single state. For example, in 2017, PSI published a comprehensive evaluation of Connecticut's four EPR laws on electronics, thermostats, paint, and mattresses, which were targeted for EPR by the state due to their toxicity, bulk, cost to manage, and ubiquitous presence in the waste stream. We found that these four programs saved municipalities, in total, over $2.6 million, diverted over 26 million pounds of waste, created more than 100 jobs, and provided additional services valued at more than $6.7 million.[11] The data was derived primarily from surveys of municipal programs and producer responsibility organizations, supplemented by PSI research. This statewide analysis provided justification for the significant efforts of many stakeholders that were instrumental in enacting and implementing the four laws. It also provided momentum to seek to enact additional EPR laws in the state.

Those supporting EPR laws need data not only to improve programs but also to prove to naysayers that EPR works and that the benefits are significant. I don't need to spend time articulating the antics of those with financial interests in maintaining the status quo, or those for whom any EPR policy should be opposed if it falls short of their dream bill. It is important to recognize, however, that if you don't do your own research, there will always be those who selectively pluck data to twist a narrative to meet their political and financial objectives.

There is no current model for how an evaluation should be conducted, thus programs are evaluated in a variety of ways that will evolve over time. Just

11. Product Stewardship Institute, "Connecticut Extended Producer Responsibility Program Evaluation: Summary and Recommendations" (January 2017).

as importantly, EPR program evaluation needs to occur on a regular basis. It is not a one-time endeavor but an iterative process. Program evaluations provide an opportunity for program managers to step back and consider whether the system has achieved the goals established when the law was created or, better yet, how it compares to global best practices. This regimen will keep a program vibrant. Static programs—like static businesses, organizations, and people—stall, wither, and become irrelevant. As additional information becomes available, a more robust picture of program success will evolve. Seeking periodic stakeholder feedback through advisory councils and program evaluation reports will surface issues that need to be addressed.

PSI has incorporated this concept of program evaluation in our model EPR bills, requiring the oversight agency to develop, or contract for, a multiyear report that includes data, summarizes results, and recommends improvements. The cost for this report should be included in the funding that producers provide to oversight agencies. Most government officials prefer that they (or their consultant) conduct the evaluation rather than the PRO, although an option can be for the PRO to choose an independent contractor approved by the agency. The requirement for such an evaluation should be included in the statute, along with the requirement for an annual report.

Of course, we all want our work to be useful. To ensure this is the case, we need to assess our results with input from multiple experts who seek the same clarity we do. No one person is a complete program expert; we each have a piece of the puzzle. Evaluating a program requires various skill sets, like those offered by economists, scientists, lawyers, educators, and marketers. Evaluation also benefits from multiple stakeholder perspectives, like those provided by producers, local governments, state governments, collectors, recyclers, and environmental groups. Facts can easily be manipulated, and we are living in unprecedented times in which, to some, facts don't matter at all. Facts, as well as the conclusions we draw from them, must be discussed, as well as debated, with others. Facts strewn on a table can become an awful mess. Assembling those facts from only one perspective will be limited to that perspective. A structured, facilitated evaluation process will allow people to contribute their unique perspectives. Discussing results together in facilitated dialogue prior to finalizing a report will produce the most useful assessment.

DATA MANAGEMENT SYSTEMS

The accumulation of massive quantities of data that needs to be organized for easy retrieval and analysis poses a challenge for all EPR programs. "Radical transparency" is a term that is often used to refer to the interest of the public and multiple stakeholders who want to know how consumer products are made and how postconsumer products are managed. We have all seen the fancy flow diagrams of materials mined and manufactured into products that are then sold and used, and at end of use are ready to be collected as scrap for reuse and recycling into new products. We put a fancy label of "circular economy" on our goal of creating value all along the supply chain so that there is an infinite loop of regeneration of materials and products.

But let's not kid ourselves. We are still trying to figure out how to do this. The only way we will know is through the use of tools that enable businesses to collect and analyze data, and provide relevant information to key stakeholders, including the government oversight agency, with the assurance that confidential information will be protected. Governments need access to data to ensure compliance, evaluate progress toward the goals, and recommend improvements. Those paying the bill want to know that their funds are being used efficiently and effectively to address the right problems in the right way. Program vendors need data to plan effective collection routes, negotiate fair contracts, and find markets for processed materials. Environmental groups want clear visibility into environmental data for assurance that people and the environment are being protected. These disparate but intertwining needs can be satisfied through data management systems that provide individualized access based on stakeholder type.

Over the past five years, PSI has been approached by well over a dozen data management companies with systems designed specifically for the burgeoning EPR field. Some have had experience managing EPR systems in other countries. Others have experience with recycling and waste systems. The pace of software development and innovation has been rapid, and there is no better confirmation that EPR is the path forward for waste management than to witness the growth of this new industry. Most of all, these systems are needed by producers themselves to ensure that their programs operate efficiently, that collectors are paid fairly for materials collected, that misreporting and fraud are identified and corrected, and that the program is meeting its collection goals. Producers will want to model the collection and processing of different material types through the system, in accordance with differential fees now being required by EPR programs, so they can reduce their costs.

These data systems have the potential not only to improve operations but also to help develop EPR policies that allow stakeholders to monitor upstream and downstream impacts, such as knowing who makes our clothing and under what labor conditions; how the cotton is grown and where other materials are sourced; what environmental and labor protections exist in the facilities that sort used clothing; how much of the clothing we buy is reused or recycled; and the cost of global textile disposal. There is so much we do not know, and we are only scratching the surface with EPR programs. Data systems will be an essential tool for all stakeholders to work together to achieve joint goals of reducing economic cost and health, environmental, and social impacts.

CHALLENGES TO EPR PROGRAM SUCCESS

Our success in implementing US EPR programs hinges on the cultural acceptance of EPR throughout our institutions, including governments, corporations, and the public. It is clear that in the past 22 years the United States has taken significant steps toward this cultural adaptation. In 2000, the term *product stewardship* was used almost exclusively by some corporations to describe internal sustainability measures. Now, it is widely accepted in the United States to refer to a corporation's role in reducing impacts from their products throughout the life cycle, most particularly at product end-of-life. The term EPR, extended producer responsibility, was even more obscure, known only to a handful of academics, federal officials, and others working at the national level. Now, everyone is complaining about the awkwardness of the term, but they know what it means.

In the past two decades, we have moved beyond education about the EPR concept to passing 133 laws on 17 products in 33 states (and counting). Although that might not be considered a saturation across the United States, it is certainly a permeation into a vast number of regions. Not surprisingly, EPR activity is concentrated in states and regions governed by those who believe government has a role to play in regulating corporate activity: the West Coast, the Northeast, and several Great Lakes states. These states are also leaders in developing strategies to mitigate climate change. Implementation of EPR laws has provided US states with broad knowledge that has helped us develop new and better policies based on patterns across products and lessons learned. These two decades of experience have also informed the creation and refinement of PSI's signature method for developing EPR bills, which engages and educates stakeholders through a central model from which new policies are then crafted with modifications for state variations.

The growth of EPR laws, along with other US policies (such as bans on certain single-use plastics) and the growing global environmental movement, have begun to change US culture toward greater societal and corporate acceptance of producer responsibility for materials management. The national and global evolution of EPR will reduce the obstacles to enacting and implementing additional EPR policies. The main challenge at this time is to achieve a balance between producers assuming responsibility for eliminating the life-cycle impacts of their products and governments allowing producers a degree of flexibility to make the transition. Other stakeholders play important roles in assisting these two central players in achieving this balance. Of course, this is not an all-or-nothing situation. State politics, culture, and individual relationships will all play a role. Overall, however, we are seeking nothing less than a cultural shift away from governments being burdened by financing and managing waste, and toward producers taking their rightful place in financing and managing their postconsumer products.

The biggest challenge in program implementation, therefore, is achieving this balance on a large scale, which will ultimately reflect a cultural transformation. Although I have perceived glimpses of this transition, there is often a tension regarding EPR program implementation between government oversight agencies and PROs. Producers all too often haggle with government oversight agencies about whether they have fulfilled their legal obligations, which presumes that, once they do, the public is safe. But how do we plan for the inevitable reality that our scientific knowledge of public safety often changes? There is a strong perception among governments that, once producers have fulfilled their legal requirements, they will resist further efforts to reduce impacts from their products and packaging. By contrast, producers often feel that no matter how well they perform, governments will always push for more, even if the additional benefits might not be proportionate to the added cost. This inevitable dynamic shows the need for regular communication between the regulator and the regulated entities to seek continuous improvement.

Like the environmental movement itself, the relationship between the regulated and the regulators is ever evolving, with both progress and setbacks. The traditional "cat and mouse" game between producers and government officials can prolong implementation, which will only increase impacts due to increased product disposal. It will also create animosity between governments and producers, which can create a downward spiral that weakens program implementation. This dynamic can also lead to a PRO and its members being in legal jeopardy as they teeter on the edge of being noncompliant. Enforcement action by a government oversight agency against a PRO tarnishes that industry's image.

CalRecycle, the state department responsible for California's solid waste management and recycling programs, took legal enforcement action against CARE in 2017, which was finalized in 2021, for repeated failure from 2013 to 2016 to meet recycling and landfill diversion goals under California's Carpet Product Stewardship Law. In three more instances since then (in 2021 and 2022), CalRecycle referred CARE for potential enforcement action for noncompliance with the law, the regulations, and/or its program plan. The California Department of Toxic Substances Control (DTSC) has issued multiple summaries of violation to noncompliant mercury thermostat manufacturers and, in 2016, entered into a consent agreement with 25 of the manufacturers represented by TRC in order to improve out-of-service, mercury-added thermostat collections while avoiding legal proceedings. However, DTSC did issue legal enforcement action against five of the manufacturers that refused to enter the consent agreement, four of which subsequently entered into a separate consent order. In 2021, California significantly amended its thermostat law with a relatively high bounty to incentivize thermostat collection.

Manufacturing and using products and resources inevitably leads to waste. But if each entity is earnest in seeking as little harm and adverse impact as possible, they can form a sustainable relationship that will change our culture and begin to actualize a circular economy. Governments try hard to avoid time-consuming enforcement action against a company or PRO. Noncompliance, therefore, is often egregious before a government will take action. Governments seek compliance and usually provide producers with many chances before pulling the trigger and taking enforcement action. They will issue notices of noncompliance, and often spend countless hours seeking to bring a PRO into compliance and making adjustments within the confines of the law. By contrast, I have also seen unreasonable government actions against companies that are sincerely expending effort to meet program goals and provide the agency what it seeks. In both cases, it is important for those with authority in the oversight agency to resolve discrepancies between their agency staff and those representing the regulated entity. The continuation of EPR programs is essential to reducing harm to people and the environment. Disagreements over implementation cause program interruption and prolong harm.

One example highlights this problem of continued product impacts over an impasse in program implementation: In eight of the 24 US states with electronics EPR laws, producers are required to meet annual recycling performance targets articulated as pounds per capita collected and recycled per year. The eight state agencies overseeing these programs intended that these goals would be a floor—a minimum collection target—that, even if reached, would not excuse producers from continuing to collect. Producers, however, interpreted the goal in these laws as a ceiling that once reached allowed them to stop collecting. This interpretation created significant unexpected costs for municipalities in these eight states that had to pay to recycle the excess amount of electronics their citizens brought to collection sites. Although the legal interpretation made by the Consumer Technology Association, the industry's lobbyist, seems strained, it held up legally. This unfortunate oversight in multiple state electronics EPR laws has become a source of bitterness and disdain for electronics EPR among many local governments, resulting in reduced interest in passing EPR laws for other waste streams in some states. These are the conflicts in implementation we want to avoid.

CONCLUSION

As producers have taken on postconsumer product management responsibility, their resistance has, at times, turned into support. The paint industry, through PaintCare, has mostly embraced their role as managers of a sustainability program that brings paint back into the circular economy. To varying degrees, so do other industries. Their challenge, however, is growing beyond initial implementation to achieve the more comprehensive goals established at the very beginning of a product stewardship dialogue, when stakeholders first gather to consider an EPR solution to the problems created by their products. No EPR program can remain static in an ever-changing cultural landscape. The expectations for EPR programs have changed, often significantly, since PSI began developing these programs in 2000. As mentioned earlier, EPR is a policy principle that seeks to lower environmental, economic, and social impacts all along a product's life cycle. This principle can be achieved through multiple strategies, including through the evolution of EPR laws.

For example, the 2007 mediated agreement with the paint industry, which led to the first paint EPR law in 2009 in Oregon, must evolve. At the time of the agreement, governments set aside many issues in the interest of establishing architectural paint management as the clear focus for a successful dialogue. Left for another day was the management of aerosol cans (with a potential to explode), empty paint cans that are recyclable yet disposed in municipal trash barrels, upstream mining of paint ingredients, maximizing paint-to-paint recycling, and full coverage of municipal paint collection costs. Periodic dialogue, institutionalized in statute, is needed to maintain the collective understanding of the challenges that each of these issues represents and to establish new goals to continually improve programs.

As a second example, the mattress industry has taken great pride in its recycling program over the past eight years. However, unlike the paint industry, which still sponsors legislation in new states without government pressure, the Mattress Recycling Council (MRC) seems to engage with stakeholders only when governments in that state show interest in introducing a bill. It is also unclear whether the mattress industry is willing to update the initial model bill on which they worked with PSI and our members nearly a decade ago. Will they be willing to include current best practices and accept responsibility for managing mattresses that are too soiled to be recycled, or continue to leave that managerial and cost burden to governments?

Without a mechanism in statute to ensure continual stakeholder discussions, these and other questions will be addressed at a pace too slow for the program improvements needed. Challenges that emerge will likely result in a tug of war between producers, governments, recyclers, and environmental groups. We need an institutional commitment to discuss and resolve issues that arise not only on a state-by-state basis, but also on a national level, since issues that impact one state impact all states. When we revisit the goals discussed at the outset of an EPR initiative and refamiliarize ourselves with the context in which a program was first created, we enable all stakeholders to have reasoned conversation, outside of a regulatory context, so that programs can continuously improve in a harmonized fashion nationwide. The best mechanism for such an ongoing review is through a multi-stakeholder advisory council, organized by the state oversight agency (or its consultant) and funded by producers.

PART II

Case Studies

Chapter 8

Paint Stewardship: A US Case Study

I stood in line at the paint store, waiting to pick up several gallons my boss had requested for the jobs facing us that day. I was the newest member of T&T Painting Contractors, and within weeks of starting, my white pants were already stiff with color—reds, yellows, blacks, browns, and blues, dripping, splattered, smeared, and smudged. I was like a walking Jackson Pollock exhibit. This was my first time painting houses. It was 1979 and I had just finished my second year at the University of Pennsylvania where I was studying English. I took the job to make some money before taking off to travel cross-country for a year.

Before I got on the road, though, I spent my time climbing high onto the rooftops of houses, churches, and other structures with a scraper, brush, and bucket to paint my way through the day. In contrast to my studies, painting houses was tangible and gratifying. Our crew would start with a structure that looked tired, faded, and peeling, and within days transform it with fresh coats into a well-dressed, cared-for property. When a job was over, we left the remaining paint with the property owner for touch-ups. It never crossed my mind that one day I would be working with manufacturers to collect, reuse, and recycle leftover paint from job sites all across the country.

This case study illustrates how one event can lead to another, and then another. My love of painting houses led me to promote reuse and recycling as part of household hazardous waste (HHW) collections while I was a founding board member and president of the North American Hazardous Materials Management Association. It also provided me with insight to develop the first HHW strategic plan for Massachusetts in the 1990s, providing collection sheds to municipalities to safely manage paint and other hazardous materials, and offering a statewide collection and processing contract that lowered costs.

My experience in Massachusetts managing leftover paint attracted the attention of the Massachusetts Paint Council, which invited me to speak at its annual dinner, and where I met the plant manager of an in-state Benjamin Moore manufacturing facility, who was also interested in recycling leftover paint. That fortuitous meeting led to the two of us partnering with a Benjamin Moore corporate executive. Together, we developed a pilot project collecting leftover Benjamin Moore paint from the sheds our agency provided to municipalities. That paint was brought back to the Benjamin Moore plant to be mixed with batches of virgin paint. In 1998, consumers never knew that they were buying Benjamin Moore product that might have contained a small percentage of recycled Benjamin Moore paint.

That pilot project led to Benjamin Moore's leadership in the Product Stewardship Institute's (PSI's) first national multi-stakeholder dialogue, which we called the Paint Product Stewardship Initiative (PPSI). The series of dialogue discussions eventually included well over 300 people, including those from manufacturers, recyclers, retailers, state and local government agencies, the US Environmental Protection Agency (US EPA), and others. The dialogue led to an agreement on a model paint extended producer responsibility (EPR) bill that has since been enacted into law in 10 states and the District of Columbia (as of March 2023). These laws are saving state and local governments millions of dollars per year, diverting tens of millions of gallons of leftover paint from disposal into recycled paint, creating hundreds of jobs, and reducing pollution costs and impacts.

Primarily, this case study is a quintessential example of an industry that got ahead of government regulation by collaborating early with agency officials and others to forge an agreement that satisfied all key stakeholders. The paint industry had been highly regulated by government agencies for its past use of lead, mercury, cadmium, and other toxic compounds in their products, as well as the use of volatile organic compounds that contribute to ozone pollution. The industry was well versed with environmental regulations, which required manufacturers to reduce or eliminate these chemicals from their products.

When PSI came knocking at the door to discuss postconsumer paint management, the paint industry got a chance to engage with government regulators in a different manner. They were offered an equal seat at the table to discuss their perspectives and to identify and solve problems together with governments and other stakeholders. Still, the industry could have resisted regulation. They could have ignored our organization or attacked me and PSI for raising concerns. Like many industries, they might have denied our research data, which showed that a significant amount of paint purchased becomes leftover and a minuscule amount of postconsumer paint was being collected and recycled. Instead, the leaders of the National Paint and Coatings Association, now the American Coatings Association (ACA), decided to give a national dialogue a try. They entered the PSI-facilitated discussions with no preconditions other than to seek a joint solution to the safe management of millions of gallons of leftover paint. They knew they could walk away from the table at any time, although they never did. The industry's willingness to engage government officials, recyclers, and others face to face was significant—and it worked.

The consensus that PSI achieved among stakeholders led to model state legislation, which was introduced jointly by the industry and other parties. In states enacting the legislation, the agreement requires producers to develop and manage a system to collect, transport, reuse, and recycle leftover paint, as well as fund it through a consumer fee. The resulting industry-run stewardship program, managed through a nonprofit organization, PaintCare, reduces a substantial cost and management burden on governments, reuses and recycles tens of millions of gallons of leftover paint, and reduces environmental impacts.

Eventually, the attention paid to the collection of leftover paint raised interest in the creation of markets for the newly increased supply of recycled paint, fostering a robust recycled paint manufacturing industry that has now become the backbone of the paint industry's success in bringing leftover paint into the circular economy. Paint recycling companies, including those providing collection services, have benefited greatly from the enactment of paint EPR laws in the United States, which vastly increased the supply of *accessible* leftover paint.

Currently, though, due to limited US markets, about 50 percent of the recycled paint manufactured in the United States is exported. The International Paint Recycling Association (IPRA), which PSI founded in 2019 with a dozen recycled paint manufacturers across North America, seeks to increase the acceptance and use of recycled paint. As consumer interest for sustainable products has grown in the past few years, IPRA members have made great strides in expanding domestic markets for recycled latex paint. Further partnering with the industry, governments, and other stakeholders will strengthen IPRA over time and enable it to continue to expand domestic markets, ensure environmentally and socially sound facility standards, and lower overall paint life-cycle impacts.

Lastly, this case study shows that EPR programs are not static, but evolve in alignment with program experience, stakeholder collaboration, and the global sentiment to minimize environmental, economic, and social impacts. Over the past decade, PaintCare has continually made changes to existing programs and incorporated best practices into new programs. What is innovative at one point will inevitably become outdated unless new ideas are incorporated into program implementation. EPR laws, therefore, need to provide flexibility in anticipation of the need for program change over time to accommodate input from a constantly changing array of stakeholders.

SCOPE: ARCHITECTURAL PAINT AND COATINGS

Early in the PPSI dialogue, which started in 2003, the multi-stakeholder group considered the full range of life-cycle impacts from paint, ranging from the upstream mining and manufacturing of raw materials to the downstream postconsumer management of leftover product. Since the most immediate impacts on state and local governments related to postconsumer paint, the stakeholder group decided that this would be the focus of the dialogue. The group also decided to limit the dialogue's scope to include only household latex and oil-based paints, which the industry calls architectural coatings. Latex paint refers to water-based paints that once contained latex from rubber trees, but now contains plastic (vinyl and acrylic). Latex paints, which comprise more than 90 percent of paint sales—up from 80 percent about 20 years ago[1]—are formulated with water, thinned with water, and cleaned up with water, although they may contain a small percentage of solvents. Oil-based paints, which comprise approximately 10 percent of sales,[2] are solvent-based; the term derives from natural oils that were originally used as binders. The oils have since been replaced by plant-derived, and later synthetically derived, alkyds. These paints are soluble in hydrocarbon and oxygenated solvents, but not water.[3] Omitted from the scope of the dialogue were marine paints, artist paints, and paints sold in aerosol containers (e.g., spray paints), which were considered important to address at some later time.

WHO ARE THE ARCHITECTURAL PAINT AND COATINGS PRODUCERS?[4]

Approximately 1.3 billion gallons of paints and coatings, valued at about $25.2 billion, were sold in the United States in 2020. These products provide two primary functions—decoration and protection. Of the three main types of paints and coatings, architectural coatings, which are used to maintain existing structures (e.g., residential homes and apartments, public buildings, offices, institutions, and factories), account for 63 percent of US coatings by volume (819 million gallons) and 51 percent of sales ($13.3 billion). Industrial and

1. American Coatings Association, email to author, October 29, 2021, with edited document attached, entitled "Market info for Paint EPR Case Study."
2. American Coatings Association.
3. Product Stewardship Institute, *Paint Product Stewardship: A Background Report for the National Dialogue on Paint Product Stewardship* (March 2004), 4, https://cdn.ymaws.com/www.productstewardship.us/resource/resmgr/imported/Background_Report_for_the_National_Dialogue_on_Paint.pdf.
4. Data in this section was provided by the American Coatings Association, email to author, October 29, 2021, with edited document attached, entitled "Market info for Paint EPR Case Study."

original equipment manufacturer finishes, which are applied to manufactured goods as part of the production process, account for 26 percent of US coatings by volume and 29 percent of sales. Finally, special purpose coatings, including traffic paints, automotive refinishing, high-performance coatings for industrial plants and equipment, and coatings for marine structures and vessels, account for the remaining 11 percent of US coatings by volume and 20 percent of sales.

Three multinational companies accounted for 81 percent of the 2019 US market share of architectural coatings: Masco, PPG Industries, and Sherwin-Williams. By contrast, three other national companies—Benjamin Moore, RPM, and True Value—represented 8 percent of the US market share for architectural coatings. The remaining 11 percent of the US market was comprised of mostly regional and local formulators, such as BLP Mobile Paint Manufacturing, California Paints, Diamond Vogel, Farrell-Calhoun, Harrison Paint, Hirshfield's, Kelly-Moore, Lanco, McCormick, Miller Paint, Muralo (Norton & Son), Vista, and Yenkin-Majestic. Over the past 20 years, the trend has been for national architectural coatings suppliers to acquire regional brands. Although sales of online and boutique brand sales (e.g., Claire, Back-Drop) are currently less than 1 percent of market share, this amount is expected to increase.

WHAT'S THE PROBLEM?[5]

Each year in the United States, about 82 million gallons of architectural paint,[6] or approximately 10 percent of what consumers purchase, go unused.[7] Those of us with paint stored in our sheds, basements, and garages know this firsthand. A conservative cost estimate for collecting and reusing, recycling, or otherwise properly disposing of this paint is about $8 per gallon.[8] If all leftover paint in the United States was properly managed, it would cost local

5. Some information in this section originally appeared in the *Product Stewardship Action Plan for Leftover Paint* (March 18, 2004), Product Stewardship Institute, Inc., https://psi.wildapricot.org/resources/Paint/2004-03-ActionPlans-Paint-Paint-Product-Stewardship-Action-Plan.pdf. Current data points have been provided by the American Coatings Association, 2021.

6. This estimate is derived from data provided by the American Coatings Association, as previously cited. Total paint and coatings sales in America equals 1.3 billion gallons, of which architectural coatings account for 63 percent by volume (819 million gallons). Ten percent of architectural coatings are left over, equal to 81.9 million gallons.

7. Abt Associates Inc., *Quantifying the Disposal of Post-Consumer Architectural Paint,* US Environmental Protection Agency (April 2007), 6, https://archive.epa.gov/sectors/web/pdf/paint_quantity_report.pdf. "The amount of architectural paint that is disposed of each year is equivalent to approximately 6 to 16 percent of sales, with a best estimate of 10 percent."

8. SCS Engineers, *Paint Product Stewardship Initiative Infrastructure Project,* Washington Department of Ecology as part of the Paint Product Stewardship Initiative facilitated by the Product Stewardship Institute (March 2007), 8–4, https://www.productstewardship.us/resource/resmgr/Paint/Paint_Infrastructure_Report_.pdf.

governments and their taxpayers more than $650 million annually.[9] Since local governments can't afford to spend that kind of money to manage all leftovers, most officials instruct residents to dry out latex (water-based) paint and place it in the trash,[10] although some residents also illegally dump it down storm drains or throw liquid latex and oil paints in the trash.

Disposal of paint in the trash can cause health and environmental impacts. Although many states prohibit liquid paint disposal, rarely is it enforced because it is hard to see paint cans that residents put in garbage bags and barrels. It is also difficult to dry out paint enough to put it in the trash, unless there is only a very small quantity remaining. Drying paint in the container is time consuming and rarely effective, and renders the container, which is often recyclable when empty, now unrecyclable. Liquid paint can splatter in garbage trucks and streets, leach from unlined landfills into waterways, and kill organisms that degrade sewage in wastewater treatment plants. Oil-based paints are particularly difficult and expensive to manage. But even household latex paints can contain low levels of volatile organic compounds, fungicides, and (in the case of very old paint) hazardous metals such as mercury, lead,[11] cadmium, and hexavalent chromium.

GOALS FOR PAINT STEWARDSHIP

When I painted houses and commercial buildings, my goals were to buy a good product, buy enough so I wouldn't have to waste time going back to the store, and do a good job. Reducing the amount of paint that became left over from a job, or finding a way to reuse or recycle what became left over, was not a consideration for me back then. Fortunately, those days have changed. For the PPSI dialogue, PSI conducted extensive research to identify the problems outlined above, the goals that stakeholders wanted to achieve, the barriers to reaching those goals, and various potential solutions. Through this research, it became evident that one of the best ways to understand the complexity of sustainable paint management was to follow the circular path of postconsumer paint (see solid-line paths in figure 8.1).

Paint management starts with buying the right amount for a job. This source reduction stage (Step 1) is, of course, the most cost-effective and

9. This estimate is derived through the following calculation: Total paint and coatings sales in the United States equals 1.3 billion gallons, of which architectural coatings account for 63 percent by volume (819 million gallons). Ten percent of architectural coatings are left over, equal to 81.9 million gallons. At $8 per liquid gallon to properly manage leftover architectural paint, the cost would be approximately $655 million per year. See related citations for other references.

10. In California, latex paint must be handled as hazardous waste, unless it is destined for reuse or recycling.

11. Lead was phased out of use by manufacturers in the mid-1950s, but EPA did not ban its use until 1978; American Coatings Association, previously cited.

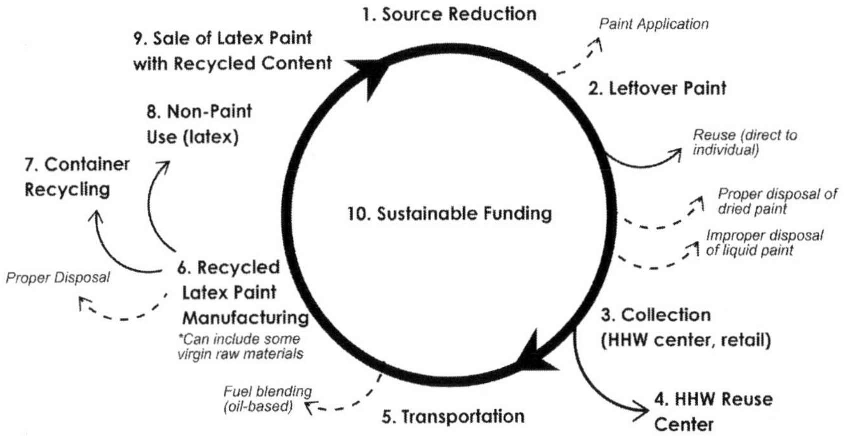

Figure 8.1 Leftover Paint Management: Solid lines represent materials staying in the circular economy (even if for one cycle only); dotted lines represent materials leaving or outside of the circular economy.

Source: © 2023 Product Stewardship Institute, Inc. This figure has been modified from a figure that appeared in *Product Stewardship Action Plan for Leftover Paint* (Product Stewardship Institute, Inc., March 2004), 6, https://www.productstewardship.us/resource/resmgr/imported/PaintProductStewardshipActionPlanFINAL3_18_04.pdf.

environmentally sound management strategy. It is challenging to buy the exact amount of product needed and use it all up, even with small sample sizes, computer-assisted technologies that help consumers buy the right amount, and retail staff advice. If the right amount is purchased for a job, there will be nothing left over to manage and no cost to manage it. But if leftover paint remains, as unfortunately it usually does, many people think they will use it for future jobs or touch-ups, but won't (Step 2). Storing this paint, with the container lid securely closed, in a location where it will not spoil in hot or cold weather, will preserve its value. If what is left over is in good condition, the most sustainable choice is to give it to a neighbor, friend, or someone else who can use it up. We call this reuse. For paint that becomes spoiled, hardened, or otherwise unusable, it can be put in the garbage if dried out (i.e., proper disposal), which is legal in all states except California.[12]

Our goal, however, is to efficiently collect, reuse, reblend, and recycle[13] as much leftover paint as possible. Convenient, well-publicized locations

12. California prohibits all paint from disposal in household garbage.

13. Paint reuse refers to paint given away to another user in its original container. Reblending refers to pouring out leftover paint from multiple containers into a larger container, mixing it together manually, and usually donating or selling it at low cost in a community. Recycling refers to reblending or remanufacturing postconsumer leftover paint, but not to in-plant recycling of manufactured materials. Reblending and recycling leftover paint into paint are higher on the paint management hierarchy than recycling into non-paint products (e.g., garden stones or cement), which is followed by fuel blending (energy value from combustion), landfill cover (as a beneficial use), and lastly landfilling as a waste.

for residents to bring back their leftover latex and oil-based paint discourage wasting resources, and instead give people opportunities to take personal responsibility for managing any leftovers (Step 3). There are two general types of leftover paint collection sites: municipal HHW centers and private retail stores (generally paint or hardware stores). Paint collected at HHW facilities and events is sometimes sorted, with full or mostly full cans pulled aside for later reuse by residents (Step 4). Some HHW facilities will go a step further and mix together multiple cans of good-condition latex paint, then pour the reblended product into one- or five-gallon containers for use by residents and painting contractors for free or at a nominal cost.

Since reused and reblended paints are not transported downstream for processing, they lower program costs. Interestingly, although reuse and reblending are higher on the paint management hierarchy, these uses can conflict with the goal of establishing a vibrant recycling industry, since reuse, reblending, and recycling all rely on having the highest quality paint, which has the most value. Striking the right balance between these competing goals is an important element of a robust paint EPR program. Empty paint containers collected at HHW centers are recycled, and spoiled or hardened paint is disposed of. Most of the paint collected at municipal HHW sites, and all leftover paint collected at retail locations, then continue the journey through the circular economy. The majority of collected paint goes to a transporter's facility for sorting and repacking (Step 5), although some is transported directly to a recycled latex paint manufacturer (Step 6).

The goal at Step 6 is to cost-effectively produce high-quality recycled latex paint that can meet market demand. To do so, this step starts by sorting all material received into latex and oil-based varieties, with most oil-based paint being repackaged and transported to cement kilns for fuel blending. The latex paint is first sorted according to its usability in making recycled paint. Latex paint deemed suitable for recycling is further sorted by color, sheen, finish, and other variables, then poured into large batches, filtered, processed, and often mixed with various amounts of virgin paint or additives to meet market specifications. The packaging in which paint is sold, in this case metal and plastic containers, and sometimes a hybrid (a plastic can with a metal lid and handle), is also recovered and recycled to the extent possible (Step 7).

Most recycled paint manufacturers make large batches of about a dozen colors that are readily available for sale. Many of these companies can also match any virgin paint color with advanced notice and a commitment to purchase a specified volume. About 71 percent of all leftover latex paint that arrives at a recycled paint manufacturing plant is in a condition favorable to being turned into recycled paint.[14] The remaining 30 percent is used to make

14. International Paint Recycling Association, *Annual Report for 2021* (August, 2022), prepared by the Product Stewardship Institute, 4, https://productstewardship.us/wp-content/uploads/2023/02/2022-08-IPRA-2021-Annual-Report.pdf.

non-paint recycled products, such as garden stones, or used as an additive in cement (Step 8). If those markets are not available, the paint is disposed of either by spraying on active landfills as daily cover in place of dirt, burned in a waste-to-energy plant, or dried and landfilled.

Leftover latex paint is a valuable resource that is highly sought by over two dozen North American companies that turn leftover paint into high-quality, low-cost recycled paint. When I was a state environmental official in Massachusetts, our office supported several paint recycling companies with significant funding, including conducting demonstration projects at public facilities to prove the efficacy of recycled paint. Back then, without EPR laws to create a large, consistent supply of available leftover paint and the funding to hire recycled paint manufacturers, our efforts did not take hold. Only in the past 15 years, with the enactment of paint EPR laws covering about 27 percent of the US population, the recycled paint industry has begun to flourish with newfound markets for the sale of recycled paint (Step 9).

Creating reliable markets for recycled paint is critical to the sustainability of the recycled paint manufacturing industry, as well as to providing a consistent industry solution for leftover paint. Achieving a stable recycled paint industry is a high-level goal of a paint EPR program. However, to achieve this goal, consistent and sufficient funding is needed to cover the cost of properly managing postconsumer leftover paint. For this reason, sustainable financing (Step 10) has been central to the successful management of leftover paint in the United States. This funding contributes to the larger local economy since paint EPR programs create jobs, not only at the recycled paint manufacturing facilities, but also in the collection and transportation of leftover paint from municipal and retail locations to paint recyclers.

These simple steps in leftover paint management—source reduction, collection, reuse, recycling, and proper disposal (as a last resort)—were developed by consensus with the full PPSI and led to a July 2004 one-page fact sheet entitled "Be Paint Wi$e, Buy the Right $ize."[15] This first PSI mediated agreement, as part of the paint dialogue, provided significant momentum for the group to work on the more difficult issues that lay ahead.

BARRIERS TO PAINT STEWARDSHIP

At some point in their lives, most homeowners stash leftover paint with the intention to use it for another job or touch-ups later. They probably stored that paint thinking it had value, and either didn't use it—instead making a

15. For updated postconsumer paint management recommendations, see PaintCare.org website: https://www.paintcare.org/paint-smarter, accessed February 16, 2023.

different color choice for their next painting job—or tried to use it for touch-ups only to realize that it was spoiled or the sun faded the color on the part of the wall not touched up. Let's say they learned their lesson and went to buy new paint, determined to purchase only what they needed. If their paint store had staff knowledgeable about buying the right amount for a job, they might have offered them small sample jars to test it on the wall. They might even have offered trained staff to help them calculate the amount needed based on the size of the areas to be painted.

Try as you might, though, you likely still bought too much paint. One barrier we all face is that paint is usually sold in limited container sizes (a quart, gallon, or five-gallon bucket) and has a pricing structure that encourages the purchase of a larger-size container. Faced with the choice to buy a quart or to buy a gallon at only slightly higher cost, people usually choose the gallon because they don't want to risk having to make an extra trip to the store to buy more. We are all trapped by the economics of paint manufacturing—the cost of the paint ingredients and container are small compared to the cost to manufacture and market paint. Manufacturing and marketing a gallon of paint costs nearly the same as a quart. And the cost to return to the store to buy more paint is perceived as much greater than the cost of buying more than needed for a job.

Once there is leftover paint, however, another challenge is maintaining its value. Paint can stay in good condition for long periods of time under the right physical conditions, but if stored at cold temperatures, latex paint can freeze and, when thawed, will be unusable. If stored in hot temperatures, latex paint can dry out if the lid is not sealed tight. Only good-condition leftover paint can be reused or recycled. Improper storage is why some leftover latex paint can only be used for lower-level uses, such as garden stones or landfill cover, or it must be disposed of. Even if leftover paint is in good condition once collected, paint color can present another challenge. Light-pigmented leftover paints are more easily remanufactured into a full array of colors than darker shades, which have limited ability to achieve the same color palette. Therefore, there is greater demand for lighter leftover paint colors since they are more versatile, and more valuable, in the secondary market.

A related barrier to returning leftover paint to the circular economy is the lack of domestic recycled paint markets, particularly for darker pigment paints. A greater percentage of darker colors get disposed of as a result. Another market barrier for recycled paint is that retailers have often established longtime, exclusive sales agreements with a virgin paint supplier that does not want competition from other brands, whether virgin or recycled paint. One regional manufacturer of virgin paint, Miller Paint, is unique in that it also sells Metro Paint, a recycled paint made by Metro, a regional government in Oregon.

Other key barriers to achieving paint stewardship goals are a lack of convenient collection sites and an accompanying lack of awareness among consumers about why it is important to recycle paint and where to bring leftover paint for reuse and recycling. We cannot tell people to recycle paint if there are no convenient locations collecting it for no charge. These two key challenges, consumer awareness and convenient collection, tie back to a lack of sustainable financing—and are most prevalent in states without an EPR law regulating the management of leftover paint.

AN INDUSTRY LEADER EMERGES

The problems and barriers to paint stewardship described above existed even more strikingly in 2002, when PSI started to engage the paint industry, than they do today. Luckily, one major factor was in our favor in seeking a solution to managing leftover paint: I had already built a trusted relationship with Benjamin Moore executives several years before, from my time as the Massachusetts waste policy director. Back in the late 1990s, the Massachusetts Paint Council had invited me to give a presentation on our agency's efforts to collect and recycle leftover paint. After my presentation, I took a round of questions. After I responded to a few easy ones, a guy from California Paints stood up and delivered a slow soliloquy blasting the practice of paint recycling. Without missing a beat while he spoke, he poured his orange juice and coffee into his water at the table, emphatically making the point that paint recycling would only lead to contaminated paint just like orange juice and coffee contaminated his water.

As the event ended, I began to shuffle out with the crowd, feeling slightly deflated. That's when Van Stogner, the plant manager from Benjamin Moore's Milford, Massachusetts, manufacturing plant, quietly took me aside. Apparently, Van was already taking back his company's paint from residents in several surrounding towns and funneling it back into the paint production process after extensive analysis and filtering. To Van and Benjamin Moore, paint recycling was not a tale of contamination and disaster. Instead, it was a way to recover valuable material to be recycled back into paint. Van asked me to meet with him to discuss his operation and explore its potential.

At our first meeting, Van invited Carl Minchew, a senior Benjamin Moore executive, and the three of us mapped out a strategy for recycling Benjamin Moore paint in Massachusetts. Early on we hit our first barrier. What I learned from my agency's regulatory staff was that collecting leftover household paint from municipal locations and transporting it to the plant was not technically legal without a special transporter's license. At the time, postconsumer paint was treated as a waste. Since no municipality in Massachusetts had collected paint before with the goal to recycle it, we needed to change our state policy to align with the new reality we wanted to create.

Fortunately, the idea of treating waste as a commodity had already been introduced by the US EPA several years earlier with a federal regulation called the Universal Waste Rule, which is one of the most innovative federal waste policies ever enacted. While the idea of relaxing a regulation to advance environmental progress may seem counterintuitive, the Universal Waste Rule relaxes regulatory constraints on designated wastes destined for recycling, which increases their recovery and reduces pollution. It also lowers management costs. Applying this concept, I worked with our agency regulatory experts to provide Benjamin Moore with regulatory relief by developing a state policy that considered the leftover paint they collected to be a material destined for recycling rather than a waste destined for disposal.

With this issue resolved, we were able to continue, and later expand, Benjamin Moore's collection and recycling operation to more communities in the state. That expansion was made possible through a state grant program that provided municipalities with special storage sheds for the collection of paint and other HHW. Since Benjamin Moore only wanted to collect its own paint (because they did not know the ingredients in other companies' paint), we instructed municipal officials to separate out Benjamin Moore latex paint. Once collected from municipal depots, the paint was brought back to the Milford plant and filtered and tested before being fed back into the virgin paint production process and sold to consumers as virgin paint.

By late 2000, I had left my state job and started PSI, but I continued this project as part of my transition. Once the program in Massachusetts was underway, we recruited government officials from Connecticut[16] and New Hampshire to replicate the program. Together, we set up a unique residential take-back of Benjamin Moore paint at the company's Massachusetts facility. Paint recycling was indeed alive in the Northeast, although in a limited manner. Most importantly, it showed that paint recycling was possible. Although this first-in-the-nation paint producer take-back program was a small pilot and limited to one company's paint, that experience forged an important path to bigger changes that lay ahead.

ORIGINS OF A DIALOGUE

At PSI's first national Product Stewardship Forum in December 2000, solid waste managers from more than 20 states discussed a preliminary *Paint*

16. The Connecticut Department of Energy and Environmental Protection was represented by Tom Metzner, who later became PSI's board president in 2018.

Stewardship Action Plan I drafted for the conference, which outlined the problem with paint disposal, goals for a paint stewardship initiative, and the need for a sustainable solution. Government officials had already identified paint as a top waste that significantly impacted the environment and their budgets. As a result, in 2002, I contacted ACA and invited them to work with us to address the environmental and economic problems associated with leftover paint. In response, a colleague and I were invited to an ACA Architectural Coatings Committee meeting held that year in New Orleans to present our case.

When we arrived at the location, we were quickly ushered through two large doors that led into a cavernous hall. The nearest person was far away, and all committee members were blurs in the distance. I felt as if we were just let into the Roman Colosseum, the stands filled with spectators, and the lions were about to be released. After being introduced by my hosts, I began my pitch. I outlined the problems with leftover paint, including the millions of gallons of paint purchased each year that went unused, resulting in hundreds of millions of dollars in municipal costs. I discussed our goals to collect, reuse, and recycle leftover paint, and the barriers we knew existed. Most importantly, I laid out a dialogue process that would allow the paint industry to voice its perspective, along with other stakeholders, on potential solutions to the problem of managing leftover paint.

After about 25 minutes, I concluded with a key question: would the paint industry join PSI to work on this problem together? By this time, I had convinced myself that the answer would certainly be a resounding "YES!" Or at least a flurry of informed questions. Instead, there was utter silence. I asked if anyone had any questions. There were none. Then, before I knew it, we were swiftly ushered out the door and into the hallway, and my colleague and I planned our trip home. Our day was done.

For two weeks I heard nothing. I kept replaying the meeting in my mind. What did I say? What *should* I have said to make it turn out better? How could I have been so off base? Then, as quickly as the door had shut behind us at the meeting, I got a phone call from the paint industry that opened another door. They accepted our invitation to engage in finding solutions together to the problems created by leftover paint. Instead of facing an utter failure and working on plan B, we were now at the starting line, ready to facilitate PSI's first national multi-stakeholder dialogue. And I was about to design my first-ever national dialogue process.

PAINT PRODUCT STEWARDSHIP INITIATIVE

In 2002, when the paint industry agreed to work with PSI to better manage leftover paint, our government members had data on the costs of collecting and recycling leftover paint through HHW events, which showed that they needed significant funding to expand those programs. After identifying stakeholders whose viewpoints were deemed critical to finding a solution, PSI began to develop a *Paint Product Stewardship Project Summary*[17] (*Paint Project Summary* for short) by facilitating multiple calls with our state and local government members to understand their perspectives on the leftover paint problem, proposed goals for a dialogue, barriers to achieving the goals, and potential solutions to overcome the barriers. (See chapter 6 for more detail about PSI's stakeholder engagement process.) This core group helped establish a foundation of knowledge and perspectives about leftover paint upon which other stakeholders could build. PSI shared the draft project summary with the paint industry, recyclers, and other stakeholders that laid out a process for stakeholder engagement, which included six basic steps outlined in textbox 8.1.

The project summary provided stakeholders with a tangible vision of what a successful dialogue could look like and how it could lead to a consensus-based agreement. PSI invited key stakeholders to provide comments on the document—including the problem, goals, barriers, and potential solutions—which were incorporated into a subsequent version that was sent out to the next stakeholder for review. This iterative process of continual review, comment, and revision showed each stakeholder that PSI was interested in their viewpoint and would incorporate it into the official dialogue document.

The project summary was the first tool PSI used to engage the paint industry and other stakeholders. Our process also included ways for potential dialogue participants to learn together through the joint development of two other documents, the first of which was a comprehensive technical background report on

TEXTBOX 8.1 PROCESS FOR STAKEHOLDER ENGAGEMENT

1. Identify and Contact Key Participants
2. Develop Paint Product Stewardship Action Plan
3. Assess Viability of a National Dialogue/Develop Strategy to Address Paint Problem
4. Convene National Dialogue or Implement Alternative Action Plan
5. Implement Consensus-Based Agreement or Alternative Plan
6. Monitor Agreement or Alternative Plan

17. Product Stewardship Institute, *Paint Product Stewardship Project Summary* (August 2, 2002), https://psi.wildapricot.org/resources/Paint/2002-08-Dialogue-Paint-Paint-Project-Summary.pdf.

paint management, entitled *Paint Product Stewardship: A Background Report for the National Dialogue on Paint Product Stewardship*[18] (shortened here to *Paint Background Report*). This report included sections on paint composition, environmental hazards of paint, paint production, leftover paint management, recycled paint markets, regulatory barriers to collecting and managing leftover paint, product stewardship examples, and major market players.

The *Paint Background Report* lay the technical foundation for a national dialogue by providing key information to enable all stakeholders to effectively participate in the dialogue. A second document, the *Product Stewardship Action Plan for Leftover Paint* (shortened here to *Paint Action Plan*) includes a spectrum of views regarding how to manage leftover paint from interviews with about 40 potential dialogue participants, including government officials, paint manufacturers, retailers, painting contractors, recyclers, and other key parties.[19] These interviews sought further input on the problem, goals, barriers, and potential solutions initially outlined in the project summary. In the midst of developing the *Paint Background Report* and *Paint Action Plan*, the dialogue group collectively assessed the viability of participating in a series of meetings and agreed to take this next big step together.

The *Paint Action Plan* laid out the vision for four face-to-face stakeholder meetings that we hoped would lead to an agreement. The group decided to begin meeting in December 2003, before the two documents were finalized in March 2004, because the collaborative process PSI used to develop those reports and the project summary provided assurance to stakeholders of the legitimacy and objectivity of the dialogue. The project summary, background report, and action plan provided stakeholders not only with critical information and a view of the way forward, but also with a fabric of familiarity with PSI and our process. Those steps ultimately allowed industry executives to become comfortable with an unfamiliar process run by an unknown organization leading down a path to an as-yet-to-be-determined destination. On one hand, it was uncomfortable for those who wanted absolute certainty and control about where the process was heading and where it would end up. But the process also allowed a great degree of freedom to create an innovative solution among those who could trust other participants, and ultimately themselves, to develop a good agreement or walk away if necessary. Most importantly, it was a process that could potentially lead to outcomes to which the entire group could agree.

---

18. Product Stewardship Institute, *Paint Product Stewardship: A Background Report for the National Dialogue on Paint Product Stewardship* (March 2004), https://psi.wildapricot.org/resources/Paint/2004-03-Dialogue-Paint-Paint-Product-Stewardship-Background-Report.pdf.

19. Product Stewardship Institute, *Product Stewardship Action Plan for Leftover Paint* (March, 2004), https://psi.wildapricot.org/resources/Paint/2004-03-ActionPlans-Paint-Paint-Product-Stewardship-Action-Plan.pdf.

Having a tangible plan was essential for participants to commit to meeting and starting down the dialogue path, which ultimately led to strong, and sustained, stakeholder engagement. The three documents provided ample research and fact-finding opportunities for dialogue participants to learn together. As explained in chapter 6, the process we developed to engage the paint industry was based on standard consensus-based techniques, which PSI modified to fit our unique role as a facilitative leader. Although we are not a completely neutral party, neither will we demand a specific solution. Instead, we advocate for product stewardship solutions that meet specific goals and seek outcomes that integrate complex interests, knowing that the best agreements are those made by all key stakeholders with strong commitments to implement them.

DIALOGUE MEETINGS

Once ACA and its paint industry executives agreed to enter a PSI-led national dialogue with government officials, paint recyclers, and other stakeholders, we began the stakeholder process in earnest. The ensuing national dialogue took shape under the moniker of the Paint Product Stewardship Initiative (PPSI). Between December 2003 and September 2004, PSI designed and facilitated four two-day stakeholder meetings to identify and prioritize solutions to the growing problem of leftover paint management. Participants represented local, state, and federal governments; recycled paint manufacturers; retailers; painting contractors; and ACA and its most prominent member companies, such as Sherwin-Williams, Benjamin Moore, PPG, ICI, Dunn-Edwards, and others. The initial meetings included about 40 participants from an initial stakeholder list of about 75 people. That list grew to over 300 by the end of the dialogue.

At the first meeting, in a sunny conference room overlooking the Boston Harbor at the University of Massachusetts President's Office in downtown Boston, I introduced participants to PSI, explaining our role in the dialogue—to provide a forum for objective discussion—and our interest to promote producer responsibility solutions, reduce product impacts, forge partnerships that assigned financial and management responsibility, and get results. I emphasized that PSI would not be advocating for a specific technical or financing solution. Even so, since California legislators had already introduced a paint recycling fee bill and Minnesota officials were considering their own bill, the industry knew that producer responsibility legislation was likely the end result. Their interest was to negotiate with government officials in a national forum through PSI to develop a model state bill for introduction in multiple states rather than conduct multiple independent state negotiations.

At the meeting, I also explained the concept of a dialogue, stakeholder responsibilities to one another, and the definition of consensus before outlining the paint stewardship viewpoints of paint manufacturers, retailers, governments, and recyclers. This technique of speaking on behalf of each stakeholder group (from extensive interviews) lets each person feel they have been heard, add to what is presented, and clarify misperceptions. It also saves significant time so not every person needs to speak on the topic.

Next on the meeting agenda was to confirm the consensus that had generally been reached on foundational items through the pre-meeting interviews and multiple drafts of the *Paint Action Plan.* These items included the dialogue mission ("to develop voluntary initiatives/agreements to enhance product stewardship in the paint industry by January 1, 2005"), dialogue focus (e.g., postconsumer leftover paint), and dialogue purpose (e.g., reduce impacts and costs of leftover paint, and develop a national solution that includes a sustainable financing system). We also walked through a road map that included four meetings and multiple workgroup calls over nine months. In addition, we gave a tangible sense as to what the outcomes might be (e.g., national standards or definitions, model state or federal legislation), and even how that agreement would be endorsed (e.g., letter of agreement, memorandum of understanding).

One key early PPSI group decision was to narrow the scope of the dialogue to "architectural coatings," which is the industry term that refers to household paints. We decided not to include specialty paints like automotive, boat, or artist paints, or paint sold as aerosols. We also decided, as a group, to tackle postconsumer paints, and not become involved in the entire life cycle of paint, which could include material sourcing for production (e.g., the mining of titanium dioxide, which is used by paint manufacturers for whiteness and opacity). Although government participants considered the wider scope of products and the entire paint life cycle important, they agreed to focus the dialogue on postconsumer household paints, which seemed more manageable to address.

Once we gained consensus on these initial items, we moved on to the problem statement and dialogue goals, which included increasing paint source reduction, reuse, and recycling, as well as creating a sustainable financing system. These two items were of critical importance to a successful program, and once participants confirmed their approval, we were ready to launch into topics not previously raised with the entire group. By this time, I felt that I had gained the group's trust and we had momentum. I then highlighted the most contentious issue (also known as the elephant in the room): whether latex paint should be dried and disposed of or reused and recycled. I knew in advance from the stakeholder interviews that this was a key issue. Since

I never want to enter a meeting and be surprised at what stakeholders might say, the only way to avoid such an instance is to speak to people in advance.

Prior to the meeting, I knew that all stakeholders agreed that oil-based paint should not be disposed of in solid waste landfills and waste-to-energy plants but should instead be collected and safely managed. However, I also knew they differed on managing latex paint, which is the vast amount of all paint sold. We designed the meeting to have a frank but respectful conversation that allowed people to speak to each other about their views and positions. As expected, most paint industry executives argued that latex paint should not be collected for recycling since it is not hazardous. Instead, they felt it was simpler to dry and dispose of it in the household trash. Government officials, by contrast, argued that latex paint has resource value no different from consumer packaging collected in curbside bins, and that their residents would be confused by a disposal message. They also said that residents do not distinguish between latex and oil-based paint and it would be futile to try to educate them about this.

With agreement on how to manage oil-based paints, it became clear that the outcome of our dialogue would hinge on the management of latex paint. Isolating this one issue was important because the options for proper management of oil-based paint were clear. With only two recyclers of oil-based paint in North America then (Société Laurentide in Québec and Hotz Environmental in Ontario, Canada) and another added today (Loop Recycled Products, also in Ontario), the main option was, and still is, to burn it for fuel value in cement kilns or other conversion processes. At our first meeting, all stakeholders agreed that consumers should buy the right amount of paint, use what they buy, and reuse what they have left over. They also agreed that oil-based paint should not go into the garbage under any circumstances and instead should be collected, reused, or transported for energy recovery. If, at that first meeting, we had also agreed on how to manage leftover latex paint, we could have signed an agreement that day and started to set up a paint stewardship program. That outcome was not to be, and the policy debate on latex paint remained. Although latex is the least problematic of the paint types, it is the largest component of the problem, and we still needed to figure out an acceptable solution.

At the conclusion of the first meeting, though, we had significant momentum. We had gotten agreement on one major point: that manufacturers were willing to take responsibility for leftover oil-based paint. We also started to explore the differences held regarding latex paint. Most importantly, we began to listen to one another and understand each other's perspectives. After that one meeting, the stakeholders sensed the progress they had made and were now fully committed to resolving more difficult issues that lay ahead.

THE DIALOGUE ROAD MAP

The progress made at this first meeting was only possible due to work conducted in advance. To prepare for it, we were clear about what would take place at the meeting and who would be participating. We provided a draft agenda that participants could review before it was finalized and advised them to review the *Paint Action Plan*, which provided essential background information so that all participants could participate equally. At the first meeting, following introductions, we used a slide deck to focus attention and started by reviewing ground rules for respectful conversation. We went through key aspects of the *Paint Action Plan*, assuming that some participants would not have read the document. We affirmed the goals for the first meeting and confirmed the problem, goals, barriers, and solutions to which they all agreed to individually prior to the meeting.

The beginning of a dialogue often feels like starting a group hike up a mountain. Meeting participants who arrive early have time to get a coffee and a bagel, chat with others who came early, and settle in at the table. This is akin to hikers exiting their cars, chatting in the parking lot, and assembling at the trailhead. Dialogue facilitators, like hiking guides, have a responsibility to keep people safe, make sure they know where they are going, and engage them in collaboratively solving problems as they arise. As the facilitator, by finalizing logistics the night before and early that day, I make time to personally welcome each participant, spending more time with those I do not know. Making a personal connection with every stakeholder is critical to getting to know them and helping them get to know me. Every communication from the facilitator to a participant is another opportunity to engage them in the difficult work that lay ahead. My aim is to assemble our group at the table for the first meeting just as a guide might assemble hikers at the trailhead. We are both about to climb a mountain with a group of people, many who do not know each other. Each group contains a mix of those who are excited to get going and those who are tentative and nervous.

At the first meeting of a dialogue, participants are in a neutral zone, waiting to see what unfolds. They likely won't know what they want to happen and will evaluate every step. During this initial meeting, stakeholders are often wary of other participants and likely somewhat uneasy about how they will respond to others' comments. They might be partly concerned about what will happen next but also partly hopeful that the facilitator can bring them beyond the current situation. From the outset, most stakeholders are willing to be led down a path that will solve their problems, but will recoil quickly if they get the sense that things are not going well. Into this space of mixed expectations, a facilitator has to quickly build rapport, confidence, and capability with meeting participants.

Those most important are individuals with the greatest ability to lead their constituency into an agreement. Facilitators will seek out key participants who have the authority and insight to convince others in their constituency to trust the process and make decisions that can lead to an agreement.

For the paint dialogue, building rapport started with our individual interviews, which fed into the *Paint Project Summary*, and then the *Paint Action Plan*. Those documents were provided to the stakeholders interviewed so they could refine what we heard their comments to be. Providing a draft document that each stakeholder could review allowed them to gain confidence that we listened to them and understood what they said. It also allowed *them* to refine the text, which provided them with authority and control over how their positions would be represented. These initial steps built trust in PSI as the facilitator. Before even coming to a meeting with others, participants got the sense that they would have an opportunity to articulate their positions, and that these positions would be considered legitimate.

To maintain the trust of stakeholders throughout a dialogue process, it is important for a facilitator to present, and follow, a predictable process. The main vehicle I used to provide continual assurance to paint dialogue stakeholders was the *Paint Dialogue Road Map* (see figure 8.2), which outlined four meetings along with the issues and solutions to be discussed *during* the meetings and *between* them. The concept of choreographing a meeting, or a

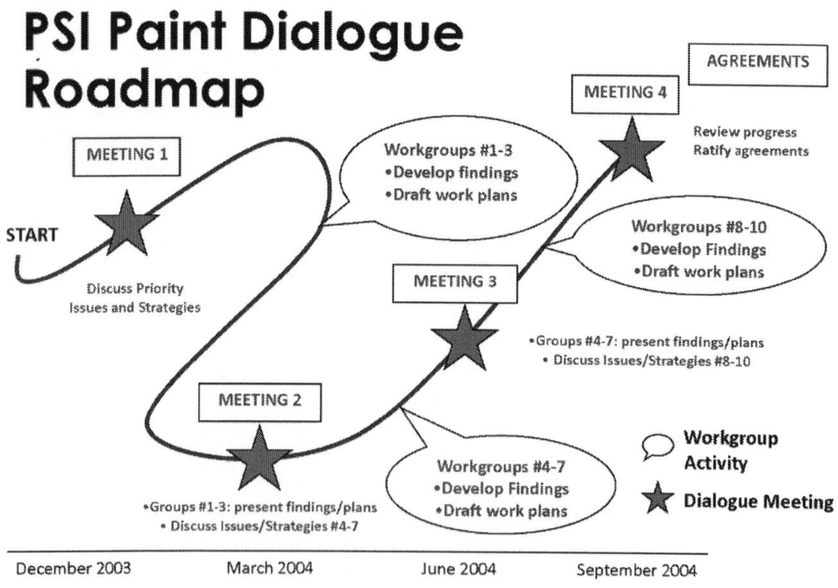

Figure 8.2 PSI Paint Dialogue Road Map
Source: © 2023 Product Stewardship Institute, Inc.

series of dialogue meetings, is outlined further in chapter 6. The road map I designed for the paint dialogue took place about 20 years ago for a specific situation. It was the first national stewardship dialogue that PSI convened, and stakeholder experience with facilitated dialogues, or even product stewardship, was almost nonexistent. The choreography used for the paint dialogue became a blueprint that PSI refined continually as the familiarity with product stewardship and EPR gradually became firmly embedded in the United States. Today, after having designed and facilitated hundreds of meetings and numerous dialogues, we have been able to adapt the process for efficiency based on stakeholder goals as well as time and funding availability. Designing meetings cannot be done solely by using a model. It is not a cookie-cutter approach. Each meeting or dialogue will have its own unique set of problems to solve, stakeholders to involve, and barriers to overcome. Even so, I still use many of the concepts from the *Paint Dialogue Road Map*, which was developed two decades ago, to illustrate basic techniques that can be used in any meeting or multi-meeting dialogue today.

Designing a series of meetings, explaining the process to stakeholders, and then following the process will give confidence to participants that their time is being used wisely and there is a vision for success. Walking paint dialogue participants through our road map, which included the sequencing of meetings, tangibly showed them what would take place at each step in the process. As long as PSI followed the process we established, and they stayed with us, no one would get lost.

During each meeting, I presented a limited number of key issues and facilitated discussion on related potential solutions. The outcome of the initial discussion on solutions at one meeting became the starting place for a work group tasked to come back to the next meeting with specific questions answered. For example, during the first meeting, we presented on three key issues—leftover paint, disposal, and collection—along with potential solutions related to the problems that pertained to each issue. At the meeting, participants volunteered for work groups to further address each of the three issues prior to the next meeting. These meetings would occur virtually, usually with PSI coordinating logistics and facilitating discussion. Since dialogue participants rarely have the time or expertise to schedule, coordinate, and facilitate such meetings, it is often best to use an internal or external professional facilitator.

For the second dialogue meeting, each work group chairperson presented to the full group about solutions discussed for each issue. PSI then facilitated discussion on that issue to refine solutions, develop next steps, and get as close as possible to an agreement. During the second meeting, PSI also introduced four new key issues—sorting (collected paint), container management, transportation, and non-paint uses (for collected paint in poor condition)—then facilitated discussion, established four new work groups (each with a chairperson), and set up discussion for the next meeting. For the third dialogue meeting, we covered the remaining three key issues: (recycled) paint manufacturing, sale of paint with recycled content (markets), and sustainable financing. After each meeting, next steps were clearly delineated, which included the development of a work group to meet through conference calls between in-person meetings. The outcomes of those work groups were presented at the next in-person meeting. At the same time, a set of new issues was introduced. The fourth and last meeting was reserved for ratifying all agreements. The 10 key issues and numerous potential solutions discussed during the dialogue meetings are those outlined in the *Paint Action Plan.* These 10 issues are also closely aligned with those presented previously in figure 8.1 above (the "Leftover Paint Management" diagram).[20]

I borrowed and adapted the technique of reviewing and refining information previously discussed before presenting new information from my first harmonica teacher, Emile D'Amico. Emile, a fine Philadelphia blues, jazz, and classical harmonica player, would review material with me from a previous lesson before launching into a new lesson. Using this technique, I have found that reviewing material with a stakeholder group at the beginning of a meeting sets a solid foundation for new material to be assimilated. It worked for learning harmonica, and it worked for our paint dialogue.

During the dialogue, I also used a mediation technique known as joint fact-finding, in which PSI, as technical researcher and facilitator, assisted stakeholders in identifying key technical questions. We then guided participants in answering their own questions through joint inquiry, often using PSI research and the inclusion of outside technical experts. For the nine months of the dialogue, we carefully gained consensus every step of the way. We planned each meeting, with agenda items placed in strategic order and participants strategically arranged around the table to foster discussion. We also developed conscious strategies for effectively handling difficult stakeholders to keep discussion moving.

Throughout the dialogue, there was ample opportunity for all stakeholders to voice their opinion, and they were kept engaged through concise meeting notes and documents that were jointly developed with participant review. Also important, but often overlooked, was the creation of informal opportunities for participants to develop personal relationships through evening receptions and dinners. Following these encounters, participants often contacted each other to clarify issues or explore new solutions. These orchestrated

20. The actual issues discussed during the dialogue were slightly different than those outlined in the *Paint Action Plan*. They were modified for figure 8.1 to emphasize source reduction and reuse, which became key aspects of the dialogue discussion.

meetings built personal rapport, which proved as valuable to our progress as understanding technical information. Personal relationships are key to deriving technical solutions. It has often been my experience that developing solutions together can also change how people feel about each other, which can often lead to future collaborations.

I would like to convey one last concept that I use as a planning tool: based on my experience, 80 percent of work for most dialogues should take place *outside* of the meetings and the other 20 percent involve activities that *require* meeting, such as developing personal rapport, making decisions, and sealing a deal. When people get together in person (or virtually), it should be to conduct business that cannot be done outside the meeting. Meetings are costly for everyone and take considerable planning for participants' time to be used efficiently. Planning also allows a facilitator to respond more effectively to inevitable unpredictable comments, questions, or events that arise during a meeting, where stakes are higher.

The dialogue proceeded as laid out in the *Paint Dialogue Road Map* (figure 8.2) and did not need to be altered. I presented the road map at the beginning of each meeting to visually depict, and reinforce, the path we were taking, which also helped participants get acclimated to my informal 80/20 rule. The road map clearly set the expectations that much work would take place in virtual work groups between in-person meetings. The work groups functioned to take key issues off-line for intensive discussion by those most passionate about that issue to allow the larger meeting to flow. We always made sure that each work group was balanced among producers, recyclers, retailers, and government stakeholders. These work groups were later transformed into management teams that conducted eight projects that became the focus of the first major agreement.

PROJECTS AND MEMORANDUMS OF UNDERSTANDING (MOUS)

The first paint agreement was captured in a memorandum of understanding (MOU)[21] that was signed and endorsed in April 2005 by over 60 entities, including ACA, state and local government agencies, the US EPA, and recycled paint manufacturers. It took nine months to develop the first agreement, but another six months to get it signed by stakeholders since it had to be approved by ACA's board of directors, the US EPA, and executives at

21. Product Stewardship Institute, *Paint Product Stewardship Initiative Memorandum of Understanding* (April 2005), https://psi.wildapricot.org/resources/Paint/2004-10-Dialogue-Paint-First-MOU.pdf.

numerous government agencies and companies. The agreement committed stakeholders to work for another two years on the eight technical projects to develop a greater understanding of the problem and solutions.

PSI assisted in raising over $2 million, jointly funded by government and industry, to design, facilitate, and complete the eight projects, with assistance from several non-PSI technical consultants. The projects followed the reduce, reuse, and recycle hierarchy. It included source reduction options; a paint reuse manual; a needs assessment identifying the need for, and cost of, leftover paint collection and processing infrastructure; a recycled latex paint standard developed with California Polytechnic State University and Green Seal; market development for recycled paint; and a life-cycle assessment (LCA) and cost-benefit analysis comparing the collection and recycling of latex paint to drying and disposing of it. All projects aimed to identify solutions to eliminate or lower leftover paint volume, management cost, and environmental impact. These projects were key to addressing vital questions asked by stakeholders, such as whether to collect latex paint and containers for recycling (yes), options for reducing the amount of leftover paint (few), areas in the country in need of processing infrastructure (many), and how to finance the system (consumer eco-fee). These projects ultimately laid the technical groundwork needed for the model agreement that became the basis for all paint EPR laws in the country today.

During the ensuing two years, PSI facilitated four additional multiday dialogue meetings to share information and data with stakeholders regarding the eight projects. Again, most of the work took place outside the in-person meetings. By participating directly in the technical projects, ACA developed a deeper understanding of the issue and how it could take a leadership role. They were also able to explore all options for minimizing and financing system costs. In addition, the agendas we developed included presentations from experts from Canadian companies, governments, and stewardship organizations who explained how paint stewardship laws operated in Canada. These experts showed that paint could be successfully collected and recycled through an industry-run program funded by purchasers of paint.

In March 2007, based on the results of the eight projects, ACA's board of directors issued a historic resolution committing to work with other stakeholders toward a national solution for the management of leftover paint. PSI followed by mediating a second paint MOU with PPSI dialogue participants that was finalized in October 2007 and signed by over 50 stakeholders and supporters. The second MOU outlined a timeline for establishing an industry-run stewardship organization that would form a nationally coordinated paint management system. As of spring 2023, this is the only current EPR agreement in the United States signed by the US EPA.

A key element of the agreement has since become known as the PaintCare fee, which legislatively establishes a fee to be paid by consumers on each can of paint purchased. During negotiations, government officials argued for full cost internalization. However, ACA made it clear that any agreement would need to contain a fee (the same funding mechanism used in Canada), and that full cost internalization would result in unequal competition among its members. ACA believed that while some member companies (e.g., Benjamin Moore and Sherwin-Williams) could pass the cost on to consumers directly since they also owned retail paint stores, other companies selling through major retailers (e.g., Walmart, Lowe's, and Home Depot) might be forced to cover the added cost. ACA also insisted that legislation include antitrust protection to allow each company to pass on the same fee amount to consumers, through the purchase price of paint, to cover the program cost.

Numerous times throughout the dialogue process the agreement seemed as if it would fall apart. However, what kept the stakeholders working together was the potential cost savings for government, the ramifications of failure and the threat of unilateral patchwork legislation for paint manufacturers, and the promise of a state-based harmonized system for all parties. I still recall the bizarre feeling I had as the group reached full agreement on the second MOU.[22] I felt as if I was bodysurfing through the dialogue crowd as the decision was made. Benjamin Moore's Carl Minchew later told me he knew exactly what I was feeling. I had, in essence, given myself up to the group, and it was making its own decision. The work, which started as a PSI initiative in 2002, had taken on the identity of the Paint Product Stewardship Initiative, which became "owned" by the group, not PSI. With the burden of steering the dialogue for four years suddenly lifted, I felt weightless as the final decision was made to adhere to all agreements.

THE OREGON MODEL PAINT EPR LAW

The second MOU included a detailed plan for creating a model paint EPR program for the United States with Minnesota as the pilot state, which enthusiastically sought pioneer status. The MOU also called for the rollout of the agreement to the other eight states listed in the MOU timeline. Despite recommendations to ACA to introduce simultaneous bills in several states, the association wanted to focus its resources on translating the MOU into legislation in Minnesota. Unfortunately, after ACA sponsored legislation

22. Product Stewardship Institute, Inc., *FINAL Paint Product Stewardship Initiative 2nd Memorandum of Understanding* (October 24, 2007), https://psi.wildapricot.org/resources/Paint/2007-10-Dialogue-Paint-Second-MOU.pdf.

that passed nearly unanimously in the Minnesota legislature, Governor Tim Pawlenty vetoed the bill, claiming inaccurately that the industry-supported eco-fee was a tax on consumers. The following year, again following near unanimous support from the legislature, Governor Pawlenty vetoed the same bill. However, this time, the PPSI group was ready, having already lined up Oregon as the fallback demonstration state, but not without wasting 18 months in Minnesota.

The Oregon bill was endorsed by ACA, the Oregon Department of Environmental Quality (DEQ), and Metro regional government (representing 1.3 million people in the Portland Metro area), among other stakeholders. The bill was enacted in July 2009, and in July 2010, Oregon launched the nation's first manufacturer-run EPR program for the end-of-life management of left-over architectural paint, nearly seven years after PSI first met with the paint industry. I still recall receiving an email from Jim Quinn, longtime architect and manager of Metro's recycled paint program, the day the Oregon program started. Jim included a photograph of the very first paint can in the United States being recycled as a result of the new EPR law. Writing to me, his email kindly referenced "the twinkle in your eye as we set out long ago to develop a national paint stewardship system, and now we have succeeded."

The Oregon law has resulted in the proper management of an estimated 768,000 gallons of leftover paint each year[23] and provides Oregon governments with direct financial savings and added service valued at over $4.9 million annually.[24] By 2021, these financial benefits for Oregon alone reached $47.7 million over the 10-year span.[25] The funding for the program comes from paint consumers, not governments, taxpayers, or ratepayers. For this reason, program costs, determined through PaintCare annual reports, are considered equivalent to government savings. It is a *direct financial savings* for governments if they are already spending money on paint management. However, it is considered an additional *financial benefit* for governments if they are not spending money to manage paint that eventually gets managed in the PaintCare program. Although an agency does not receive a direct financial benefit in all cases (since they are not spending money in some cases), they do get additional benefits, such as having residents provided with convenient access to leftover paint collection sites, reduced pollution from paint properly managed, and fewer waste management operational costs.

---

23. The annual gallons processed figure is calculated as the five-year average of the gallons processed as reported on PaintCare Oregon annual reports from 2016 through 2020, available at www.paintcare.org.

24. The annual program *value* to municipalities is calculated as the five-year average of the total program *costs* reported on PaintCare Oregon annual reports between 2016 and 2020, available at www.paintcare.org.

25. The sum of total program cost savings as reported on PaintCare Oregon annual reports from 2011 through 2020, available at www.paintcare.org.

Oregon was the first state to translate the model paint agreement into law and provided a legislative template for other states that followed. Each of the laws contains the same basic requirements to which manufacturers, retailers, state agencies, and others adhere. For example, all brands must pay an architectural paint stewardship fee (i.e., PaintCare fee) for each container of paint sold into the state, which they recover from the retailer, which in turn recovers its cost by including the fee in the sale price of new paint. Retailers have the option of informing consumers on the receipt that a paint stewardship fee is included in the cost, and almost all do. This type of financing system was unique at the time, and was later used to finance the first carpet EPR law and four mattress EPR laws in the United States.

Under each of the 11 paint EPR laws (as of March 2023), producers manage stewardship programs funded by consumers with government oversight. Three states currently require retailers to inform consumers, through educational materials provided by producers, about how to reduce their paint waste and where to bring their leftover paint for reuse or recycling. To ensure proper implementation and monitoring of the program, producers in almost all states[26] must pay an administrative fee to the designated state agency for plan approval, oversight, and enforcement and submit an annual report to the agency. Retailers are not allowed to sell products from producers that do not have an approved product stewardship plan indicating their participation in the program. The agency also must report to the state legislature on the outcomes of the paint stewardship program and recommend if any changes should be made to the program. Although some minor variations of the laws exist, the key program elements of each law follow the 2007 mediated agreement (see table 8.1).

STATUS OF US PAINT EPR LAWS[27]

The consensus-based process that led to the development of a model legislated paint stewardship program that multiple stakeholders supported laid the groundwork for harmonization in program design across the United States. Following Oregon's 2009 enactment of the nation's first EPR paint program, nine additional states and the District of Columbia have now enacted EPR laws based on the same model, all with the support of ACA's legal team and lobbyists, led first by Alison Keane and now by Heidi McAuliffe. All of

26. The paint EPR laws in Rhode Island and New York only require manufacturers to pay for plan approval but not additional oversight and administration costs.

27. Part of this section is an edited version of an excerpt taken from: Product Stewardship Institute, "EPR Achievements: The Case of Paint in the US," in *Coatings World* (November 2019), 40–43. Much of the information in the article was derived from published and unpublished Product Stewardship Institute sources, as well as PaintCare annual reports.

Table 8.1 Key Elements of the Paint Stewardship Program

Key Element	*Explanation*
Stewardship Plan Contents	Producers must submit a stewardship plan to the state for review and approval that includes a description of how producers will collect, transport, reuse, and recycle leftover paint according to the waste management hierarchy and environmentally sound management practices.
Governance	Producers have the ability to organize under a producer responsibility organization (PRO)—in this case, PaintCare, a nonprofit organization—which submits a plan to the state and manages the program. Brands not registered with the state under an approved stewardship plan are banned from selling paint. A third-party financial audit is required, in part to ensure that the fees cover the cost of the paint stewardship program only.
Funding Inputs	A fee is remitted by producers to the PRO for the operation of the stewardship program. The cost is passed to dealers and retailers, who add it to the price consumers pay on each can of paint they purchase (currently \$0.30–\$0.49 for cans larger than a half pint and smaller than one gallon; \$0.65–\$0.99 for one to two gallons; and \$1.50–\$1.99 for larger than two gallons up to five gallons).
Collection and Convenience	Producers must meet convenience requirements based on distance and population (e.g., 95 percent of residents must be within 15 miles of a drop-off site, and one additional site must be added per 30,000 residents). Paint retailers can choose to serve voluntarily as a drop-off site to collect leftover paint.
Outreach and Education Requirements	Producers are required to educate consumers about the program, the fee, and the importance of leftover paint management.
Annual Report Contents	Producers are required to submit an annual report detailing paint collection and processing, consumer education, convenience, and financials.

Note: Fee ranges were provided by PaintCare in February 2023.

these programs are in operation (see figure 8.3 and table 8.2). What makes the mediated agreement on paint stewardship unique is that the paint industry sponsors state legislation and lobbies for bill enactment. No other industry group in the United States has taken responsibility for reducing the postconsumer impacts of their products like the American Coatings Association. PSI, state and local government agencies, and recycled paint manufacturers have been integrally involved in enacting these laws and introducing new bills, all in close collaboration with ACA.

Even with a supportive industry and a clear model, passing and enacting legislation require considerable education, persistence, and prolonged stakeholder

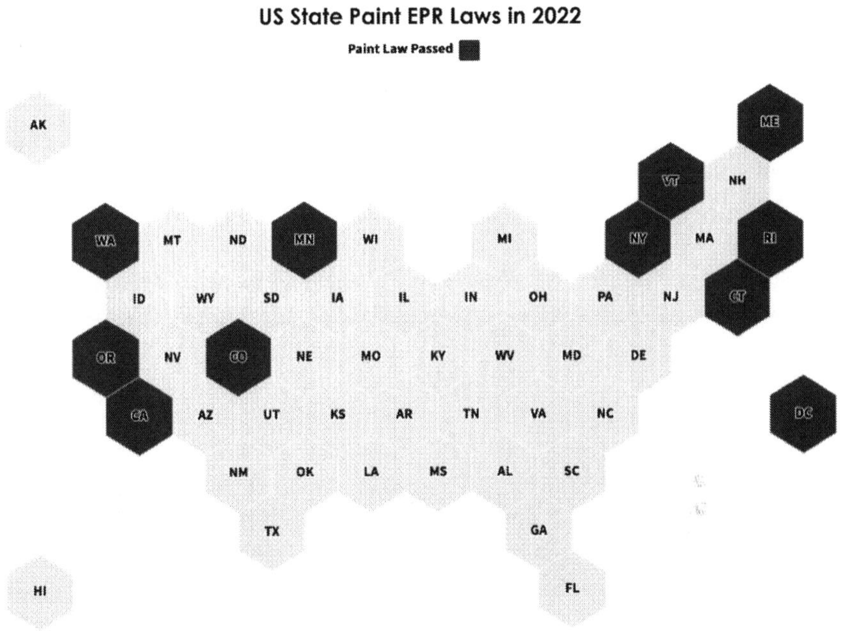

Figure 8.3 US State Paint EPR Laws in 2022
Source: © 2023 Product Stewardship Institute, Inc.

engagement. Bills introduced in New York and Washington State took seven years to get through the legislature to the governor's desk. New Jersey is the only state with a bill to have passed the legislature four times only to be vetoed by a Republican governor twice and a Democratic governor twice. As of spring 2023, states in which paint EPR bills have been introduced but not yet enacted include New Jersey, Massachusetts, Illinois, Maryland, Missouri, Texas, and

Table 8.2 US Paint EPR Laws: Dates Enacted and Implemented

State	*Law Enacted*	*Program Implemented*
Oregon	2009	2010
California	2010	2012
Connecticut	2011	2013
Rhode Island	2012	2014
Minnesota	2013	2014
Vermont	2013	2014
Maine	2013	2015
Colorado	2014	2015
District of Columbia	2015	2016
Washington	2019	2021
New York	2019	2022

New Hampshire.[28] Some of these bills have lingered for up to eight years or more. The long road to enacting legislation, as well as continued interest in the introduction of new bills, points to the need for long-term engagement and collaboration on paint stewardship in the United States.

PAINT STEWARDSHIP PROGRAM PERFORMANCE

By spring 2023, 11 paint EPR laws have been enacted, covering approximately 27 percent of the US population, or 90 million people, and all programs have begun collecting leftover paint. Since data reporting typically begins at least one year after a program is first implemented, data was available for 10 programs as of March 2023. By that time, the PaintCare program had collected more than 62 million gallons of paint, established more than 2,400 voluntary collection sites (76 percent of which are retail sites),[29] and saved local governments and taxpayers over $253 million.[30] Reporting on state paint programs is largely harmonized. In 2014, PSI led a facilitated process in conjunction with PaintCare and state and local governments from six states to create a report template with an outline and key information to be included in each section. The consistent template harmonizes reporting and simplifies comparison among programs. This template is used for all programs with the exception of those in Oregon and California, which had established their reporting requirements prior to the development of the template.

The PaintCare program has achieved considerable success, in part because of the process by which it was developed and in large part because of the strong leadership of its first and current president, Marjaneh Zarrehparvar, who had previously participated in the PPSI dialogue process as a representative of San Francisco's waste management program. PaintCare's main role is to develop the collection site network and facilitate the transport and processing of leftover paint, which also ensures that paint companies are in compliance with state laws. PaintCare performs these functions through staff who manage numerous service contracts for leftover paint management, as well as consumer education. These staff also work closely with local governments to ensure that the program is convenient for residents across the state, in rural areas as well as in urban settings.

Table 8.3 summarizes the most recent PaintCare program performance in nine states with programs that collect leftover paint as of March 2023. The

28. Illinois's legislature passed its paint EPR bill in the House and Senate in 2023; it is expected to be signed by Governor J. B. Pritzger.

29. Most recent program data available from PaintCare annual reports and press releases, available at www.paintcare.org.

30. Product Stewardship analysis of the transportation and processing costs reported in PaintCare program annual reports, and thus avoided by local governments.

Table 8.3 US Paint Stewardship Program Performance by State

	CA	*CO*	*CT*	*DC*	*ME*	*MN*	*NY*	*OR*	*RI*	*VT*	*WA*	*Totals*
State Characteristics												
Population (in millions)	39.5	5.7	3.6	0.7	1.3	5.6	8.5	4.2	1	0.6	7.7	78.4
Urbanization rate	95%	86%	88%	100%	39%	73%	88%	81%	91%	39%	84%	
Year-Round Drop-Off Sites*												
Retail store	652	166	101	8	81	209	244	135	23	68	191	1,878
HHW and Other	206	29	56	1	47	58	30	49	4	11	64	555
Totals	858	195	157	9	128	267	274	184	27	79	255	2,433
Convenience*												
Percent of residents within 15 miles of a drop-off site	99.0%	97.6%	100%	100%	95.3%	94.4%	N/A	96.6%	99.9%	99.5%	95.7	
Paint Processing**												
Annual gallons processed*	3,580,449	806,605	437,799	36,903	136,818	1,007,909	N/A	740,102	92,811	111,847	581,363	7,532,606
Percent of latex paint recycled of amount collected***	67%	85%	82%	81%	82.5%	49%		78%	83%	77%	88.1%	71%

* PaintCare, email communication, April 10, 2023. Based on most recent available PaintCare annual program report data through calendar year 2021 (CA data through June 2022), https://www.paintcare.org/paintcare-states. Annual reports for each state can be accessed from this website.

** PaintCare, email communication, Apri 10, 2023. PaintCare fact sheets by state. Available at https://www.paintcare.org/paintcare-states.

*** Does not include reuse, which ranges from 3 percent to 9 percent of amount collected.

states vary in terms of population size, population density, and land area, all of which can impact the ease with which the industry can meet convenience standards. Programs that are more convenient for community members are expected to recover more leftover paint. California processes nearly four million gallons of paint annually for a population of about 40 million people. The next largest program is Minnesota, at nearly one million gallons of paint for a population of 5.6 million people. The smallest program is in the District of Columbia, which processes 36,000 gallons for a population of about 700,000 people.

Several other factors can influence paint recycling rates, including the presence of a strong reuse network and local climate (which may impact the condition of leftover paint stored by residents).[31] About 20 percent of paint processed through the state programs is oil-based, while 80 percent is latex. Currently, approximately 71 percent of collected latex paint is reused or recycled back into paint.[32] The remainder is recycled into non-paint recycled products, solidified and landfilled, used as fuel, or mixed with a thickening material and applied as daily landfill cover. Most oil-based paint is used as fuel or incinerated; a small amount (less than 5 percent) is reused.[33]

Most year-round collections take place at retail stores, with 1,878 voluntary drop-off sites at paint and hardware stores. An additional 555 year-round drop-off sites are at HHW facilities, landfills, transfer stations, recycled paint manufacturers, reuse stores, and other locations. Each PaintCare program provides significant consumer convenience, ranging from 94 percent to 100 percent of households and businesses having access to a year-round drop-off site within 15 miles of their home or business. By January 2022, the PaintCare program has also provided 7,600 large-volume pickups (direct from painting contractors, other businesses, or households with more than 100 gallons of paint), 6,777 periodic HHW events, and another 319 PaintCare-run paint collection events in less populated underserved areas.

RECYCLED PAINT MANUFACTURE

A successful EPR program for paint relies heavily on a healthy paint recycling industry to accept, sort, remanufacture, and market recycled paint. By 2017,

31. Product Stewardship Institute analysis of data from PaintCare annual reports, accessed June 1, 2021, https://www.paintcare.org/paintcare-states/. Annual reports for each state can be accessed from this website.

32. International Paint Recycling Association, *Annual Report for 2021* (August, 2022), prepared by the Product Stewardship Institute, 4, https://productstewardship.us/wp-content/uploads/2023/02/2022-08-IPRA-2021-Annual-Report.pdf.

33. Product Stewardship Institute analysis of data from PaintCare annual reports, https://www.paintcare.org/paintcare-states/. Annual reports for each state can be accessed from this website.

PSI began organizing paint recyclers in North America, eventually bringing them together to identify common challenges and opportunities. In 2019, the International Paint Recycling Association (IPRA) was conceived and developed by PSI in collaboration with a dozen recycled paint manufacturers to promote the quality, availability, and value of recycled latex paint.[34] IPRA is the first organization in the world to represent the recycled paint industry, and its founding members operate in 17 states and two Canadian provinces. Collectively, in 2021, IPRA member companies achieved the following:[35]

- Recycled almost 4.8 million gallons of paint
- Reached a 71 percent recycling rate for latex paint
- Created 577 direct jobs and more than 2,700 indirect jobs in the recycling industry

With 11 programs operating, new programs being established, and new bills being introduced, paint stewardship and the recycled paint industry are positioned for considerable growth.

PAINT DIALOGUE LESSONS LEARNED[36]

PSI amassed a wealth of experience in the development of the paint stewardship program. The process designed for the stakeholder engagement that led to the national agreement was extensive, detailed, and costly. Since this was PSI's first national dialogue, the process had not been tried before, and we were deliberate in our approach. We were starting from a place where stakeholder engagement was rare and few people even understood the concept of producer responsibility.

The paint dialogue process became the blueprint for PSI dialogues on other products. Each iteration refined our approach, increased its efficiency, and reduced its cost. What stands out most, however, is that no other US industry has followed the paint industry and engaged with government agencies as they have. For this reason, the way the program was built, through

34. IPRA founding members (see https://recycledpaint.org/) include: Acrylatex Coatings and Recycling, Amazon Environmental Inc., American Paint Recyclers, GDB International, Local Color Paints, Loop Paint, MetroPaint, MXI Environmental Services, Recolor Paints, Re-coat Recycled Paint, Société Laurentide/Boomerang Paint, and Visions Quality Coatings.

35. International Paint Recycling Association, *Annual Report for 2021* (August, 2022), prepared by the Product Stewardship Institute, 4, https://productstewardship.us/wp-content/uploads/2023/02/2022-08-IPRA-2021-Annual-Report.pdf.

36. This section adapted from: Scott Cassel, *The Dynamics of Dialogue: Lessons Learned from the US Product Stewardship Movement*, Product Stewardship Institute (July 22, 2012), https://psi.wildapricot.org/resources/Paint/2012-04-Report-Paint-The-Dynamics-of-Dialogue-Lessons-From-US-Product-Stewardship-Movement.pdf.

continual collaboration, remains unique. What is more striking is that, among industries required to take responsibility under EPR laws, the paint industry is the most visible in touting its accomplishments and taking pride in its achievements. The industry has assumed ownership of the program that reflects its past efforts with others to jointly develop a system that works for each stakeholder. There may well be a connection between effective stakeholder engagement early in the process, before legislative sausage making, and corporate satisfaction with the result of a law or regulation. The paint stewardship program is not perfect, and it will certainly need to be adjusted periodically. Even so, it has a stable foundation, and this is due in large part to the hard work by many people over many years to develop a model that has produced significant results.

Below are the lessons learned that most stand out for me. While these experiences are vast, they are by no means complete. Even after more than a decade of operation, new ideas are being implemented and new lessons are being learned. What is most important is that opportunities to collaborate present themselves on a regular basis to those most involved in working on these programs: ACA, PaintCare, recycled paint manufacturers, waste management companies, state and local governments, and nonprofit organizations. It is critical to have ongoing communication among stakeholders, not just appearing together at state and national conferences, but also stepping back periodically to assess the program from a national view and, when programs take hold internationally, from a global perspective as well. Integrating sustainability principles into an industry sector, upstream and downstream, is a continual process that relies on innovation and commitment.

EPR Produces Results

For more than 40 years, local governments throughout the country have set up programs to collect, reuse, and recycle leftover paint. A few of these programs have been very successful in capturing much of the available leftover paint in their jurisdiction, but only after spending millions of dollars of public funds. Most other municipal paint collection programs capture only a fraction of the amount of leftover paint available, with the remainder disposed of in landfills and waste-to-energy plants. Without the EPR program model developed through the PPSI group, governments would be faced with the choice of either spending millions of taxpayer dollars to recover leftover paint or reluctantly allowing that paint to be wasted. States now have a choice to pursue legislation, based on a proven model that recovers a valuable paint commodity, reduces pollution, and shifts the financial and management burden of leftover paint from government and taxpayers to producers and consumers.

The Power of Champions

There were many individuals who were instrumental in developing the paint agreement, including those from both government and industry. However, a few key individuals took risks, significantly influencing the group and contributing to its success. Carl Minchew from Benjamin Moore was an early proponent of product take-back options, and he developed a company-specific take-back program that demonstrated that leftover paint from consumers could be recycled back into new paint at a reasonable cost. Another key champion is Robert Wendoll of Dunn-Edwards, a California company that manufactured and sold virgin paints, and also sold a private label recycled paint manufactured for Dunn-Edwards by Amazon Environmental, one of the country's first recycled paint manufacturers. Minchew and Wendoll were well respected by their companies, as well as by the ACA board of directors. Their support for the PSI-facilitated effort helped persuade other companies to engage with PSI and ultimately reach an agreement.

ACA's Alison Keane had the legal skills, access to the ACA president and board, and an approach that was consistent, confident, and innovative. Keane led ACA's members into an agreement that broke new ground for the industry, with ongoing technical support from ACA's David Darling and Steve Sides. She also had the courage to stand up to her members if they were not being constructive. I recall one time during a meeting when one of ACA's members from a large paint company was spouting unreasonable comments that threatened to sour the entire meeting. Keane politely interrupted this person while they were speaking, making it clear that their view was contrary to that of ACA. This action was significant in allowing the meeting to proceed and stay on track. More importantly, it set a standard for discussion during the entire dialogue, letting industry colleagues know that their comments should seek to solve the problems we had all identified. It also sent a message to government officials that the industry was serious about an agreement. Keane's leadership was on display at other critical times as well. After becoming more involved with the ACA-funded life-cycle assessment and cost-benefit analysis that was pivotal to determining whether latex paint should be recycled or disposed, she ended it, reasoning that the studies were inconclusive and that consumers wanted both latex and oil-based paint to be collected and recycled, so that is what the program should provide.

Another champion who was instrumental to the success of the agreement was US EPA's Barry Elman. Elman had the analytical capability to participate in the LCA and other technical projects, the authority to represent the agency, the willingness to cajole the industry to take responsibility for their leftover paint, and the knowledge of the federal bureaucracy to weave through channels to gain the agency's support. Although the US EPA might not have had the clear legal authority to require companies to take responsibility for their postconsumer products, Elman was extremely effective at persuading the industry to make commitments to assume that responsibility.

Lastly, there were many state and local government champions who stepped up repeatedly to strongly and consistently promote the interests of not only their own constituents, but also other agency officials across the country who did not participate in the dialogue. They are largely responsible for creating a new paradigm in the United States for leftover paint management. These officials had a great depth of experience in managing latex and oil-based paints, and they educated the paint industry about the postconsumer problems caused by their products. Since so many state and local government officials engaged in the dialogue, the problems were irrefutable. Further, these same agency officials knew which solutions were needed. They were patient and took the time to explain to a receptive industry how problems were created and what the industry could do to help solve them. The credibility of these government officials was unshakable and provided the basis for reasoned dialogue and ultimate agreements. At the risk of missing worthy individuals, those who come to mind include Jen Holliday, Sego Jackson, Jim Quinn, Theresa Stiner, Dave Nightingale, Abby Boudouris, Tom Metzner, Leslie Wilson, Garth Hinkle, Scott Mouw, and Shirley Willd-Wagner.

The Need for Participation from Key Member Companies and Their Association

It was important for both ACA and key members to participate in the dialogue. Senior representatives for companies such as Sherwin-Williams, Valspar, Benjamin Moore, ICI, PPG, Ace, True Value, Dunn-Edwards, and others were critical to gaining internal company support. Having one without the other would not have sufficed. By participating directly in the dialogue, company representatives could speak at industry committee and board meetings about their involvement in the dialogue. These firsthand accounts of progress supported ACA staff during board decisions, including whether to sign the two MOUs. Without this support, ACA staff would have been left to convince company CEOs on their own, which would have been a daunting task. ACA's role, on the other hand, was critical in leveling the playing field for all company members and speaking on behalf of the entire industry. While each company is limited by their own interests, ACA understands the range of industry interests and can put each company's position in context with those of other companies.

The Importance of Legislative Drivers

As mentioned, the paint industry was unique among industries in its early engagement with government officials. Even so, paint companies were motivated to engage due to past negative legislative experience with lead paint and volatile organic compounds. In addition, several states were intent on introducing paint recycling legislation. When PSI invited ACA to participate in a collaborative dialogue, California had already introduced paint recycling legislation that the industry opposed. Shortly after, Minnesota also expressed interest in developing a bill. Legislation got the industry's attention. It showed clear government intent, and if the industry did not engage, they could have been stuck with a law they did not like. The paint industry knew that, if it did not engage, government was well organized and would move unilaterally in a way that would likely not take into consideration many of their interests. Engaging with PSI provided the potential for greater regulatory certainty and gave the industry an equal voice with government agencies and recyclers.

The Influence of Results from Existing Programs

Ideas can easily be dismissed, but results from existing programs are hard to ignore. By the time PSI engaged the US paint industry in finding a solution to leftover paint, several Canadian provinces, most notably British Columbia, Quebec, and Ontario, had already enacted EPR laws that required paint manufacturers to manage leftover paint in those provinces through a consumer-funded eco-fee. Mark Kurschner, president of Product Care, presented multiple times to our dialogue group about how the system functioned in Canada, answering questions, providing results, and revealing few problems. After the third time presenting—remotely—to our group, Kurschner told me he could sense that the group finally grasped what he was saying. I felt the same illuminated discovery on this third call. Not only was it important to have experts present on how a similar program could function in the United States, but it took three presentations for the message to sink in that this same system could succeed in our country. Kurschner and Product Care were also instrumental in getting the Oregon, California, and Connecticut programs ramped up while PaintCare was building its own staff.

Highlight Facts and Information

Dialogue is most fruitful when robust and balanced information is available. Without data that is accepted by key stakeholders, solutions will rest on faulty assumptions. For the PPSI, presentations based on credible data were consistently woven into the discussion and framed so that participants could make clear decisions efficiently. When government agencies repeatedly emphasized that, on average, 10 percent of paint purchased becomes left over, and that this would cost over half a billion dollars for local governments to manage, it got the industry's attention. Negotiations and decisions throughout the PPSI dialogue process were built on data, developed by either a trusted source or by the multi-stakeholder group itself within the dialogue context. The PPSI, as a group, designed much of the research conducted by PSI and other consultants. The group did not take the research at face value, but instead spent time analyzing the data, participating in work-group meetings, asking questions, and challenging results. This "joint fact-finding" approach, facilitated by PSI, ensured that any data developed for the group went through an expert multifaceted analysis, and that results were derived through consensus from multiple stakeholder perspectives.

Choose the Right Messenger

We all can sense when a presenter grabs the attention of the people they want to influence. At times it is the passionate person who can articulately engage others through emotion. Another time it will be the person with the most experience. Other times it is the quiet participant who gains attention through few, but powerful, words. In the paint dialogue, presentations on government costs from paint management by three well-respected officials were a turning point for the dialogue. Jen Holliday, of the Chittenden County Solid Waste District in Vermont, and Sego Jackson, of Snohomish County in Washington, spoke first to set the context. Theresa Stiner of the Iowa Department of Natural Resources spoke next. Iowa is a centrist state, and Stiner—a quiet, reserved government official—nevertheless strongly and clearly articulated the need for a paint stewardship program. Once these three officials spoke, it was as if all governments were speaking about the need for change. It became a critical moment in the understanding by industry of core government concerns.

Build on the Problem

Without key stakeholders agreeing to a problem statement, it is futile to go the next step. Solutions cannot be built on a faulty foundation, and acknowledging the various aspects of a problem, even if one doesn't fully identify with each aspect, is an essential element of gaining agreement. The key is that each stakeholder needs to acknowledge the problem as viewed by other stakeholders. Once a complex problem is accepted and goals established, participants can move ahead to develop solutions. Regarding the paint dialogue, once all participants acknowledged that both latex and oil-based paints presented significant financial and management problems for governments, as well as environmental impacts, they could move on to considering joint solutions.

Put Legislative Eggs in Multiple Baskets

PSI coordinated the national paint dialogue for about four years before a consensus bill was introduced in the first state, Minnesota. The PPSI planned for Minnesota—the selected pilot state—to operate the paint EPR program for one year, after which it would be evaluated by the group, with lessons learned being incorporated into future state programs. As a result of former Minnesota governor Tim Pawlenty's two consecutive vetoes, the original idea of learning from a pilot was helpful to Oregon but became less of a driver for other states as the whole process was set back 18 months. Once a law was finally enacted in Oregon, the new pilot state, there was little time to evaluate the program before paint EPR legislation was also enacted in California and Connecticut, which did not want to wait any longer.

States not part of the MOU were also eager to enact paint stewardship legislation owing to the significant costs of managing paint. ACA spent considerable funds and staff time lobbying *against* some state bills because its resources could not keep up with the backlog of interest in the industry-sponsored program. Pawlenty's vetoes also caused increased costs for governments and others, stakeholder hesitation, and a contorted process that was akin to a train pileup when a locomotive suddenly comes to a grinding halt. This was a costly lesson for all stakeholders. In subsequent years, the model paint bill was introduced in multiple state legislatures as were EPR bills on other products developed as a result of other PSI dialogues. Politics is too unpredictable to plan a state-by-state approach with any degree of certainty. Pilot legislation in a single state is an attractive idea, but in practice it is better to spread the possibilities for enacting legislation among a group of states, with lessons shared among them.

Attitudes Can Change

Stakeholder attitudes changed throughout the paint dialogue, sometimes drastically and many times even personally. As with most initial meetings, the first meeting of the multi-stakeholder group was tense. It was as if company representatives believed the government officials across the table would tie them up in regulatory knots, and government officials believed that the industry representatives would dismiss their concerns. As the meetings unfolded, one by one, and space was provided for honest discussion, each person felt empowered since they had the opportunity to provide their own perspective.

Good ideas rose to the top as priorities to pursue, and bad ideas sank to the bottom and were not carried over to the next meeting. By the end of the dialogue, views on all sides changed, allowing consensus on major solutions to emerge. The personal and professional changes that took place during the paint dialogue were profound. Adversaries became allies, and the concerted effort of multiple individuals seeking the same result enabled that result to become a reality. In the end, fear of the other was replaced by a willingness to withhold judgment until facts were presented and reasoned discussion took place. The participants came to trust a process of which they had not previously been aware.

Trust the Process

Stakeholder dialogues are needed because people have strong divergent viewpoints that have been allowed to linger unresolved. Few stakeholders want to relinquish control of a process, and most would prefer that the outcome be the one they establish. The paint dialogue was no different. It took time to build rapport with each stakeholder and among stakeholders. Over time, however, participants began to trust the process, and I, as facilitator, learned to step back and let the group momentum lead the way. Admittedly, there were two times during the dialogue when I needed to pay attention to my own lesson of trusting the process, and I tried to force a direction when the group was not ready. I recall feeling that, during these two instances, I was perceived by the paint industry as advocating for a position rather than letting each stakeholder advocate on their own behalf. Each time this happened an industry representative called me on it, and I backed off to allow the group to move on its own. By the end of the dialogue process, it reaped dividends. One gratifying moment, as I explained earlier, was when I tossed control of the process to the group during a critical final decision and, almost miraculously, it became a key turning point that led to the second MOU. During the dialogue, there were many times when no one knew what would happen next and everyone was open to new ideas. At these moments, the vast majority of stakeholders trusted the process, and each other, to come up with the right next step.

Have Authority to Represent Your Constituency

Some discussions suffer because those at the table do not have the authority to represent their agency, company, or organization. I have been at meetings where a participant says, "I am only providing my own opinion." One of the ground rules in the paint dialogue was that each person needed to have the authority to represent the interest of their agency, company, or organization. Without that authority, an agreement would have been impossible. The PPSI had the right people who were entrusted to represent their constituencies. It was crucial during the paint negotiations that each key stakeholder could effectively represent their affiliation's interests. It was understood that each representative would have to check back with upper management before the MOU language was finalized. But effective participants will know what they can and can't agree to during a meeting, and when to check in along the way with those in their internal management chain. The authority of each participant to speak for their constituency was a key element of the success of the paint dialogue.

Social Change Requires Consistency and Perseverance

Changing an industry sector through multi-stakeholder collaboration requires a continual forum for discussion, which can only endure if stakeholders believe that progress is being made. Such fundamental change will not happen quickly, and there is a cost to coordinating stakeholders to keep a dialogue group together. People change jobs and new staff need to be acclimated to the process. Group dynamics shift as new leadership takes over a company or agency, or new people change the group chemistry that was slowly formed through formal and informal meetings. Even active participants must juggle numerous other priorities and be reengaged if a project slows. Stakeholders also forget decisions previously made if the lag time is too long, as it was for paint, leading to the need to reeducate stakeholders several times. The lag also caused some stakeholders to second-guess key elements of the original MOU. Keeping the PPSI group together, particularly during and after the 18-month veto period, was a formidable challenge.

PSI's main role in the paint dialogue was its continual presence and our ability to create and maintain momentum, reach closure on issues, and prepare the group to make decisions. PSI became the glue that held the pieces together, including the eight projects developed after the first MOU, the ever-changing database of contacts, and bits of news that affect each stakeholder and the dialogue itself. As agreements were reached, PSI became one of many advocates for that agreement. Over time, PSI's process became so familiar to the group that it developed its own identity (the Paint Product Stewardship Initiative). Ownership of the dialogue by participants drove their joint interest in attaining a solution.

The Need for Federal Involvement

US EPA staff were instrumental in helping the dialogue achieve a mediated solution. Barry Elman, Rebecca Smith, and Matt Keene served as key technical resources that brought with them the authority of the agency. Their high-level expertise in policy, program evaluation, life-cycle assessment, and recycled paint procurement, among other areas, provided the group with facts and insights that kept the group on track. The involvement of US EPA lent a credibility that elevated the importance of ACA taking responsibility and the need to reach an agreement. EPA staff never overstepped their waste management authority, which, as emphasized earlier in this book, rests largely with state and local governments. Instead, they expressed a sincere interest in solving the economic, environmental, and human health problems created by leftover paint. The signature on the paint agreement from Deputy Assistant Administrator Thomas Dunne was significant in providing federal government leadership to a paint industry searching for a stable path forward.

Complexity of Life-Cycle Assessment (LCA)

The paint dialogue provided PSI members and me with an insider's view of the potential and challenges of LCA, along with equally complex cost-benefit analysis (CBA). The paint industry hired top-notch LCA and CBA consultants, Franklin Associates and Eastern Research Group, to attempt to provide policy answers to the key question of whether the industry should be held responsible for managing oil-based paint only, or if their responsibility should also include latex paint. The LCA isolated and compared innumerable variables to help one of the paint dialogue work groups make a policy judgment on whether there would be fewer net environmental impacts if latex paint was disposed in a landfill or waste-to-energy plant as compared to it being recycled into paint. After more than a year of analysis and more than $300,000 spent, the results were clearly contingent on modeling assumptions that the work group members had difficulty agreeing to.

What I learned is that LCA and CBA are fascinating approaches that hold tremendous promise for understanding the interrelated and variable factors that are relevant if one wants to reduce life-cycle environmental impacts while understanding the costs and benefits of policy decisions. A key limitation, however, is that LCA and CBA can be extremely expensive and, as analytical tools, are easier to apply to simple comparisons (e.g., individual packaging choices) and more difficult to apply to complex policy decisions. I also learned that the intense concentration needed to follow each aspect of a calculation was one of the most mentally taxing analyses I have conducted, even as I found the conversations with colleagues thoroughly engaging. A problem we encountered is that only two people in our dialogue, David Allaway of the Oregon Department of Environmental Quality and Barry Elman of US EPA, understood all aspects of the calculations. Unfortunately, the highly capable paint industry technical staff did not have the same level of experience with LCA and CBA as the government officials.

As facilitator, I helped the PPSI group reach resolution on innumerable aspects of the calculations. In the process, I came to understand that one tiny assumption could skew the outcome of an entire LCA. For example, one of the most influential assumptions was whether a calculation assumed that recycled paint would only be used on a structure that would be painted with virgin paint if recycled paint was unavailable, or whether the structure would go unpainted if recycled paint was not available. This assumption had significant impacts on both environmental outcomes (whether or not virgin paint was displaced) and cost-benefit outcomes (the value of painting unpainted structures). Another influential variable was the extent to which the calculation assumed that people drop off leftover paint while performing other errands as compared to making special trips to get rid of paint. Perhaps not surprisingly, government officials had one perspective and producers another. My key takeaway is that LCA and CBA are valuable as a way of thinking, but for public policy decisions, they must be conducted by a publicly funded neutral party, often with the assistance of a facilitator or mediator. Even academic institutions are often funded by corporations and/or other private entities, even if not directly for a project, and are not always neutral venues for such a stakeholder-driven process.

THE FUTURE OF EPR FOR POSTCONSUMER PAINT MANAGEMENT

The management of leftover paint is one of the best-performing US EPR programs due to the strong implementation by PaintCare and the collaborative process that allowed all stakeholders to participate in the development of a model program. PaintCare has refined its program over time to respond to issues raised by stakeholders, including developing large-volume pickups, door-to-door collection when needed, customized outreach materials for national retail chains, refinement of its bidding process for transporters, and planning for state regulatory adjustments for the collection of paint at retail locations. It is always a challenge for newly developing EPR programs to introduce a model bill in additional states while also incorporating lessons learned from current programs so that new programs display "next generation" best practices. Otherwise, new programs just duplicate a program developed a decade before.

Continuous Improvement

As mentioned before, the paint industry is at the top of the class of producers taking responsibility for their postconsumer products. A key consideration, though, for all stakeholders—the paint industry, government officials, recycled paint manufacturers, and collectors—is how to incorporate shared best practices into new programs while revising existing programs over time to include those same best practices. These programs need to continually determine how best to reduce the overall life-cycle impacts from paint, both upstream during mining and manufacturing, and downstream in reuse and recycling. And lastly, stakeholders need to consider, from a regulatory and industry perspective, how best to manage other paint and paint-related products that continue to be disposed of by households and small businesses, such as spray paints (aerosols), marine paints, artist paints, wood furniture stains, thinners, and strippers. Should the scope of products in paint EPR programs be expanded to include these additional materials, or would it be more effective for those paint products to be managed with other HHW in a separate EPR program?

The paint stewardship program is also seeking new ways to minimize the portion of leftover latex paint that cannot be recycled into new paint. Although PaintCare provides information on proper storage, many people are still unaware that how paint is stored is critical to whether that paint can be turned into recycled paint. There is also a need to determine the long-term environmental impacts of using latex paint for cement additives, garden stones, or other such non-paint recycled products.

Another issue that is beginning to be resolved is how to maximize the collection, reuse, or recycling of all paint containers. The PaintCare program was not initially set up to accept empty paint containers, and many material recycling facilities are reluctant to accept these containers from municipal recycling programs. If 10 percent of paint is left over, how many paint cans are completely used up and need to be recycled, but are disposed of instead? How many of these containers are generated by painting contractors compared to households? Should these paint containers be included in the PaintCare program, or managed under a separate packaging EPR program that could collect the containers in existing municipal curbside and drop-off recycling programs? Now that four state packaging EPR laws have been enacted and about a dozen such bills were introduced in 2023, the policy that has emerged is for paint cans to be included in the packaging EPR bill unless handled through a separate stewardship program (i.e., PaintCare).

These and other issues are typically raised by state and local government officials, recyclers, and other stakeholders as EPR programs mature. One new key element of state EPR legislation that has emerged for other products is a multi-stakeholder advisory council that provides non-binding recommendations to a producer responsibility organization and the government oversight agency. These councils, which have become standard in most PSI model EPR bills, provide an opportunity for ongoing input from an expert group of key stakeholder representatives (often 9 to 15 people) to address issues and advance a stewardship program. Stakeholders may want to assess whether new and existing paint stewardship programs could be further strengthened through more structured, periodic input that an advisory council can provide. In the case of paint, a council might include representatives from local governments, recycled paint manufacturers, paint collectors, retailers, environmental groups, and others. Although an advisory council can provide an important opportunity for continuous improvement of state programs, a program like PaintCare that is now national in scope might also benefit from annual or biennial gatherings of stakeholders across states to share experiences and address emerging issues from multiple perspectives.

Recycled Paint Markets

As the recycled paint industry has matured, recycled paint quality has risen to equal that of virgin paint, and its availability has vastly expanded to include online sales. Although the reputation of recycled paint has improved, the industry still faces obstacles in achieving broader market acceptance. Currently, much recycled paint is sold to bargain-conscious consumers due to its lower price compared to virgin paint. Even so, as homeowners, painting contractors, building maintenance companies, architects, and others have learned about the value and quality of recycled paint, the market has begun to expand. For example, several recycled paint brands are carried by Walmart and other brands have opened their own retail paint outlets.

By enacting EPR laws, government agencies create an increased supply of leftover paint; they need to buy that paint through their procurement programs for use by government officials and contractors. Creating more supply of leftover paint without helping to establish a domestic market has forced recycled paint manufacturers to export about half of the paint collected in PaintCare programs. Creating more domestic markets will keep those jobs in the United States.

A related challenge is the complex relationship between the virgin paint industry and their leftover product. While consumer product manufacturers have made increasingly bold pledges to use recycled content in their packaging, paint manufacturers have made no such claims. Perhaps this is because leftover postconsumer paint material is recovered and used by companies comprising a different industry to make a competing product. The recycled paint manufacturers comprising IPRA compete with virgin paint manufacturers that belong to ACA. Although recycled paint manufacturers can also belong to ACA, the interest of virgin paint manufacturers in selling recycled paint is dwarfed by the market sale of virgin paint. Currently, while Unilever, Procter & Gamble, General Mills, and other brand companies are being forced through EPR laws to create the postconsumer markets for the packaging in which they sell their products, EPR laws have not yet required virgin paint manufacturers to do the same for leftover paint.

Further complicating this dynamic is that, over the past 100 years of paint sales in our country, virgin paint manufacturers have established exclusive sales relationships with retailers. Some companies, such as Benjamin Moore and Sherwin-Williams, have their own branded retail outlets, while Home Depot, Lowe's, Walmart, and most Ace, True Value, and independent hardware stores have exclusive relationships with virgin paint brands. In addition, even when there are not exclusivity agreements, recycled paint is not typically ordered or sold in large quantities at present, so shipping is expensive, and retailers hesitate to give floor space to what is seen as a niche product. It remains to be seen whether retailers will begin to carry recycled paint to meet an increased consumer demand for low-cost, high-quality paint, or whether recycled paint manufacturers will continue a trend to establish their own retail stores. Another possibility is whether virgin paint companies will acquire recycled paint manufacturers and add a line of recycled paint to their virgin paint sales.

The establishment and growth of IPRA will bolster the recycled paint industry and support continued expansion of EPR programs for paint in the United States. These recycled paint manufacturers are the backbone of the paint recycling infrastructure, and ACA members rely on these companies for their burnished image as a green industry. At the same time, ACA has been reluctant to promote recycled paint since, to the association, that would be akin to promoting one paint type over another. How the relationship unfolds between virgin paint and recycled paint companies will be a key factor in the ability of both related industries to meet public expectations for a sustainable future.

Tracking Leftover Paint

Regulators are increasingly interested in tracking the flow of leftover paint from the point of generation to its ultimate disposition, whether to domestic recycling facilities, overseas markets, non-paint recycled markets, or disposal facilities. This interest is not unique to paint but part of the increasing attention to how materials used in products are managed following consumer use. The degree of transparency now expected of companies has already created a burgeoning data management industry that will provide paint companies with information to reduce environmental and social impacts across the full life cycle of paint manufacture and remanufacture. New data tools will also help to increase the efficiency of EPR programs by tracking costs and materials along each step of the postconsumer management process.

Reducing Life-Cycle Impacts

Not all leftover paint has the same value, costs the same to manage, or could result in the same impacts. "Eco-modulated fees," which have now been incorporated into packaging EPR laws and bills, incentivize producers to reduce the end-of-life impacts of their packaging by using more recyclable or reusable materials, and to consider other environmental factors such as greenhouse gas emissions in packaging design. These producer payments seek to influence upstream decisions to optimize environmental outcomes and minimize impacts by imposing lower fees on packaging that are less costly to manage and have fewer environmental impacts.[37] As the concept of eco-modulated fees gains acceptance, there may be interest in considering how this concept might apply to paint.

Currently, oil-based paint costs more to manage than latex paint, and darker color paints have lower recycling value than lighter colors. Will there be an interest to discuss whether to set higher fees for paints that cost more to recycle and manage? Another step toward sustainable paint management will be to assess the full life-cycle impact of paint manufacture, including which toxic materials are still used in paint products and which of these cause health and environmental impacts. Can LCA be a tool to evaluate the policies needed to reduce the use of these toxic materials? Finally, what steps can the industry take to reduce the impacts of material extraction, and how can LCAs help to establish criteria to develop eco-modulated fees for the sale of new paint?

Paint Source Reduction

As with other products, reducing the amount of leftover paint that needs to be managed has, unfortunately, become a secondary goal. Passing along the cost to consumers through the eco-fee has reduced the motivation of industry and government to reduce the amount of paint that becomes left over. Those working on paint stewardship will need to consider what policies can incentivize the reduction of leftover paint. One step is to revisit the source reduction strategies recommended in the paint dialogue that have yet to be piloted and evaluated. Another approach is to consider having the paint industry internalize postconsumer paint management costs by covering those costs themselves, replacing the current consumer fee, or would such a significant change jeopardize the current successful program? The question remains, what policies and programs can incentivize new approaches to reduce leftover paint, which will always lead to lower environmental and economic impacts?

---

37. Product Stewardship Institute, *Extended Producer Responsibility for Packaging and Paper Products: Policies, Practices, and Performance*, published March 2020, updated September 2020.

Chapter 9

Battery Stewardship: A US Case Study[1]

Batteries are sources of energy that are ever present in our lives. They allowed me to listen to late-night baseball games under my pillow and fool my parents into thinking I was asleep. They powered my radio/cassette tape player so I could travel cross-country while jamming out on my favorite music. They were also one of the main components of household hazardous waste (HHW) about which I received calls on an HHW consumer hotline at the start of my career in 1983. Due to their growing volume in the waste stream, batteries were the sole focus of a daylong summit I attended in the early 1990s convened by Dana Duxbury, an early HHW program champion, along with representatives of all major household battery brands.

A battery is defined in the *Oxford English Dictionary* as "a container consisting of one or more cells, in which chemical energy is converted into electricity and used as a source of power."[2] Advancements in battery technologies have continually powered consumer and industrial product innovations and are now an essential strategy for reducing greenhouse gases by transitioning to a renewable energy economy. The European Green Deal seeks to significantly increase the use of electric vehicles, e-bikes, and other "green" products, as well as increase solar and wind energy production and foster a networked renewable energy grid. All of these initiatives will require increasingly powerful and effective batteries and the ability to capture and recycle used batteries. Thus, building European infrastructure for manufacturing and recycling batteries is also a key strategy of the European Green Deal.

Many US governments and forward-thinking companies are fast following this European global climate stabilization blueprint by investing in green

1. This case study was written with input from Call2Recycle, Inc.

2. Also see the federal definition of "battery" in universal waste regulations: "Batteries," US Environmental Protection Agency, accessed February 12, 2023, https://archive.epa.gov/epawaste/hazard/web/html/batteries.html.

energy projects that require batteries for energy storage and distribution. US auto manufacturers have set ambitious goals to transition entirely to electric vehicle production, some as soon as 2035,[3] and the US federal government is investing heavily in infrastructure to support the transition to a carbon-neutral economy, including electric vehicle charging stations. More and more consumer products previously powered by gasoline or other means, from lawn mowers to bicycles, are increasingly powered by batteries.

Our growing reliance on batteries and the proliferation of new battery types and technologies make the recovery and recycling of spent batteries ever more significant. The mounting recognition of the importance of battery recycling and the economic, strategic, and environmental opportunities that battery recycling offers are finally shifting the mindset of both policymakers and the industry to ensure effective battery recovery systems. Battery extended producer responsibility (EPR) laws that are flexible enough to accommodate continual advancements in technology are considered the best way to ensure that valuable materials are returned to the circular economy.

This case study illustrates the US household battery stewardship journey, from removing batteries with toxic materials from the waste stream to managing batteries as resources for a circular economy. It includes some of the key successes and challenges faced by policymakers, manufacturers, and recycling program operators along the way. It also references how the materials used in portable batteries have changed significantly over time, often influenced by product policy. These changes in battery chemistry illustrate the need for EPR policies to be adaptable enough to continually incorporate new product types.

This overview also emphasizes the need for ongoing stakeholder engagement to develop collaborative solutions. While a handful of early state battery stewardship laws in the 1990s motivated the industry to reduce toxics used in batteries (e.g., lead, mercury, and cadmium) and recycle them, PSI and its government members have worked with the battery industry and other stakeholders over the past 15 years, during which battery technology has significantly evolved. From this engagement, several iterations of PSI model battery EPR bills were developed. These model bills, starting in 2010, were part of a mix of policy developments across the country that led to the subsequent introduction of multiple state, and sometimes local, battery EPR bills. While some of the bills covered single-use batteries only, others focused solely on rechargeable batteries, and still others proposed to regulate both types of batteries. This consistent, and persistent, drumbeat of interest in battery EPR

3. Michael Wayland, "General Motors plans to exclusively offer electric vehicles by 2035," CNBC (January 28, 2021), https://www.cnbc.com/2021/01/28/general-motors-plans-to-exclusively-offer-electric-vehicles-by-2035.html.

laws ultimately led to the country's first single-use, multi-chemistry household battery EPR law in Vermont in 2014, as well as the first EPR law that includes both single-use and rechargeable multi-chemistry consumer batteries in the District of Columbia in 2021[4] and California in 2022.

Household batteries are an example of a product type that has caused continuous impacts on the environment if not managed properly. While limited government resources have often been shifted to higher priority products, the impacts from household battery disposal have never gone away. Instead, they have rested just beneath the surface of conscious concern for regulators, ready to pop up once again when staff resources become available.

This continuum is now changing. With the advent of a new generation growing up with climate change and the concept of a circular economy, these battery EPR policies are destined to play a much more prominent role in the years ahead. Coupled with significant disposal impacts from lithium-ion (Li-Ion) batteries and greater experience among government regulators with EPR policies, increased attention to reducing household battery impacts through EPR is likely to emerge in the years ahead.

SCOPE: HOUSEHOLD BATTERIES

PSI's work has been focused on consumer batteries, including rechargeable, which represent an estimated 40 percent of the market, and single-use (also known as "primary" batteries), an estimated 60 percent of the market.[5] These battery types are also known as "portable" batteries since many of them are used not only in households but also in commercial settings. Single-use and rechargeable batteries vary by chemistry, size, shape, and voltage, and are used in a wide range of portable devices and household items, including toys, clocks, flashlights, laptop computers, cameras, watches, and cell phones. Increasingly, batteries are being used in hand power tools (e.g., drills), landscaping equipment (e.g., lawn mowers, leaf blowers), "e-mobility devices" (e.g., wheelchairs, bicycles, hoverboards, scooters, cars), home battery storage units, and other consumer products. Many of these new types of batteries are currently being considered for inclusion in the next generation of battery EPR bills. These bills, however, do not yet include those larger batteries

4. The District of Columbia's legislative body passed the bill, and the mayor signed it, in December 2020. However, the bill did not become law until March 2021, following the required period of congressional review.

5. "US Consumer Battery Sales & Available for Collection," prepared by Kelleher Environmental in association with SAMI Environmental, commissioned by Call2Recycle, Inc., May 2016, accessed March 25, 2023, https://www.call2recycle.org/battery-sales-available-for-collection.

used in vehicles or commercial, industrial, or military applications if they are already recovered through existing networks.

WHO ARE THE HOUSEHOLD BATTERY PRODUCERS?[6]

The world battery market is dominated by a small number of companies, shown in table 9.1. Roughly 80 percent of *single-use* batteries sold in North America are made by Duracell, Energizer,[7] Panasonic, and Kodak, while roughly 90 percent of *rechargeable* batteries manufactured for the North American market are made by Panasonic, Sony, LG, and Samsung. Many companies make and/or market both single-use and rechargeable batteries, but one type of battery usually dominates their total battery sales. For example, although Duracell and Energizer sell both types of batteries, approximately 80 percent of each company's battery sales are derived from single-use batteries.[8] In the United States, the National Electrical Manufacturers Association (NEMA)

Table 9.1 Companies Manufacturing or Branding Single-Use and Rechargeable Batteries

Company	*Single Use*	*Rechargeable*
Duracell	✓+	✓
Energizer/Rayovac	✓+	✓
Sony		✓
BYD		✓
Panasonic	✓	✓
Maxell	✓	✓
Varta	✓	✓
LG Chem		✓
Samsung		✓
Lishen		✓
NEC		✓

✓+ = dominant battery type manufactured by these companies. Panasonic, Maxell, and Varta manufacture both battery types but neither is dominant. Six companies manufacture only one battery type.

Source: Battery Summit 2011, Briefing Paper Factbase, April 5–6, 2011, provided by Call2Recycle, March 1, 2023.

6. This section is excerpted (and significantly updated) from *Battery Stewardship Briefing Document*, Product Stewardship Institute (July 2, 2014). The document was developed in preparation for a PSI-facilitated meeting held in Hartford, Connecticut, on June 11–12, 2014. The briefing document was funded by the Connecticut Department of Energy and Environmental Protection, although it incorporated parts of a 2010 PSI document by the same name funded by Call2Recycle.

7. Including Rayovac, which Energizer acquired in 2019.

8. Call2Recycle, email communication with attached edited document, March 1, 2023.

represents single-use battery manufacturers and PRBA—The Rechargeable Battery Association—represents the rechargeable battery manufacturers. Some companies are members of both associations. In 1994, PRBA established an industry-funded take-back and recycling program for rechargeable batteries, the Call2Recycle® program, administered by Call2Recycle, Inc., a nonprofit public service organization (formerly called the Rechargeable Battery Recycling Corporation—RBRC). Under EPR laws, Call2Recycle is known as a producer responsibility organization (PRO). Another industry association, the Corporation for Battery Recycling, comprised of Energizer, Panasonic, Duracell, Kodak, and Spectrum Brands (Rayovac),[9] was active in EPR discussions for several years, although it has been inactive since 2020.

WHAT'S THE PROBLEM?

Each year, millions of single-use and rechargeable household batteries are purchased in the United States. Only about 12 to 15 percent of rechargeable batteries, and a much smaller percentage of single-use batteries (80 percent of the market), are recycled.[10] The remainder are disposed of in the trash. Certain rechargeable batteries contain toxic materials, such as cadmium and lead, which can pose environmental and public health risks when disposed of improperly. Li-Ion rechargeable batteries can generate large amounts of energy and pose a greater safety risk than other types. Since these batteries contain significant residual energy even when they are no longer powering products, they can smoke or ignite when shredded, punctured, or shipped (if their terminals are not insulated). They have been implicated in an increasing number of fires that have caused worker injuries and millions of dollars in damage to trucks, recycling plants, landfills, and other waste management facilities.[11] In addition, toxic gases are released during Li-Ion battery fires.[12]

Although federal law prohibits the sale of single-use batteries containing added mercury (except very small button cells), a small percentage of these batteries manufactured overseas (including those shipped in products) may contain mercury. Some of the non-domestic batteries are also designed to have short lives, just enough to power products (e.g., toys) for a brief time after purchase. Batteries that are thrown away squander resources that could

9. Spectrum Brands (Rayovac) dropped out of the association several years into the effort.

10. Call2Recycle, email communication with attached edited document, March 1, 2023.

11. US Environmental Protection Agency, Office of Resource Conservation and Recovery, *An Analysis of Lithium-ion Battery Fires in Waste Management and Recycling* (July 2021), https://vlsrs.com/wp-content/uploads/2021/08/EPA-lithium-ion-battery-report-update-7.01_508.pdf.

12. Fredrik Larsson, Petra Andersson, Per Blomqvist, and Bengt-Erik Mellander, "Toxic fluoride gas emissions from lithium-ion battery fires," *Scientific Reports* 7, 10018 (2017). https://doi.org/10.1038/s41598-017-09784-z.

be reclaimed through recycling and thus require raw materials to be mined, thereby creating additional environmental impacts—including greenhouse gas emissions. Although the cost for managing household rechargeable batteries collected by Call2Recycle is covered by the industry program, government agencies must pay for and manage single-use batteries, as well as rechargeable batteries that enter the waste stream. The risk inherent in using Li-Ion batteries has caused confusion for community battery recycling programs, with some curtailing collection out of fear of handling, storing, and shipping the batteries, while others seek mandatory collections (e.g., EPR) to remove them from the waste stream to reduce safety incidents.

WHAT ARE THE GOALS?

Many consumers would like to recycle all batteries if convenient, and they generally find it difficult to distinguish between single-use and rechargeable batteries. Municipalities seek to eliminate their cost to manage all household battery types. They also want to maximize battery reuse and recycling to save resources and reduce greenhouse gas impacts by curtailing the mining and manufacturing of new materials. In addition, they want to collect single-use and rechargeable batteries together, to eliminate the need to instruct residents on managing various battery types. All stakeholders want to ensure the safe collection, transportation, and recycling of batteries to minimize potential fire risk. Making the collection of spent batteries convenient is a key aspect in recovery, and this most often takes place at retail locations where batteries are sold and at municipal depots, which collect additional materials.[13] Another goal of managing batteries is to ensure that these efforts have a net environmental and economic benefit, which can be calculated through life-cycle assessment (LCA) and cost-benefit analysis (CBA). Finally, an important goal for battery manufacturers is that all state or local policies are consistent nationally.

WHAT ARE THE BARRIERS?

As is typical of managing many waste products, there is a lack of consumer awareness about how to recycle batteries, as well as a lack of collection opportunities for consumers, particularly for single-use batteries. Further,

---

13. In 2019, 23 percent of batteries collected in the Call2Recycle program were through municipal programs, and 42 percent were from retailers. Call2Recycle, email communication with attached edited document, March 1, 2023.

consumers do not generally distinguish between batteries that are rechargeable (for which there are many industry-sponsored opportunities to recycle), and single-use batteries (for which there are rarely recycling opportunities). In some parts of the country, this difficulty is compounded by an overall low motivation for recycling. Another barrier is that consumers mistakenly place Li-Ion batteries in household recycling bins, and they are difficult to identify when comingled with other materials. To reduce costly fires and increase the safe collection and transport of batteries, regulators have required consumers and recycling managers to tape battery terminals and/or place them in a bag prior to collection, making recycling less convenient and more costly by requiring additional outreach and education.[14]

Increasingly, to protect consumers, rechargeable batteries have been embedded in products, preventing consumers from easily removing them. This has increased the difficulty of recycling, resulting in the disposal of both the product and the batteries inside. These battery-containing products also make it difficult to calculate an accurate recycling rate, which is based on the number of batteries available for collection as well as the number collected. For example, some batteries are collected and recycled with the host product in electronics recycling programs and are not reported. The fact that some battery-only collection programs do not publicly report their collection results further exacerbates accurate data gathering.

A significant barrier to determining whether efforts to recycle household batteries have a net environmental and economic benefit is that life-cycle assessments (LCAs) and cost benefit analyses (CBAs) are costly, time consuming, and complex. A further challenge is that LCAs are often undertaken by those with a particular interest in the outcome rather than by integrating multiple stakeholder perspectives and mediating differences to get a result trusted by all. Finally, LCAs are only a snapshot in time and must be revised as significant changes occur, such as the addition of battery processing facilities that can reduce transportation costs and greenhouse gas impacts, or advanced battery recycling technologies that can recover a greater amount of material in batteries that are collected.

THE EPR SOLUTION

Battery EPR laws are considered best practice because these policies are one of the most effective approaches to capturing valuable materials, which are increasingly strategic from a national perspective. Battery stewardship in

14. "Used Lithium-Ion Batteries," US Environmental Protection Agency, last updated May 24, 2022, https://www.epa.gov/recycle/used-lithium-ion-batteries.

the United States contrasts significantly with the European Union (EU) and Canada. In the EU, the European Commission (EC) enacted a directive in 2006[15] mandating that EU member states ensure that a 25 percent collection rate for all batteries (single use and rechargeable) be attained by September 30, 2012, and that a 45 percent collection rate be attained by September 30, 2016. While data collection methodologies are inconsistent, most member states have attained this performance standard, and the EU has attained an average collection rate of 47 percent, with six countries reporting rates of over 60 percent.[16] (See chapter 5 for a more detailed discussion of the EC battery directive, which covers both primary and rechargeable batteries.) The EC battery directive also imposes a standard for "recycling efficiency," which is the percentage of battery materials entering a recycling facility that is reclaimed for a secondary use. For example, if 100 pounds of batteries enter a recycling facility and 75 pounds of material are shipped to end markets, that recycling facility has a 75-percent recycling efficiency, with the rest of the material being disposed.

Battery EPR policies in the United States have taken time to emerge, but we are on the verge of a widespread breakout as the move toward a circular economy has taken hold. To understand the trajectory of battery stewardship, however, we need to go back to early concern over toxic materials used in batteries.

MANAGING BATTERIES IN THE UNITED STATES: HISTORICAL OVERVIEW

Historically, public policy addressing battery recycling has been most concerned with material toxicity. Lead-acid batteries used in automobiles have the highest recycling rate of any product in the country, diverting harmful lead from disposal facilities. These recycling rates were fueled by early public concern over lead, as well as a positive market value[17] for lead that makes collection and processing profitable. The high rates are also due to a "core

15. European Parliament and Council of the European Union, "Directive 2006/66/EC of the European Parliament and of the Council of 6 September 2006 on batteries and accumulators and waste batteries and accumulators and repealing Directive 91/157/EEC," *Office Journal of the European Union* (2006).

16. "Waste statistics—recycling of batteries and accumulators," Eurostat Statistics Explained, data extracted January 2023, https://ec.europa.eu/eurostat/statistics-explained/index.php?title=Waste_statistics_-_recycling_of_batteries_and_accumulators.

17. Global Newswire, October 13, 2022. The global lead acid battery market value is projected to rise from USD 26.07 billion to over USD 63.44 billion by 2028, Vantage Market Research, accessed February 12, 2023, https://www.globenewswire.com/en/news-release/2022/10/13/2533714/0/en/Lead-Acid-Battery-Market-Size-to-Hit-63-44-Bn-by-2028-Manufactures-CAGR-Opportunities-Obstacles.html.

charge" mandated in 30 states that requires the consumer to pay a deposit when purchasing a battery that is refunded when a used battery is returned in exchange.[18] Based on the enactment of the federal Resource Conservation and Recovery Act (RCRA) in 1976, states also began to require recycling of nickel-cadmium (Ni-Cd) rechargeable batteries, small sealed lead-acid (SSLA) rechargeable batteries, and mercury-containing, single-use button cell batteries because of the harmful effects of cadmium and mercury.

In 1991, Minnesota enacted a producer responsibility law that required manufacturers to design, finance, and operate a program to collect and recycle all rechargeable battery chemistries, and to label Ni-Cd batteries. Several states also followed with EPR laws, but with less ambitious requirements on manufacturers, limiting their recycling obligation to Ni-Cd and SSLA rechargeable batteries. Those states included Iowa, Florida, Maryland, New Jersey, Vermont, and Maine. The rapid succession of these diverse state laws was the impetus behind the creation in 1994 of the Rechargeable Battery Recycling Corporation (RBRC), now known as Call2Recycle, Inc. Five manufacturers of Ni-Cd batteries—Energizer,[19] Panasonic, Saft, Sanyo, and Varta—founded RBRC to both comply with newly enacted state laws and create a consistent approach to handle this battery chemistry nationwide.[20]

With its creation, RBRC became one of the first product stewardship programs in the United States, and the rechargeable battery industry became the first product sector to assume responsibility for financing and managing the collection and recycling of the products they sold into the US market. Unlike most current US product stewardship organizations, RBRC did not only operate programs in states with laws that obligate manufacturers to take back their products; the organization was set up as a national entity to harmonize the free, convenient collection of rechargeable batteries through a common strategy of mostly retail and municipal HHW collection sites connected to a network of nationwide processing facilities. RBRC also offered consistent education and outreach programs for any interested retailer or municipality.

Seeking consistency throughout North America, RBRC created the Rechargeable Battery Recycling Corporation of Canada (RBRCC) in 1997, which quickly grew to manage rechargeable battery collection programs in all provinces and territories. RBRCC later managed the collection and recycling of both single-use and rechargeable batteries in the EPR-regulated provinces

18. "State Recycling Laws," Battery Council International, accessed February 12, 2023, https://batterycouncil.org/Education-Research/State-Recycling-Laws.

19. Energizer, through the sale of its Gates subsidiary, left the Ni-Cd business in the late 1990s, withdrawing as a share member of RBRC. At the same time, Sony entered the Ni-Cd business and became an RBRC share member. When Call2Recycle later expanded into primary batteries, Energizer was invited back as a share member.

20. Call2Recycle, email communication with attached edited document, March 1, 2023.

of British Columbia (2010), Manitoba (2011), Québec (2014), Prince Edward Island (2019), and Ontario (2020). Given its differing audience and regulatory mandates, the Canadian affiliate was spun off to be a separately governed organization in 2017.

BATTERY PROGRAM FINANCING

Call2Recycle is funded by over 300 manufacturers and marketers of rechargeable batteries and rechargeable battery-powered products (representing over 90 percent of the rechargeable battery market).[21] The remaining 10 percent of companies that do not pay into the recycling system still benefit by having their batteries recycled (known as "free riders"). The organization has established over 17,000 active collection sites, including retailers, municipalities, manufacturers, and other businesses, across the United States.[22] From the outset, RBRC (and later Call2Recycle, Inc.) based the fees it charged to member companies to run the recycling program on the weight of covered batteries sold into the marketplace, as well as the costs involved in battery handling, sorting, and processing. This type of funding mechanism is called "cost internalization" because the fees are invisible to consumers and absorbed by companies into their cost of doing business. The fee system is an early form of what we now call "eco-modulated fees," which have become popular in funding packaging EPR programs.

To finance growth in its collections, Call2Recycle has since developed value-added, battery-related fee-based services that create an alternative revenue stream to fund its stewardship efforts. Those services include handling large-format batteries, product recalls, warranty returns, and other challenges often facing battery and product manufacturers. This business model enables Call2Recycle to continue to grow battery collections under a voluntary stewardship scheme.

EXPANSION OF BATTERY INDUSTRY RECYCLING PROGRAM

Over time, the widespread commercialization of lithium-ion (Li-Ion) batteries transformed consumer technology. It also changed the nature of battery collections. With their light weight and energy density, Li-Ion batteries virtually replaced most other rechargeable chemistries in the marketplace, including

21. Call2Recycle, email.
22. Call2Recycle, email.

Ni-Cd and nickel metal hydride (Ni-MH) batteries. Given the technology advancement of Li-Ion rechargeable batteries, these lighter and more powerful batteries outpaced sales of Ni-Cd batteries by the early 2000s. In response, RBRC expanded its mission to include the collection and recycling of all consumer rechargeable batteries under 5 kilograms (about 11 pounds) that are sold in North America. As RBRC grew to accept more chemistries and more products (cell phones and all batteries, including single-use, in Canada), the scope of its work rapidly expanded. As a result, RBRC and RBRCC formally changed their names to Call2Recycle, Inc., and Call2Recycle Canada, Inc., respectively, in 2013. The name was more relatable to consumers and aligned more closely with the organization's mission.[23]

EXPANDED US BATTERY STEWARDSHIP POLICIES AND KEY INITIATIVES

Soon after the creation of RBRC in 1994, the public policy landscape for batteries changed dramatically. Formalized by the federal Mercury-Containing and Rechargeable Battery Act of 1996, manufacturers no longer created consumer batteries containing mercury, except for a small portion of button cells, some of which still contain a minimal amount of mercury. The legislation also allowed small quantities of batteries to be transported as "universal" waste, a regulatory designation that more easily permitted retailers and other entities to participate in battery collections. Additionally, the law required that consumer nickel-based batteries be easily removable by consumers, facilitating their ability to be recycled. Since early battery laws were enacted to keep toxic materials out of disposal facilities, the phaseout of toxic materials from consumer batteries (along with Call2Recycle's efforts to divert the more toxic batteries from disposal) quelled most efforts to enact further battery stewardship legislation.

Although state legislation in the 1990s sparked the creation of Call2Recycle and its industry-run voluntary national battery recycling program, few other US initiatives attempted to increase battery recycling. One exception was a 2006 California law that banned the disposal of consumer batteries and required retailers that sell them to take them back upon request, with funding provided by producers and consumers. Following the California battery disposal ban, two California counties—San Luis Obispo and Santa Clara—imposed requirements on retailers to take back consumer batteries for recycling. Also, in 2005, New York City imposed a retailer take-back requirement

23. Consumers can find opportunities to recycle their batteries by going to the website Call2Recycle.org.

for rechargeable batteries, although this ordinance was later preempted by a statewide law passed in 2010.

For much of the decade (2000–2010), however, most EPR legislative activity in the United States was focused on developing and enacting bills on other products, such as electronics, paint, mercury-containing products, and pharmaceuticals. By 2010, an increased number of states finally had the staff capacity to address other priority products, such as batteries.

2010 PSI NATIONAL DIALOGUE

In response to a growing interest by governments to increase the recycling of all household batteries, PSI conducted research and convened a national stakeholder dialogue, both sponsored by Call2Recycle, to identify the problems and potential solutions for battery stewardship. The research was synthesized in a report (the *Battery Stewardship Briefing Document*[24]) that prepared participants for the PSI-facilitated discussion that took place at our 2010 National Product Stewardship Forum.

The briefing document was the first step in PSI's multi-stakeholder battery dialogue process, outlined in chapter 6. The report includes background information on battery composition, markets, and life-cycle management. It also proposes a project focus, issue (problem) statement, project goals, and meeting outcomes. Finally, it presents potential solutions pertaining to each of the project goals. The information in this report was derived through interviews and discussions with more than 40 key stakeholders, as well as a review of available literature. Following the dialogue meeting, PSI revised the document based on stakeholder review and comment to ensure that the information presented was as comprehensive and accurate as possible. It was the first multi-stakeholder conversation about battery EPR in the United States and provided a foundation for subsequent discussions.

MIT LIFE-CYCLE ASSESSMENTS

One participant at the 2010 PSI National Dialogue was the lead MIT researcher who presented preliminary findings of a life-cycle assessment (LCA) that compared landfilling of single-use batteries to several collection and recycling scenarios using existing North American battery processing facilities. The study, which was published in 2011 by MIT and updated in 2018 by

24. Product Stewardship Institute, *Battery Stewardship Briefing Document* (September 8, 2010).

Camanoe Associates,[25] was funded by the National Electrical Manufacturers Association (NEMA). The study concluded that "alkaline battery recycling has a net environmental burden . . . higher than landfill or incineration impacts for the majority of indicators used in the LCA," although two indicators (human health carcinogens and human health non-carcinogens) show a net environmental benefit.[26] This means that, except for the two human health indicators, the industry-funded study found that there are more impacts from recycling single-use batteries than from throwing them in the trash for disposal (from the limited recycling scenarios studied).

As a result, the LCA has been used as a tool to decide on battery recycling strategies that will maximize environmental benefits and minimize adverse impacts. The study found that the production of raw materials dominates the battery life-cycle impacts, with the transport of those raw materials to manufacturing having only a minimal environmental effect. The highest impacts during production come from only a few materials: manganese dioxide, zinc, and steel. Recycling can offset the need to mine raw materials and can reduce overall impacts. However, recycling has its own impacts, which the study found derive mainly from the energy consumed in the actual recycling process, the number of dedicated trips consumers take to return used batteries, and (to a lesser extent) the impact of transporting the batteries to be processed.[27] The study opened the possibility that, under the right circumstances, recycling single-use batteries can have a positive impact on the environment. The MIT studies played a significant role in spurring the battery industry to explore a voluntary collection and recycling solution for all batteries.

2014 PSI NATIONAL DIALOGUE

The 2010 PSI research and national stakeholder meeting fed into a 2014 national stakeholder meeting facilitated by PSI and sponsored by the Connecticut Department of Energy and Environmental Protection (CT DEEP). The agency's EPR lead and environmental analyst, Tom Metzner, was also a PSI board member and one of the founders, along with PSI, of the Connecticut Product Stewardship Council. The 2014 meeting, which was held for two days in Hartford, Connecticut, was attended by more than 130 stakeholders (in person and online), including government officials from 23 states. PSI updated its 2010 background briefing document for the

25. Elsa Olivetti and Jeremy Gregory, Camanoe Associates, "Life Cycle Assessment of Alkaline Battery Recycling" (March 2018), accessed March 26, 2023, https://www.nema.org/docs/default-source/products-document-library/Life-Cycle-Analysis-of-Alkaline-Battery-Recycling---2018.pdf.

26. Olivetti and Gregory, 18.

27. Olivetti and Gregory.

two-day stakeholder meeting to lay out the current understanding of the problem, goals, barriers, and potential solutions for increasing the recycling of single-use and rechargeable batteries. At the same time, CT DEEP officials began working with a key legislative sponsor who agreed to introduce a battery EPR bill in the 2015 legislative session.

To prepare for the Connecticut dialogue meeting, PSI also developed a document, *Elements of Battery Legislation for Discussion*, which included specific legislative language offered as best practices from battery EPR bills introduced in multiple states, along with government agency input. This document also proposed a starting position for discussions with the battery industry. The battery elements document, which represented the joint position of state and local government battery recycling experts across the country, defined 23 key elements needed to increase battery collection and recycling performance.[28] It also laid out a preferred government agency approach, along with options for consideration. The battery elements document was an application of the basic conceptual tool that PSI routinely uses (discussed in chapter 6) to gain consensus on numerous important technical issues prior to considering legislative language.

The battery elements document spurred the battery industry to unveil their own draft all-battery recycling bill at the start of the Connecticut stakeholder meeting. The industry bill was the first time that all associations representing battery manufacturers agreed on the need for both single-use and rechargeable batteries to be collected, recycled, funded, and managed through an EPR framework. The 2014 national dialogue represented an ideal opportunity to develop a consensus between key state and local government officials and battery manufacturers. Legislative efforts in Minnesota (2013) and California (2014) on single-use batteries, and in Washington and Oregon on rechargeable batteries (both in 2013), ended without bill passage. At the Connecticut dialogue meeting, key differences in perspectives and approaches were debated, and these talks continued through multiple PSI-facilitated stakeholder calls over the next few months. As differences on issues narrowed significantly and consensus was within reach, stakeholders agreed on a final meeting to hammer out remaining issues.

Unaware of our efforts, however, the Connecticut legislative sponsor requested another meeting with battery industry representatives, who promptly left PSI's negotiating table. Thinking they had a direct line to the bill sponsor, manufacturers promoted their own original bill. In response, Connecticut state officials, feeling compelled to follow the lead of the bill sponsor, also promoted their own agency bill before the legislature. Ultimately, the 2015 battery bill that was introduced in Connecticut did not

28. PSI later refined its elements documents to about 16 elements.

pass. Even so, momentum for managing household batteries under EPR policy had already been created in other states.

STATE BATTERY BILLS

Although Connecticut did not become the first state with a next-generation battery EPR bill, the national dialogue included representatives from numerous states. Jen Holliday, director of public policy at the Chittenden County Solid Waste District (Vermont), was one of the government officials who participated in the national battery recycling dialogue. Holliday, a founding member and chair of the Vermont Product Stewardship Council and PSI president, was one of the government leaders who helped develop PSI's *Battery EPR Elements Document* and participated in PSI-convened policy meetings. Sensing a political opportunity, she obtained a bill sponsor in Vermont, negotiated with battery manufacturers, and helped pass the nation's first single-use, multi-chemistry battery bill one month prior to the Connecticut dialogue meeting, in May 2014.

Vermont local governments focused only on single-use batteries because they did not think they had the votes to pass an all-battery bill that included batteries embedded in products, a main issue of contention among manufacturers of toys and medical devices. These manufacturers have increasingly included batteries that are difficult to remove, or not designed to be removed, from their products. The Vermont law covers approximately 81 percent of primary batteries sold in the state.[29] From 2015 to 2020, the law resulted in the collection of nearly 455,000 pounds of single-use batteries, with an additional 290,500 pounds of rechargeable batteries being collected during this same time period.[30]

The momentum for additional state battery legislation grew through the passage of the Vermont bill, the Connecticut dialogue, and subsequent bill discussions. Trying once again to pass its bill first unveiled in Connecticut, the battery industry now sensed an opportunity in Maine. By this time, the industry bill had incorporated many of the elements in PSI's *Battery EPR Elements Document*. Ironically, however, the tables had turned, with industry taking the lead and government following. Unfortunately, the battery industry sought support and input from PSI and its government agency members very late in the process. Such earlier discussions might have resulted in strong government support to counter strong opposition from the toy industry. With

29. Call2Recycle, email communication with attached edited document, March 1, 2023.

30. Mia Roethlein, Vermont Agency of Natural Resources, "Combined Battery Recycling Collections," email to author, August 19, 2021.

only tepid support among Maine local governments, the bill eventually died. The industry's 2016 effort in Maine became a turning point for the battery industry. Sensing that national priorities of government officials had shifted to other, more pressing, product stewardship issues, the battery industry quietly receded in its support for battery EPR legislation.

Fortunately, the groundwork had been laid for battery EPR in the United States. From the 2010 PSI briefing document and dialogue meeting sponsored by Call2Recycle, to the 2014 PSI-facilitated national dialogue meeting sponsored by CT DEEP, multiple iterations of state battery EPR bills were developed by PSI and its government members. These discussions, at various times including industry partners, contributed either directly or indirectly to the introduction of all-battery bills in Texas (2015 and 2017), Maine (2016, 2017, 2019, and 2020), Washington (2020, 2021, 2022), District of Columbia (2020 and 2021), and California (2021 and 2022); rechargeable battery bills introduced in Washington (2013) and Oregon (2013); and single-use battery bills introduced in Minnesota (2013), Vermont (2014), California (2014), Connecticut (2015), and multiple times in New York (since a 2010 law covering rechargeable batteries already exists).

In 2019, Rachel Clark, legislative counsel for District of Columbia councilmember Mary Cheh, sought help from PSI to introduce an all-battery bill. Suna Bayrakal, PSI's battery lead and director of policy and programs, and I provided technical advice using our *Battery EPR Elements Document* and connected the office to PSI members, who were also battery experts in Vermont and Connecticut. Councilmember Cheh subsequently introduced and eventually passed the nation's first bill covering multi-chemistry, single-use and rechargeable batteries, as well as battery-containing products, which was enacted into law in 2021. The law also references recycling efficiency targets, highlighting the importance of maximizing the recovery of material that enters a recycling facility. In 2022, California also enacted a similar single-use and rechargeable battery law, as did Washington in 2023, which became the first state to include e-mobility batteries and a study on embedded batteries.

BATTERY STEWARDSHIP: LESSONS LEARNED AND THE ROAD AHEAD

Since the 1990s, only six battery EPR laws have been enacted: California's 2006 retail take-back law for rechargeable batteries funded by producers, a rechargeable battery law in New York in 2010, a single-use battery EPR law in Vermont in 2014, and all-battery (rechargeable and single-use) laws in the District of Columbia in 2021, California in 2022, and Washington in 2023.

Numerous other battery bills have been introduced, but not passed, in multiple states, as outlined above. Even so, the past decade of battery EPR development has laid the groundwork for what is likely to be a significant increase in the enactment of battery EPR laws due to the prominent role that batteries play in developing a circular economy and in reducing climate change. The need for these laws has also increased due to the significant number of fires caused by Li-Ion batteries. Below are some of the lessons learned for battery EPR advocates. There are many important policy questions to be answered, even as new policy questions will undoubtedly emerge.

Local Government Support Is Critical to Bill Passage

Coalitions start by building support among those sharing a common vision. Local governments are natural allies of product stewardship efforts, even if not fully aware of ways they can benefit. Their support should be sought as an initial step in any EPR product policy effort. Since they often finance and manage waste systems, municipalities tend to collectively benefit from systems that require manufacturer or consumer funding. Providing more infrastructure to collect and process products, along with consumer education, relieves a significant and growing burden for local governments.

For these reasons, state municipal associations have become important allies that have, over time, taken formal positions in support of EPR bills. As municipal officials and their associations gain experience with EPR, they begin to realize that this policy approach, which seeks to reduce one of their largest annual budget items, is a key municipal financing strategy. By allowing their residents to conveniently recycle batteries and other consumer products, municipal officials provide important avenues for residents to take responsibility for safely managing the products they consume.

For state legislation to pass, it will need the support of a significant number of local governments, particularly those represented by influential legislators on important committees. In Vermont, battery bill passage was driven by a local government leader backed by other important local government supporters. Without such local interest, the Maine bill did not generate the groundswell of support that gets legislator attention and results in action.

Stakeholder Unity Increases the Success of Bill Passage

The legislative process is often chaotic and unpredictable, with myriad bills competing for a legislator's attention. Bills have a greater chance of success if there are few apparent differences among stakeholder positions, which obviates the need for a legislator to take a political stance that favors one constituency over another. Multiple stakeholders who channel their collective energies into a clear harmonious position create clarity and legislative momentum. As occurred with some of the battery bills introduced, stakeholders were not coordinated, and thus lobbied legislators independently on multiple fronts. Seeking a positional advantage, government and industry players sought to meet their own interests. The different positions taken were not resolved either before or after legislative hearings, leading legislators to take no position. Although this coordination role is often undertaken by a legislative sponsor or committee chairperson (or their aide), other parties can assume this role as well, such as agency officials or those perceived as having the ability to bring stakeholders together and provide a forum for resolving differences. PSI has often played this role.

Timing and Messaging

Once stakeholders develop a solid policy and assemble a strong coalition, there is a legislative rhythm that is similar in all states, but also unique in key aspects. Being a policy expert is not enough to pass legislation. One needs to have skills within their coalition to be capable of successfully lobbying the legislature. A bill needs to be framed in a way that resonates with legislators, as well as with those the bill is intended to help—municipal officials who make decisions to pay for recycling and disposal, recyclers and waste management companies that seek to reduce impacts from Li-Ion battery fires, community members who want to recycle, environmental groups that want to reduce the toxicity of all products, and battery manufacturers that want to equitably fund the reduction of impacts from their products.

Regarding battery bills, EPR policies are primed for a boost due to the greater awareness of battery technology in reducing climate change. Also, the increased risk of Li-Ion battery use and disposal has heightened the importance of proper postconsumer management. Another factor in boosting the visibility of battery EPR is that batteries also contain raw earth materials that have strategic and economic importance.[31] In addition, there is increased awareness of the need to recover materials in a circular economy to create jobs and economic opportunities all along a product's life cycle. Messaging around these aspects will ensure battery EPR bills get the attention they deserve.

To make certain that the hard work in messaging a bill and gaining coalition support is effective, bill leaders usually start discussions and build coalition support in the summer and into the fall. Most legislative sessions start in

---

31. Daniel R Simmons, "Advancing US Battery Manufacturing and Domestic Critical Minerals Supply Chains," US Office of Energy Efficiency & Renewable Energy (October 4, 2019), https://www.energy.gov/eere/articles/advancing-us-battery-manufacturing-and-domestic-critical-minerals-supply-chains.

January. Some allow bills from the previous session to be carried over, and others have short and long sessions that alternate year to year. No two states are exactly the same in their legislative rhythm. In any case, the best time to develop a unified position among stakeholders is during the fall, so when legislatures are in full swing, a bill can be introduced and further negotiated. In most cases, the greater the support for a bill among diverse stakeholders—those representing industry, government, and environmental interests—the greater the chances of bill passage.

Due to varied state and municipal product stewardship priorities, batteries have not quite made it into the collective government crosshairs long enough to develop the consistent message to the battery industry that the risk of inaction might lead to laws they won't like. The lack of prolonged government attention has led the battery industry to take a step back and reassess its commitment to promoting recycling programs that will require additional industry resources. Even if legislative intent becomes more aligned, many items still need to be negotiated to develop a unified model battery EPR bill.

National Legislative Consistency

As with all industries, battery manufacturers are extremely concerned with having to comply with a patchwork of state and local legislation across the country; it adds complexity, increases regulatory and administrative costs, and decreases efficiency. Although state battery laws that passed in the 1990s are considered the first EPR laws in the United States, they are splintered, with some states regulating rechargeable batteries only, while others regulate just single-use batteries, and some only specific battery chemistries. As of June 2023, only Washington, DC, California, and Washington State have enacted laws requiring producers to collect both single-use and rechargeable batteries of a wide range of chemistries.

There are two ways to achieve national policy consistency—federal legislation or model state legislation. The industry's interest in regulatory consistency drove it to develop an all-battery EPR bill for the first time in 2014, but only after legislation was introduced in Connecticut and California, and after government officials from 23 states attended a national stakeholder meeting. The threat of bill enactment in multiple states often needs to be great enough to keep industry interested in developing a model bill, which also reduces the risk of objectionable legislation. PSI's state battery EPR model, which is similar to the industry's own model, can help drive greater national consistency.

One of the main reasons why battery recycling rates are higher in EU member states than in the United States is that the European Commission (EC) mandates the collection of single-use and rechargeable batteries through its EU-wide EPR directive, which ensures greater consistency and effectiveness across the continent. By contrast, the United States does not have a federal battery recycling mandate that covers all states and requires producers to fund and manage single-use and rechargeable batteries. Only 11 US states have battery EPR laws, each with different requirements, and mostly passed years ago without collection rate or recycling efficiency mandates.

Fortunately, there has been recent federal legislative interest in battery EPR. The federal Infrastructure Investment and Jobs Act,[32] signed into law in 2021, requires the convening of a task force to develop a federal framework for battery EPR. The multi-stakeholder task force is expected to produce a framework and implementation recommendations within a year of convening. This initiative has the potential to spur greater policy consistency across the country.

VOLUNTARY BATTERY RECYCLING AND FREE RIDERS

The Call2Recycle program collects and recycles both rechargeable and single-use batteries. Even so, producers only fund single-use battery collection and recycling in jurisdictions where EPR laws have passed for single-use batteries, currently only in Vermont, California, Washington State, and the District of Columbia. They also cover the cost for rechargeable batteries (about 20 percent of the battery market) in all states on a voluntary basis, and in the ten states and the District of Columbia that have EPR laws for rechargeable batteries. Since the national rechargeable battery program is voluntary for producers, about 10 percent of rechargeable batteries are not funded by the companies that sell them into the market. This inequity results in a free rider problem; companies making those batteries are not charged for their postconsumer management. In addition, due largely to the lack of battery recycling requirements and the cost of battery recycling, there is a dearth of US consumer battery sorting and recycling facilities. EPR laws can provide sustainable funding to recycle batteries, which can supply a continuous flow of quality materials to recyclers, encouraging long-term investments in recycling facilities. Working together, stakeholders can help to achieve common goals of efficiently increasing battery recycling, decreasing externalities imposed by Li-Ion and other batteries, and reducing battery life-cycle costs and impacts.

The attention increasingly paid to the safe management of Li-Ion batteries provides an opportunity to realign battery manufacturers, government officials, waste and recycling companies, and other key stakeholders. The

32. The Federal Infrastructure Investment and Jobs Act, H.R. 3684, https://www.congress.gov/bill/117th-congress/house-bill/3684/text.

resources now available for safely collecting and managing used Li-Ion batteries should be used to develop a federal all-battery EPR bill or a model for states to introduce.

RETAILER REQUIREMENTS

In 2005, New York City imposed a retailer take-back requirement for rechargeable batteries, which the city aggressively enforced against Times Square bodegas as well as large electronics stores. This city ordinance was later preempted by New York's statewide law passed in 2010. Two California counties, San Luis Obispo and Santa Clara, also passed retailer requirements following the 2006 California statewide ban on battery disposal.

Based on its collection data, Call2Recycle has concluded that retail requirements have yielded mixed results, regardless of whether they are imposed by state, county, or city government.[33] Despite aggressive city enforcement, Call2Recycle believes it is not clear that mandates for New York City retailers attained any greater collections than if they hadn't been required. In fact, retailers learned quickly that compliance meant having a visible collection box, even if they had no intention of using it. By contrast, battery collections in California were boosted most through a cooperative effort between Call2Recycle and county governments that identified and engaged collection sites proactively. Although the retail mandate provided context and incentive, the results were further improved as a result of the collaborative effort among Call2Recycle, governments, and retailers. But even with extensive outreach, Call2Recycle has found it difficult to compel a business or another entity to collect consumer batteries, particularly with lax agency enforcement.

Although some retailers continue to host collection boxes in their stores to bring in more foot traffic, many retailers have been reluctant to divert staff resources from other important tasks to recycling programs. In line with PSI's *Battery EPR Elements Document*, producers will need to meet a convenience standard to ensure that the collection of batteries from consumers is convenient. The standard can be met by a combination of methods, including collections at retailers and HHW depots. The more convenient battery collection is for consumers, the more likely it is that batteries will be collected in greater numbers. The role of retail collection will continue to be a prime consideration in future battery EPR programs, and all stakeholders will need to jointly establish incentives and methods for effective retailer participation.

33. Call2Recycle, email communication with attached edited document, March 1, 2023.

INCREASING PRODUCT SCOPE ENLARGES THE CIRCULAR ECONOMY

As we plan to capture more of the familiar single-use and rechargeable battery chemistries, we will need to plan for recovering other battery types that have already begun to enter the market, such as those used in e-cigarettes, vapes, e-bikes, motorized scooters, electric vehicles, and other consumer products. Further, we need to craft current battery EPR policies to incorporate future battery chemistries. Each state will decide how prescriptive its legislation will be and how much flexibility to provide its oversight agency on decisions for managing new battery types under its EPR law.

The proliferation of different battery technologies (Li-Ion vs. other rechargeable technologies) combined with increased use of batteries across product categories (bikes vs. power tools vs. computers) can be confusing to consumers in terms of how and where to recycle them. Call2Recycle realized early on that, to maximize collection of rechargeable batteries, they needed to collect all batteries, since consumers rarely distinguish among them. This factor motivated program designers to include the collection of multiple types of rechargeable batteries and single-use batteries. Laws that are designed to accommodate new technologies and new battery uses will best position programs for long-term efficacy and a level playing field for producers. Reducing consumer confusion over which batteries can be collected will ultimately increase battery collection.

THE ROLE OF LIFE-CYCLE ASSESSMENT

A key goal of all product stewardship initiatives is to reduce environmental impacts across a product's life cycle. The battery industry has already taken the step to develop a detailed life-cycle assessment (LCA), which they first conducted in 2011 and revised in 2018. This LCA showed that one of the key opportunities for reducing energy use (and greenhouse gas emissions) in a battery's life cycle is to recycle, thus decreasing the need to mine raw earth materials. The industry's LCA showed that the *disposal* of single-use batteries, for the limited recycling scenarios studied, results in *fewer* environmental impacts than *recycling* those batteries. However, the results of an LCA can shift dramatically, based on assumptions made in the analysis about current conditions (e.g., distance the average person will travel to a battery collection site). LCA results are likely to change with new policies that increase the number of convenient battery collection locations and domestic battery processing facilities.

Enacting US EPR battery laws will result in the collection of more used batteries, which is a signal to investors to develop additional battery recycling capacity to accommodate an increase in battery material supply. It is highly possible that increased enactment of battery EPR laws will result in more efficient transport of batteries from collection locations to recyclers, which will be closer in proximity and process greater quantities of batteries. These changes, among others, can swing an LCA so that single-use battery recycling shows a net positive environmental impact. There is an important opportunity for all stakeholders to analyze the current battery LCA and consider steps that might result in both the recycling of more single-use batteries and reduced environmental impact from single-use battery recycling.

CONVENIENCE, OUTREACH, AND AWARENESS

To maximize battery collections, consumers need to be aware of the problem with batteries, particularly the increased safety issues caused by the mishandling of Li-Ion batteries. Once aware, consumers will often seek a convenient place to recycle their spent batteries. Education that heightens consumer awareness and knowledge, along with available collection infrastructure, are critical, along with frequent outreach to collection site managers and the public. These efforts are costly, but necessary, elements to reach program performance targets. Other strategies will also be needed to convert awareness into action, and these are best discussed jointly among Call2Recycle, state oversight agencies, and other strategic partners. For example, the District of Columbia law, which bans consumer disposal of batteries beginning 2023, will create a strong incentive for people to bring batteries to collection points.

INTERNALIZING EXTERNAL COSTS FROM BATTERY MANAGEMENT

Lithium-ion batteries have caused millions of dollars of damage to recycling and waste management facilities, including lost wages, facility downtime, and plant repairs. Who pays these costs now, and what are the best policies to reduce the impacts from Li-Ion batteries in the future? Should EPR laws require battery producers to pay fees into the recycling system based on the full cost to manage each battery type they sell into the market? These enhanced "eco-modulated fees" would, therefore, consider multiple factors based on externalized costs, such as Li-Ion battery fires, as well as the full life-cycle costs to mine and manufacture new batteries and recycle postconsumer batteries.

Would it be effective and feasible to implement a ban on the disposal of Li-Ion batteries, on all rechargeable batteries, or on those that are *both* single-use and rechargeable? How would municipalities enforce such a ban? And how is it possible to determine which products contain embedded batteries? Or should battery brand owners be required to assess the feasibility and potential benefits of a deposit return system for batteries, just as the European Commission has required of the battery industry in its updated battery directive? These are but a few of the many policy questions that stakeholders will need to grapple with in the coming years to bring us to a more sustainable battery materials management future.

RECOVERING BATTERIES IN PRODUCTS

Batteries are unique in that they exist to power other products. Thus, batteries are either sold in stand-alone packs as replacements or with products. Some products are designed to make batteries difficult to remove as a safety feature. To retrieve these batteries means collecting the product as well. This was the logic behind Call2Recycle's decision to collect cell phones in 2004. By doing so, they increased the collection of rechargeable batteries, as well as products that were low cost to repurpose or recycle. However, other products, such as toys and medical devices, often contain batteries that are difficult to remove and equally difficult to collect with the product.

Is the best policy for maximizing battery recovery to require all batteries to be easily removable from products so consumers can recycle them and, if so, should certain products be exempt? Should each battery manufacturer be held responsible for knowing whether their batteries are sold individually to consumers or are embedded in products? Should battery EPR laws hold brand owners (e.g., Hasbro, Fisher-Price, Mattel) responsible for the batteries it embeds in the products they sell?

The battery industry wants batteries sold with products to be included in all EPR laws so the brand owners pay for their recycling; otherwise, these batteries will become a free rider in the recycling system. In contrast, toy and medical device companies have sought an exemption from regulation. Many government officials take a third approach and want to recycle both the batteries and the products in which they are contained. The difference in perspectives among these three key stakeholder groups presents a current challenge in need of a resolution. The increase in Li-Ion battery fires, the increased use of batteries as we race to slow climate change, the importance to US national security of recovering critical earth minerals contained in batteries, and the acceptance of producer responsibility policies might just be the catalysts needed to enact an increased number of effective battery EPR laws in the coming decade.

Chapter 10

Packaging EPR in the United States

Imagine a product delivered to your home wrapped in a 14-karat gold package. Needless to say, you'd be quite happy. That package has obvious value. You could keep that package or sell it to someone who will turn it into a ring, other jewelry, or another product. You would receive money for your gold package because, after all, it's valuable stuff.

Now, what if the product you received was delivered in plastic, paper, cardboard, metal, glass, or a combination of materials? You would likely find little direct value in the package. You might check to make sure it is recyclable in your municipal program and, if it is, put it into your curbside bin or bring it to your drop-off recycling center. At that point, it will be transported to a recycling facility and become the responsibility of those who process materials and find recycling markets. Unfortunately, if your package is not recyclable, you would have to put it into the garbage, destined for a landfill or waste-to-energy plant. But if you try to recycle that unrecyclable package, unbeknown to you, it will add cost to your local recycling program since your package would become a contaminant at the recycling facility.

In the United States, consumers are expected to figure out what is recyclable and what is garbage, even amid inconsistent recycling programs, constantly changing packaging materials, and product labels that often conflict with municipal recycling instructions. Municipalities are expected to continually educate their residents about how and what to recycle, and to budget for recycling costs during frequently fluctuating markets. Recycling facility operators are expected to process materials that enter their plant and maximize material value by meeting end market specifications. The actor that is least visible in the picture is the one most responsible for deciding what materials are used in packaging: the brand name companies that sell consumer products.

Extended producer responsibility (EPR) laws for packaging change this inequity so that the responsibility for funding and managing packaging waste shifts to product brand owners. These companies choose the type of packaging in which to sell their consumer goods based on a myriad of factors, including material cost, product protection and safety, transportation cost, sustainability criteria, consumer convenience, and marketing pizzazz. EPR laws add another factor for brand owners to consider: the cost to manage packaging once a consumer has a product in their hands and no longer needs the packaging. Currently, local governments, taxpayers, and ratepayers in the United States are responsible for financing and managing postconsumer packaging through thousands of distinct recycling programs. Under EPR laws, this system becomes the main responsibility of brand owners, while other stakeholders, including local governments, state governments, collectors, processors, and consumers, will continue to play important roles in the system. Together, they will provide a more consistent recycling experience for consumers.

Packaging has a long life cycle, from material extraction and manufacture to conversion into a package used to fill, protect, and ship products to a retail store or direct to a consumer. Under an EPR system, after a package has fulfilled its function for the consumer—often in a matter of minutes, like a candy wrapper—the responsibility for that package stays with the company that made the decision to use it. That company is now obligated to manage the discarded packaging, removing the financial burden from municipalities, taxpayers, or individual ratepayers. In other words, the municipality or household that had no role in choosing the packaging material is no longer stuck with the cost and primary responsibility for managing it. As a consequence, this major shift in responsibility will now be part of the calculation that brand owners make in choosing materials for their packaging in the first place. Under packaging EPR systems, companies pay less for using materials that cost less to manage after consumer use and have lower environmental impacts. Conversely, they will pay more for using materials that are not reusable or recyclable, cost more to recycle, or have greater negative social and environmental impacts.

Under an EPR system, consumers will still be expected to use convenient reuse and recycling systems that companies and governments provide. However, the success of recycling programs will become the main responsibility of brand-name companies that are ever present in our lives, including Procter & Gamble, Unilever, General Mills, Amazon, Walmart, Costco, SC Johnson, Coca-Cola, PepsiCo, and numerous others. A primary goal of EPR systems is to expand access to consumers and make packaging reuse and recycling more convenient and understandable to them. It is also to fund the modernization of the US recycling system by collecting a larger and cleaner amount of packaging materials, including those not collected and recycled under current systems.

This case study provides the context for the significant and rapid change taking place in managing packaging waste in the United States. With the first four US packaging EPR laws having passed in 2021 in Maine[1] and Oregon,[2] and in 2022 in Colorado[3] and California,[4] and with over 30 individual packaging EPR bills introduced in 16 states in 2022,[5] the entire EPR movement in the United States has been elevated to a new level. The case study also provides insights into initiatives driven by PSI, state and local government members, industry, environmental advocates, and others that set the stage for the passage of these laws. It outlines the problems, goals, barriers, and solutions considered in managing packaging waste, as well as the long journey still underway among those of us advocating for EPR systems.

For packaging EPR to have taken hold in the United States, we first had to educate government officials, legislators, advocates, companies, and organizations about the general concept of EPR. PSI began that educational process in 2000 by defining the terms *product stewardship* and *extended producer responsibility*, then traveling around the country to meet with government officials to explain these terms and lay out a new vision for managing municipal solid waste. It then took over 20 years for the experience of developing, enacting, and implementing numerous EPR laws in many industry sectors[6] to provide a firm foundation from which to begin to change packaging recycling systems. But it was the "China National Sword Policy," China's severe restriction of recyclable material imports, that led to skyrocketing municipal recycling costs and the urgent need for a paradigm shift in managing waste. This exposure, coupled with a global backlash against single-use plastic products and packaging, forced political action in support of packaging EPR in the United States.

Ultimately, the story of packaging EPR in the United States provides a strong lesson about the need to lay extensive groundwork so that, when an opportunity presents itself, advocacy can succeed. We are now in the midst of significant positive change regarding the management of municipal garbage in the United States, transitioning from a linear make-use-dispose economy

---

1. State of Maine Legislature, Summary of LD 1541, https://legislature.maine.gov/LawMakerWeb/summary.asp?ID=280080518.
2. State of Oregon Legislature, Senate Bill 582, https://olis.oregonlegislature.gov/liz/2021R1/Downloads/MeasureDocument/SB582.
3. State of Colorado Legislature, HB22–1355, https://leg.colorado.gov/bills/hb22-1355.
4. State of California Legislature, SB54, https://leginfo.legislature.ca.gov/faces/billTextClient.xhtml?bill_id=202120220SB54.
5. Product Stewardship Institute (2022), www.productstewardship.us.
6. As of March 2023, there were 131 US EPR laws enacted in 16 industry sectors in 33 states. For an up-to-date list, visit the Product Stewardship Institute website, accessed at https://www.productstewardship.us.

to one that views product and packaging design, material selection and use, and viable end markets for recyclable materials, as part of a circular economic system that can restore our natural resources rather than deplete them. Let's now step back and consider the context for this momentous change that is taking place right before our eyes.

SCOPE: PACKAGING AND PAPER PRODUCTS

Packaging, which represents about 28 percent of the US municipal waste stream,[7] includes materials that brand owners use to package everything from cereal, bread, vegetables, and cleaning supplies to water, juice, soda, shampoo, and other products. It includes plastic film and containers, steel and aluminum cans, glass bottles, cardboard, and a combination of these materials. Paper products (e.g., marketing flyers, magazines, office paper) contribute another 9 percent to the waste stream.[8] Together, roughly 110 million tons of packaging and paper products (PPP) are used each year in the United States (about 37 percent of annual municipal solid waste generation).[9] EPR bills in the United States target PPP materials because they represent a significant part of the waste stream from homes and businesses; these materials are collected in our curbside and drop-off recycling programs. Additional products often included in US packaging EPR laws and bills include single-use plastic products such as straws, cutlery, cups, plates, and grocery bags, which also contribute to litter and waste.

While the Maine packaging EPR law focuses solely on packaging, Colorado's also includes paper products, since these materials are collected with packaging in recycling systems. Oregon's law, by contrast, includes packaging, paper products, and foodservice ware (e.g., cutlery, cups, plates), and California's covers packaging and foodservice ware, but not paper products. In this regard, some bills and laws often refer to their mission as

7. US Environmental Protection Agency, "Advancing Sustainable Materials Management: 2018 Tables and Figures" (December 2020), table 22: 33–34, accessed January 2, 2022, https://www.epa.gov/sites/default/files/2021-01/documents/2018_tables_and_figures_dec_2020_fnl_508.pdf. Total containers and packaging (82.2 million tons) is roughly 28 percent of total municipal solid waste generated (292.4 million tons).

8. US Environmental Protection Agency, "Advancing Sustainable Materials Management: 2018 Tables and Figures" (December 2020), table 5: 6, accessed January 2, 2022, https://www.epa.gov/sites/default/files/2021-01/documents/2018_tables_and_figures_dec_2020_fnl_508.pdf. Total paper and paperboard non-durable goods (25.5 million tons) is roughly 9 percent of total municipal solid waste generated (292.4 million tons).

9. US Environmental Protection Agency, "Advancing Sustainable Materials Management: 2018 Tables and Figures," table 5: 6. Total paper and paperboard non-durable goods (25.5 million tons) plus total containers and packaging (82.2 million tons) is roughly 110 million tons, or about 37 percent of total municipal solid waste generated (292 million tons).

"modernizing recycling" or use other terms that reflect their expanded focus beyond packaging.

Beverage containers are also considered packaging. The key question is how they are treated in a packaging EPR bill and whether the state in which the bill is introduced has already enacted a deposit return system (DRS, also called a "bottle bill"). There are 10 states that have enacted DRSs, including Oregon (in 1971), Maine (in 1976), and California (in 1986).[10] These require consumers who purchase designated beverages to pay a container deposit (usually five or 10 cents), which they get back upon return of the container. Beverage containers included in the DRS laws for Oregon, Maine, and California are *excluded* from their packaging EPR laws, since these containers are already being managed. Including them in the EPR law would result in producers of those containers being charged twice to manage the same containers. Beverage containers not included in their DRS laws are *included* in the EPR law. Otherwise, those containers would become free riders in the recycling system, whereby producers of other packaging materials would subsidize their collection and processing.

WHO ARE PACKAGING PRODUCERS?

Many of us are familiar with at least some of the brand name companies that sell paint, batteries, electronics, mattresses, and other consumer products (e.g., Sherwin-Williams, Duracell, Apple, and Sealy). Under EPR systems, these companies are held responsible for managing their postconsumer products because they make the decisions about which materials to use when manufacturing them. Typically, these companies are both the brand owner and the product manufacturer. By contrast, defining the entity responsible for managing packaging under an EPR law is more complicated. Is it the manufacturer of the packaging, the company using the packaging to sell their product, or others along the supply chain?

Under the four US state packaging EPR laws and most bills, there is a standard three-tiered definition that holds a company selling a product into a state for the first time responsible for its packaging. This entity is most often the brand owner of the product rather than the packaging manufacturer, but also could be the importer/distributor of the product. Retailers that sell their

---

10. For a full list of US states with deposit return systems (bottle bills), see "State Beverage Container Deposit Laws," National Conference of State Legislatures, accessed October 12, 2022, https://www.ncsl.org/research/environment-and-natural-resources/state-beverage-container-laws.aspx. Also see Bottle Bill Resource Guide, Container Recycling Institute, accessed October 12, 2022, https://www.bottlebill.org/index.php, and https://www.bottlebill.org/images/Allstates/10_States_Summary_and_Notes_July_2022.pdf.

products as store brands (e.g., Best Buy, Target, Walmart, Whole Foods) are also considered the responsible party under packaging EPR laws, as are online sellers (e.g., Amazon) if they add their own packaging to the product. This same EPR producer definition is also used in the United States for consumer product laws and bills and is common practice in existing EPR programs across Europe and Canada.

Consumer brand companies are held responsible under EPR systems because they make the decision about what packaging to use for their products and thus have the greatest financial incentive to change their packaging. They are often the only entities that can track sales of their products and packaging, which is needed to comply with EPR laws. For this reason, brand owners are held primarily responsible for packaging under nearly all EPR systems globally. Moreover, packaging manufacturers have far less ability than consumer brands to make packaging more sustainable, especially if brands want to switch to a different material or away from single-use packaging to reusable packaging, or even to no packaging at all.

Placing legal responsibility on brand companies is also the most effective way to manage the system, since that one brand company is held responsible for all the packaging they use. For example, holding Chobani responsible for all packaging needed to ship their yogurt to retail outlets is simpler for regulators, producers, and supply chain partners than placing responsibility on each individual manufacturer of the plastic cup, foil lid, and cardboard boxes that Chobani uses to package, seal, and ship its yogurt. Holding Chobani responsible for paying for the reuse, recycling, or disposal of its packaging materials also provides a more direct financial incentive to that one company rather than holding responsible the three manufacturers of each of the three distinct packaging types.

Let's now look at which consumer brands sell the most products, thus using large quantities of packaging. The following 10 companies generated $467 billion in global revenues in 2021: Nestlé, PepsiCo Inc., Anheuser-Busch InBev SA/NV (AB InBev), JBS, Tyson Foods, Mars International, Coca-Cola, Cargill, Danone, and Mondelēz International.[11] Fast-moving consumer goods (FMCG), other heavy users of packaging, are also tracked as a business sector: The top 10 global FMCG companies by sales are Nestlé, Procter & Gamble (P&G), PepsiCo, Unilever, JBS S.A., British American Tobacco, Coca-Cola, L'Oréal, Philip Morris International, and Danone.[12]

11. Kristen Kazarian, Packaging Strategies, "2019 Top 100 Food & Beverage Packaging Companies" (August 9, 2019), accessed March 26, 2023, https://www.packagingstrategies.com/2021-top-100-food-and-beverage-packaging-companies.

12. Raveendran, Firmsworld, "Top Ten Biggest FMCG Companies in World," last updated September 7, 2022, accessed March 26, 2023, https://firmsworld.com/top-10-fmcg-companies-in-world/.

Consumer brands held legally responsible for managing postconsumer packaging are not the only companies impacted by EPR laws. Certainly, the companies that manufacture materials used in packaging have an interest, as do the companies converting these materials into packages. The top ten packaging companies in the world by sales, with combined global revenues of about $128 billion in 2020, are International Paper Company, WestRock, Tetra Laval, Amcor, Oji Holdings, Crown Holdings, Ball Corporation, Smurfit Kappa Group, Reynolds Group Holdings, and Stora Enso.[13]

In at least one US state, some retailers have sought to put the responsibility on packaging manufacturers and not on retail brands. That producer definition is used in Italy, where packaging manufacturers, as well as importers, are held responsible under their EPR law. In practice, manufacturers are obligated for the packaging made and used *in* Italy while the importer is responsible for packaging made *outside* Italy but sold into the country. Most manufacturers, however, have informal agreements whereby brand owners reimburse them for the fees they pay to CONAI, the PRO. Since there are fewer manufacturers than brands, some claim that this arrangement has led to tracking fewer obligated companies. Since Italy is the only country with this system in Europe, companies that sell products in Italy have a more complex tracking system to apportion responsibility.[14]

In Canada, all provinces with packaging EPR systems hold brand owners, rather than packaging manufacturers, responsible for the packaging that they use because brands hold the decision-making power in the marketplace. Retailers that are brand owners of their own private-label products are also held accountable for the same reason. In Canada, as in the United States, if the brand owner does not have legal residency in the legislated jurisdiction (e.g., sells into the province from another province or country), the retailer is typically considered the first importer of those products into the province. There are exceptions to this approach, however. For example, new regulations in Ontario expand the scope of obligated parties to those who have legal residency anywhere in Canada, an approach likely to be adopted by other provinces. In addition, Ontario obligates online sellers even if they do not have legal residency in the legislated jurisdiction.[15]

Companies that manufacture and use plastic packaging and products have been criticized because plastic production relies heavily on oil and gas

13. Hemanth Kumar, "Top ten packaging companies in 2020," Packaging Gateway (September 14, 2020; updated October 25, 2021), accessed December 29, 2021, https://www.packaging-gateway.com/features/top-ten-biggest-packaging-companies-in-2020/.

14. Joachim Quoden, Managing Director, Extended Producer Responsibility Alliance, email communication, January 8, 2023.

15. Catherine Abel, Vice President, Strategic Initiatives and Policy, Resource Recovery Alliance (formerly Canadian Stewardship Services Alliance [CSSA]), email communication, July 28, 2022.

feedstocks, which are significant contributors to greenhouse gases (GHGs). A 2021 report by the Minderoo Foundation[16] found that the top 20 polymer producers by weight generated 55 percent of global single-use plastic waste. The top three were ExxonMobil (US petrochemicals company), Sinopec (China petrochemicals company), and Dow (US chemicals company). The study showed that the top three institutional asset managers investing in single-use plastics were Vanguard Group, BlackRock, and Capital Group.

These large companies contribute a significant amount to the global packaging problem because of the sheer amount of material that is generated. As such, they represent an opportunity to positively impact the environment if they adopt more sustainable practices and packaging, especially if these actions are part of an EPR system. At the same time, there are numerous, relatively small regional consumer product brands and packaging manufacturers that also contribute to packaging waste. Even so, many companies have already taken steps to transition to a circular economy. With change comes an opportunity for market leadership and financial growth for companies that understand the changes that are now being required by government agencies under rapidly evolving EPR policies. Change can also bring uncertainty for companies and the risk that they might incur significant financial loss if their business model is not in synch with new regulatory requirements. Let's explore why these vital changes are needed on a massive scale.

WHAT'S THE PROBLEM WITH PACKAGING?

The United States generates more than 292 million tons of municipal solid waste (MSW) annually, which is nearly one ton per person per year.[17] As mentioned above, about 37 percent of that waste is comprised of PPP. The sheer volume of waste this represents is a big part of the problem. While the generation of paper and paperboard products declined from 87.7 million tons in 2000 to 67.4 million tons in 2018, plastic products (e.g., durable goods, containers, and packaging) rose during this same time period from 25.6

16. Dominic Charles, Laurent Kimman, and Nakul Saran, *The Plastic Waste Makers Index* (Minderoo Foundation, 2021), accessed March 26, 2023, https://cdn.minderoo.org/content/uploads/2021/05/27094234/20211105-Plastic-Waste-Makers-Index.pdf.

17. In 2018 (the most recent national data available), 292 million tons of municipal solid waste were generated. The US population in 2018 was roughly 327 million people, which is 0.9 tons per person per year. See US Environmental Protection Agency, "National Overview: Facts and Figures on Materials, Wastes, and Recycling," accessed March 26, 2023, https://www.epa.gov/facts-and-figures-about-materials-waste-and-recycling/national-overview-facts-and-figures-materials#NationalPicture.

million tons to 35.7 million tons.[18] Not only are plastics one of the fastest-growing sectors of the waste stream, but they contribute more significantly to visible and invasive pollution. The Ocean Conservancy reported that at least five of the top 10 ocean pollutants included single-use plastics such as bags, bottles (and caps), stirrers, straws, and containers,[19] while Break Free from Plastic identified the following brand owners of the greatest number of plastic items collected in beach cleanups: The Coca-Cola Company, PepsiCo, Unilever, Nestlé, Procter & Gamble, Mondelēz International, Philip Morris International, Danone, Mars, Inc., and Colgate-Palmolive.[20] More recently, other researchers have detected plastic compounds inside our bodies.[21]

About 50 percent of residential PPP is recycled in the United States[22] and this rate has been stagnant for over a decade.[23] The US PPP recycling rate is now far lower than in many other nations that have implemented EPR. For example, in 2020 (latest data available), Belgium recycled 80 percent of its packaging, Germany and Spain 68 percent, and Italy 73 percent. The average packaging recycling rate for the 27 countries in the European Union was 64 percent.[24] In the EU, only packaging is mandated under packaging EPR laws, although paper products are still collected and recycled, sometimes together with paper packaging. Since paper products are recycled in the United States at a higher rate than packaging, the US recycling rate for packaging and paper products is skewed higher than the EU recycling rate for packaging alone.

---

18. US EPA, "National Overview: Facts and Figures on Materials, Wastes and Recycling," Total MSW Generated by Material, 2018 (latest data available), accessed March 26, 2023, https://www.epa.gov/facts-and-figures-about-materials-waste-and-recycling/national-overview-facts-and-figures-materials#Generation.

19. Ocean Conservancy, "We Clean On: International Coastal Cleanup 2021 Report," 14, accessed March 26, 2023, https://oceanconservancy.org/wp-content/uploads/2021/09/2020-ICC-Report_Web_FINAL-0909.pdf.

20. Break Free from Plastic, Brand Audit Report 2021, accessed March 26, 2023, https://www.breakfreefromplastic.org/wp-content/uploads/2021/10/BRAND-AUDIT-REPORT-2021.pdf.

21. World Wide Fund for Nature, Dalberg Advisors, and The University of New Castle, "No Plastic in Nature: Assessing Plastic Ingestion from Nature To People" (June 2019), accessed March 26, 2023, https://wwfint.awsassets.panda.org/downloads/plastic_ingestion_web_spreads.pdf; Damian Carrington, "Microplastics found deep in lungs of living people for first time," *The Guardian* (April 6, 2022), accessed March 26, 2023, https://www.theguardian.com/environment/2022/apr/06/microplastics-found-deep-in-lungs-of-living-people-for-first-time?CMP=oth_b-aplnews_d-1.

22. Product Stewardship Institute analysis of data from US EPA, *National Overview: Facts and Figures on Materials, Wastes and Recycling* (December 2020), "Table 4. Generation, Recycling, Composting, Other Food Management Pathways, Combustion with Energy Recovery and Landfilling of Products in MSW, 2018," accessed March 28, 2022, https://www.epa.gov/sites/default/files/2020-11/documents/2018_ff_fact_sheet.pdf.

23. Product Stewardship Institute analysis of data from US EPA, "Sustainable Materials Management—Materials and Waste Management in the United States: Key Facts and Figures" (2020), accessed April 29, 2022, https://edg.epa.gov/metadata/catalog/search/resource/details.page?uuid=C9310A59-16D2-4002-B36B-2B0A1C637D4E.

24. Eurostat, "Recycling rates for packaging waste (2020 data)," last updated March 21, 2023, accessed March 26, 2023, https://ec.europa.eu/eurostat/databrowser/view/ten00063/default/table?lang=en.

Adding paper to EU packaging recycling rates would raise those rates further, making the comparison even *more* in favor of EU programs with EPR. EU recycling rates have also been calculated differently as of 2020 (see chapter 5), measuring recycling as only what is sent to viable end markets following processing at a recycling facility, which removes contaminated material from being counted as recycling, as it will ultimately be disposed. Most US state recycling rates currently count recycling as what arrives at a recycling facility *prior to processing*, which includes material eventually disposed.[25] Eurostat recycling rates reflected this change for the first time when 2020 data were reported, thus further inflating the US recycling rate in comparison to the EU.[26]

US recycling programs typically face cyclical challenges. When recycling markets are restricted, there is less demand for recycled material, recycling costs rise, and local leaders are forced to make difficult choices on where to spend scarce public resources. In the wake of market setbacks, communities have often had to decide whether to stockpile recyclable materials, reduce the number of materials their recycling programs accept, raise taxes and fees, or suspend recycling altogether. Such changes have threatened the public's already fragile understanding of, and confidence in, recycling and have eroded progress made over the last half century.

The problems with the US recycling system, though, are more fundamental than the loss of export markets. The system has always placed the burden on individual municipalities to collect and manage whatever materials companies sold into the marketplace, regardless of whether they had postconsumer value. To receive the commodity price in their recycling contract, a municipality must provide a consistent, high-quality supply of materials to end markets that have strict quality standards and specifications. Loads with low-value materials that have few markets will fetch a lower price than high-value materials that have robust markets. Contaminated loads occur when residents try to recycle packaging not allowed by their municipal program, which is often due to confusion about what materials they can recycle. Contamination can also occur in many other ways; for example, when glass shards from broken containers mix with paper or when materials get wet, moldy, or contain food waste. These loads result in an additional cost for municipalities managing the recycling program with limited tax dollars. Since producers continually make new types of packaging, governments and waste management companies must constantly react, never knowing which new materials will show up next and when.

25. PSI state government member responses to PSI survey, September 2022.

26. Joachim Quoden, Managing Director, Extended Producer Responsibility Alliance, email communication, July 18, 2022.

Over time the recycling system has become highly fragmented; communities across the country, even next to one another, collect different sets of materials for recycling and communicate different recycling messages to their residents. The result is that the system lacks adequate infrastructure investment, confuses consumers with highly variable messages about what materials are recyclable, provides little consistency from place to place, and often fails to reach economies of scale that makes recycling cost effective.

With governments and taxpayers picking up the cost to manage packaging after consumers are done with it, consumer goods companies have designed packaging to meet their own interests, such as reduced transportation costs, branding, perceived customer preference, and material cost and availability. Without a financial connection between brand owners and the postconsumer management of their materials, companies have little incentive to reduce the amount of packaging they use or to use materials that are easy to reuse or recycle. In failing to recover a large portion of packaging materials, the public incurs lost economic value and recycling jobs, as well as an enormous cost to human and environmental health. Also losing out are brand owners that seek recycled material as raw feedstock for new packaging and durable products to meet postconsumer content goals.

The coronavirus (COVID-19) pandemic, which began in 2020, further challenged municipal recycling programs across the United States. With a sudden shift in the workplace from offices to homes, there was an unprecedented decline in commercially generated waste and significant increases in residential trash and recycling volumes in nearly every state. As people reduced dining in restaurants and ate more take-out foods, single-use plastics, paper, and other materials flooded the residential waste stream. Although many states considered recycling an essential service during the pandemic, numerous curbside and drop-off recycling programs were suspended, further impacting already stressed supply chains that rely on recycled materials. In some large cities, recyclables were even mixed with trash and disposed, due to staff shortages.[27]

Even before the viral epidemic, starting in 2018, the US recycling system had begun to face unprecedented challenges. Recycling costs skyrocketed due to the China National Sword[28] and inherent flaws revealed a disjointed recycling system that long needed updating. In some cases, cities and towns faced increased costs in the hundreds of thousands, and even millions, of dollars to maintain their recycling programs.

---

27. Sydney Harris, "COVID-19 Impacts US Recycling Programs" (June 12, 2020), The PSI Blog, Product Stewardship Institute, https://productstewardshipinstitute.wordpress.com/2020/06/12/covid-19-impacts-u-s-recycling-programs.

28. *New York Times*, "As Costs Skyrocket, More US Cities Stop Recycling" (March 2019). https://www.nytimes.com/2019/03/16/business/local-recycling-costs.html.

GOALS FOR PACKAGING STEWARDSHIP

Manufacturing and consuming products and packaging have used earth's resources in such a way that we have threatened our own survival.[29] Although technological innovation repeatedly unlocks new resources, the way we have developed and used them has overwhelmed the planet's capacity to absorb harmful impacts. Countering these unsustainable practices will require a revamping of the US reuse, recycling, and composting infrastructure. By setting goals and taking the necessary steps, we can reverse the imbalance in materials extraction, waste generation, and GHG emissions. The result will be a system in which the need to extract raw materials and expend the associated energy to manufacture new packaging is dramatically reduced.

A main goal, of course, is to reduce the generation of waste that causes adverse environmental, economic, and social impacts. Many seek to promote the source reduction and reuse of resources by refilling reusable bottles from water fountains instead of buying single-use plastic water bottles, or refilling reusable containers from bulk bins at grocery stores instead of repeatedly buying single-use plastic packages. These examples are only the beginning, as new reuse businesses continually emerge.[30] And, of course, packaging that cannot be reduced or reused should be sustainably recycled or composted.

To many, these statements by now are commonplace. Many of us have said them hundreds of times, but still find the need to say them again only because they have not yet been actualized by society. As consumers, we want recycling to be convenient. For those who still must transport their recyclables to a neighborhood drop-off location even when their trash is collected outside their front door, recycling is not convenient. But recycling rates will shoot up once curbside recycling is provided because that *is* convenient. Many suggest that recycling is convenient if it is as easy as throwing out the trash, whether collected at the curb or a drop-off location. Consumers also need educational materials that are simple and clear, instructing them about what, how, and when to recycle in a standardized way across the country. We also want packaging to be consistently labeled, and to be reusable, recyclable, or compostable.

Extended Producer Responsibility is a policy that helps us to map out intentional goals, design a path to reach them, and follow that path until our goals are achieved, redefined, or expanded. It creates defined roles for all key

29. United Nations, *UN News* (February 18, 2021), United Nations Secretary-General Antonio Guterres, virtual press briefing on the UN Environment Programme report, Making Peace with Nature. The Secretary-General said that "unsustainable production and consumption" leads to the "interlinked environmental crises" of climate disruption, biodiversity loss, and pollution that "threaten our viability as a species," accessed March 26, 2023, https://news.un.org/en/story/2021/02/1085092.

30. Upstream website: https://upstreamsolutions.org/the-new-reuse-economy.

stakeholders, forming a clear network of accountability. EPR laws, regulations, and stewardship plans generally require producers to meet specific waste reduction, reuse, recycling, composting, and recycled content targets by specific dates. To meet these targets, sustainable funding is needed to pay for programs to reduce waste and to collect, reuse, recycle, and compost packaging materials. This includes funding to develop new infrastructure and recover packaging that is not currently recyclable. For materials to be truly recyclable and compostable, comprehensive collection systems, efficient processing facilities, and viable end markets are all required. Funding is also needed for programs that will produce a clean, consistent, and reliable supply of recycled feedstock so that brand owners will be able to reach their recycled-content goals.

Lastly, another aspect of being sustainable is that the system needs to be fair. Brand owners of packaging that is truly reusable, recyclable, compostable, contains recycled content, and does not contain toxic chemicals should pay less into the system than those that use packaging that does not meet these criteria and results in more negative environmental, social, and economic impacts.

If we succeed in achieving all of the goals listed above, we will also have reduced negative environmental externalities, including reduced litter, pollution, and mismanagement of waste. And we will be on our way to a more habitable planet.

BARRIERS TO PACKAGING STEWARDSHIP

Unlike toxic and bulky products that have been covered by product stewardship programs in the United States for the past two decades, packaging recycling has a longer history. In fact, when we mention the term *recycling*, we imagine curbside or drop-off bins filled with packaging and paper products. One of the strongest barriers to packaging EPR may be the lingering perception that individuals, not companies, cause pollution. I still vividly recall the image from a 1970s television advertisement by Keep America Beautiful, a campaign funded by the bottling industry, of an actor playing a Native American who sheds a single teardrop as a bag of trash is tossed at his feet from a passing car. This staged ad was successful in highlighting the growing problem of litter on our streets and in the environment. Unfortunately, it also put the entire responsibility for controlling pollution on individuals.[31] This sentiment remained a subtle, but powerful, barrier to holding producers responsible for packaging waste, particularly as local governments stepped up over time to take on the responsibility for recycling. Below are some additional challenges faced by those advocating for EPR for packaging over the past two decades, many of which have been addressed at least partially.

Complexity of Packaging

In the past, packaging was simpler, or nonexistent. Over time, though, it has evolved to meet the needs of, and perhaps define, our modern lifestyles. As technological innovations advanced, our habits changed and packaging kept up with those demands. We can eat what we want, when we want, because packaging can now protect and preserve foods shipped around the world. To meet these consumer demands, packaging materials have become more complex. For example, some of the fastest-growing packages are flexible pouches, which comprise an estimated 51 percent[32] of the total plastic packaging market. There are numerous industry studies supporting the claim that the production of flexible packaging has lower GHG emissions, energy consumption, resource depletion, and other impacts compared to rigid *recyclable* packaging options. One non-industry analysis, conducted by the Oregon Department of Environmental Quality, found that the production and disposal of flexible coffee packages had lower life-cycle impacts in most (though not all) cases, even as compared to the production and recycling of recyclable coffee packages at an aspirational rate of 100 percent recycling.[33] However, due to the complexity of flexible pouches, which are often made with multiple materials and layers, no technology is yet capable of recycling multi-layer or multi-material flexible pouches to the point of having viable end markets for these materials.[34]

Complex packages introduced into the market by brand owners can become a challenge for municipally run recycling programs, even if some of that packaging has lower overall environmental impacts compared to recyclable packaging. The Consumer Goods Forum, a coalition of global

31. Finis Dunaway, "The 'Crying Indian' ad that fooled the environmental movement," *Chicago Tribune* (November 21, 2017), https://www.chicagotribune.com/opinion/commentary/ct-perspec-indian-crying-environment-ads-pollution-1123-20171113-story.html; the famed "Crying Indian" advertisement by Keep America Beautiful, featuring actor Chief Iron Eyes Cody (he wasn't really a chief or Native American) from 1971, available at https://youtu.be/lmyq3gwZcPs.

32. "The Global Commitment 2020 Progress Report" (2020), Ellen MacArthur Foundation, 42, https://emf.thirdlight.com/link/il0mcm1dqjtn-knjubr/@/preview/1?o.

33. Minal Mistry, David Allaway, Peter Canepa, and Jonathan Rivin, "Putting Beliefs to the Test," *Resource Recycling* (September 9, 2019; updated November 20, 2020), accessed January 5, 2022, https://resource-recycling.com/recycling/2019/09/09/putting-beliefs-to-the-test/.

34. Alison Keane, Flexible Packaging Association, email communication, October 1, 2022.

consumer packaged goods manufacturers and retailers, has tried to address this challenge by developing "Golden Design Rules" to increase the circularity of plastic and other packaging.[35] These rules commit members to eliminate polystyrene and polyvinylidene chloride, and use transparent, uncolored, and mono-material plastics in PET thermoformed packaging.

Lack of Consistent Education

Further challenges result from the difficulty consumers have in distinguishing whether packaging is truly recyclable or compostable. Let's start with the fact that the "chasing arrows" recycling symbol on plastic packaging was developed by the Plastics Industry Association (formerly the Society of the Plastics Industry) "to develop consistency in plastics manufacturing and recycled plastics reprocessing."[36] These Resin Identification Codes, which are required by numerous states, only identify the type of plastic resin by using a number and an acronym (e.g., HDPE for high density polyethylene). They do not mean that the plastic is recyclable or that it is recyclable in all municipal programs. If a person places that material in their recycling bin when their community does not accept it in its program, it will become a contaminant (e.g., waste) and an added cost.

Municipalities across the country instruct their residents to recycle different types of packaging. As a result, several states have developed labeling or materials acceptance laws that provide consistency about what is recycled within their states, making recycling less confusing for consumers. Unfortunately, each state law is different and provides no national consistency for companies that sell their products regionally or nationally. A company meeting the labeling law in one state might be in violation of the labeling laws in other states. Although two organizations have developed nationally standardized recycling labels—the Sustainable Packaging Coalition's How2Recycle label and Recycle Across America's label—there is still a need for federal guidelines or legislation to ensure consistency across all states for the labeling of recyclable and compostable materials.

Contaminated Recyclables

Just as people cannot readily distinguish between latex and oil-based paint or between single-use and rechargeable batteries, residents have a heck of

35. The Consumer Goods Forum, Golden Design Rules, Rule 6: Increase Recycling Value in Flexible Consumer Packaging (2021), https://www.theconsumergoodsforum.com/wp-content/uploads/2021/07/2021-Plastics-All-Golden-Design-Rules-One-Pager.pdf.

36. Kelly Cramer, "101: Resin Identification Codes," How2Recycle (July 20, 2017), https://sustainablepackaging.org/101-resin-identification-codes.

a time distinguishing non-recyclable or non-compostable packaging from similar packaging that can be recycled or composted, particularly when labels on packages are unclear or untruthful, and education is inconsistent. This is the case with flexible packaging, compostable or biodegradable packaging, and other packaging types. Putting non-recyclable packaging in recycling bins causes contamination. Recycling facilities cannot separate out all non-recyclable packages from recyclables, so they become contaminants in mixed loads that lower the value of those materials and increase the cost that local governments and taxpayers pay for recycling. The lack of feedback to residents who put contaminants in recycling bins compounds the problem. However, correcting it is also time consuming and costly. Finally, if a municipality responds to market changes by adjusting its program to shed the burden of costly items that lack markets, their residents' heads spin even more.

One major contribution to our wishcycling mindset in the United States is that many of us now have been trained to put all recyclables in one cart, known as single-stream recycling. Two decades ago, the waste management industry and many governments changed the municipal recycling system by exchanging separate bins—one for paper and another for other recyclables such as bottles and cans—for one bigger container. This switch to single stream was easier for residents, since they could put all recyclable materials in one bin, and it increased the amount of materials collected. Unfortunately, residents also took it as a sign to put everything they wanted to recycle into the bin, regardless of whether it was accepted by their recycling programs. While the amount of recyclables collected increased, so did contamination.

Reversing this contamination trend has been a challenge, although the increased cost of recycling due to China's National Sword policy has provided an incentive for some communities to invest in dual-stream recycling systems, with some reporting savings.[37] In Europe, the Waste Framework Directive (see chapter 5) requires that plastics, metal, paper, and glass packaging *each* have to be kept separate for collection, unless commingling is proven to maintain the value of the recyclables.[38]

37. Jacob Wallace, "Dual-stream recycling proponents feel vindicated after converted communities see financial benefits," *Waste Dive* (December 15, 2021), accessed March 26, 2023, https://www.wastedive.com/news/dual-stream-recycling-wilkes-barre-lake-worth-beach/611493/.

38. European Commission, "Guidance for Separate Collection of Municipal Waste" (April 2020), http://publications.europa.eu/resource/cellar/bb444830-94bf-11ea-aac4-01aa75ed71a1.0001.01/DOC_1.

Recycling Economics

Consumer confusion about recycling has long been fueled by those who maintain the myth that recycling makes money for municipalities and, when it costs more, disposal is the best option. Perhaps this myth exists because the packaging materials collected and processed often have value: they are sold to end markets to be remanufactured into new products. Unfortunately, many do not bother to factor in the cost of collecting, sorting, and processing recyclables. They assume that if you make money on the *sale* of recyclables, you make money on recycling. And if you don't make money, "the market" dictates that the material should be thrown away.

The reality is that the cost to collect, transport, and process postconsumer packaging material is almost always greater than the value of the processed material when sold to end markets. That is certainly the case with garbage: People must pay someone to take it "away." The key question, however, is *how the cost to collect and process recyclables in a jurisdiction compares to the cost to collect and dispose of waste.*

In regions where open space abounds, disposal in landfills is often much cheaper than recycling and, consequently, recycling rates are usually much lower than the national average. In other locations, such as the Northeast, where there is limited landfill and waste-to-energy facility capacity, garbage disposal costs are often higher than the national average. As a result, the cost to collect and process recyclables, until recently, has typically been *lower* than the cost to collect and dispose of that same material. In these areas, recycling has saved money for municipalities since *recycling costs are less* than disposal. Governments in these regions tend to promote recycling more heavily and recycling rates are usually higher than the national average.

The cost disparity between recycling and disposal has always been a significant challenge for municipalities. Due to the China National Sword policy, however, recycling costs have increased dramatically nationwide, often even *higher* than disposal costs in the Northeast. Although recycling markets are expected to rebound over time, particularly as domestic recycling capacity returns, the National Sword has become a painful lesson for municipalities, which have little control over the current recycling system even though most have the responsibility for funding and managing the system.

Keep in mind, however, that recycling cost calculations in all municipalities still do not adequately account for the external costs of pollution from waste collection, transportation, and disposal or from the upstream impacts of material extraction and remanufacturing, all of which increase GHG emissions. Market economics, which drives decisions on recycling and trash disposal, does not fully account for the costs of litter or aquatic impacts of waste, the environmental and social impacts of landfills or incinerators, the extraction of additional resources to make new packaging, the creation of recycling jobs, and other factors that are included in a standard life-cycle and cost benefit analysis.

These variables are important to factor into policy decisions aimed at reducing, reusing, and recycling materials so that we truly lower environmental, economic, and social impacts. Unfortunately, instead of presenting the public with viable policy options, some have pieced together incongruent information to present a skewed narrative, casting doubt and derision on the benefits of recycling.[39] These viewpoints have often confused municipal recyclers and the public, resulting in the doubt and inaction that has contributed to the waste crisis we are faced with today.

Meeting Stakeholder Interests

Of all the products and materials that PSI has addressed, packaging has met with the most industry resistance, in large part due to the number, size, and political power of consumer-packaged goods companies (i.e., brand owners) and the local and regional influence of waste management companies. Until about 2020, most US brand companies and their associations opposed packaging EPR systems even as their counterparts in Canada, Europe, and other parts of the world spent years implementing these same systems. Although they knew they needed to change their packaging to meet a growing public demand for recyclability and sustainability, they wanted to make changes on their own terms, voluntarily, and as part of competitive branding. They also had a steep learning curve to overcome regarding how their companies operated within EPR systems abroad. Over time, as the enactment of packaging EPR laws became a reality, producers have sought to incorporate their interests into US legislation. They generally believe that, if they must pay for the system, they want significant control over how it is managed to increase efficiency and lower costs.

Recycling and waste management companies, for their part, have felt particularly vulnerable to changes in recycling systems that EPR laws would bring. After all, they have played a significant role for more than 50 years in creating an extensive recycling network throughout the country. They have invested billions of dollars in collection equipment and processing facilities. They have developed enduring contractual relationships with their clients, whether an individual household, business, small town, or large city. They fear that they will lose business through the replacement or disruption of the existing recycling infrastructure if EPR programs require brand owners to manage the system.

39. John Tierney, "The Reign of Recycling," *New York Times* (October 3, 2015).

Local government officials also don't want to lose control over recycling systems that provide a valuable service for their residents. They want that system to be flexible and sustainable, and one that offers options based on municipal preferences. While municipal officials seek a significant change in the way that recycling has been financed and managed, they generally do not want to disrupt their positive, long-standing relationships with recycling companies.

Other key stakeholders encompass an array of local, state, national, and international environmental organizations. While some of these groups have experience with EPR systems abroad, most are specialists on topics such as source reduction, reuse, recycled content, recycling, plastics pollution, and toxics reduction. These organizations have sought to include their interests in packaging EPR bills, particularly advocating for strong goals to be set in statute. Many have propelled the global focus on plastics pollution and contributed to the interest in requiring producers to take responsibility for the packaging waste generated from products they sell. Distrusting corporations, they have also sought a high degree of government oversight, control, and enforcement.

Despite the many challenges of integrating these diverse stakeholder interests, four packaging EPR laws have been enacted in the United States by spring 2023 and many other states have developed bills. US producers and waste management companies have begun to engage with government officials to develop EPR systems for packaging. Although there is still resistance among companies, opposition has slowed, and a growing number support EPR as they learn how to operate under these new systems. One concept that may have contributed to the gradual acceptance of EPR in the United States has been the realization that recycling system changes take time to unfold. Systems can be designed for flexibility with a transition from one system to another, over time, and not all at once. Thankfully, we have many packaging EPR systems around the world from which to learn.

In Ontario and Québec, Canada, for example, packaging EPR programs were initially established as "shared responsibility" systems in which brand owners and municipalities typically split costs, with local governments maintaining greater system control. Over time, as stakeholders gained experience and relationships deepened, there has been greater acceptance of producers paying the full system cost in exchange for gaining greater management control over the system. In fact, the global trend is for brand owners to fully pay for the recycling system and be provided significant control over its management.

In the United States, the main challenge faced in enacting a packaging EPR law is how to balance the complex and varied interests of brand owners, recycling companies, municipalities, environmental groups, and others such as small businesses. The modernized recycling system that results from an EPR law must reflect the interests of each of these stakeholders. The complexity of these interests will always be influenced by the varying degrees of stakeholder experience with EPR, as well as different methods of influencing policy.

Together, stakeholders face a unique challenge that will require the melding of interests in a thoughtful process. Ideally, the more consensus achieved in the development of a bill, the greater the prospects for smooth program implementation of a modernized US recycling infrastructure, with stakeholders working together, rather than in conflict, to meet their unique interests. Before we explore that path, let's step back and consider how the capacity for packaging EPR was built in the United States.

ORIGINS OF US PACKAGING EPR

The first packaging EPR bill in the United States was federal legislation introduced in 1992 by Senator Max Baucus (D-MT), but it "faced stiff industry opposition and was never enacted."[40] The introduction of this novel policy concept was accompanied by an article advocating for packaging EPR, articulately penned by Allen Hershkowitz of the Natural Resources Defense Council (NRDC) for *The Atlantic*.[41] According to Hershkowitz, who worked extensively on the federal bill, this was likely one of the first times that a US publication covered the policy concept.[42] At that time, however, the knowledge and capacity to support such legislation did not widely exist in America. The concept of EPR had not yet taken hold in this country, even though those working internationally, with access to Europe's burgeoning packaging EPR programs, had already begun to embrace the concept. This early stage of the US EPR movement was marked by those like Betty Fishbein (INFORM, Inc.), Reid Lifset (Yale University), Allen Hershkowitz (NRDC), Clare Lindsay (US EPA), and Catherine Wilt and Gary Davis (University of Tennessee). These and other "pollinators" tracked the early packaging EPR systems that emerged in Europe and recognized their promise for the United States.

One year after the introduction of the federal EPR packaging bill, I started a new position as the director of waste policy for the Massachusetts Executive

40. "The Polluter Pays," Knowledge@Wharton, University of Pennsylvania (April 4, 2017), https://knowledge.wharton.upenn.edu/article/the-producer-pays/. Quotation and information provided in article by Reid Lifset, associate director of the industrial environmental management program at Yale University.

41. Allen Hershkowitz, "How Garbage Could Meet Its Maker," *The Atlantic Monthly* (June 1993).

42. Allen Hershkowitz, email communication, March 6, 2023.

Office of Energy and Environmental Affairs (EEA). Our main approach to reduce waste and increase reuse and recycling was to offer annual grants as incentives to local governments to take the many small steps needed to reach the 46 percent statewide recycling goal that was included in our state solid waste master plan. Working in collaboration with my boss, Environmental Undersecretary Leo Roy, as well as Robin Ingenthron, Greg Cooper, Brooke Nash, and others at the Massachusetts Department of Environmental Protection (MassDEP), we dangled $15 million of grant funds each year to heavily promote volume-based pricing (i.e., pay-as-you-throw),[43] education, enforcement, equipment, and technical assistance to enhance municipal recycling programs. The more programs that municipalities committed to implement, the more funds we provided to enable them to carry them out. We also annually evaluated the recycling rates of each of 351 municipalities in the state, publicly praising the high performers while shaming those scraping the bottom of the barrel. I even conducted recycling publicity campaigns across the state with our agency's environmental secretary, Trudy Coxe, appearing on local radio, television, and other outlets to build broad awareness. In addition, MassDEP enacted waste bans and hired inspectors to enforce against waste management companies that disposed of designated recyclable materials collected from residents.

Despite our efforts in Massachusetts to increase the recycling rate, it inched slowly upward by one percentage point per year, until it stagnated altogether. The same phenomenon was happening all around the country. No matter how hard they tried, local governments were not able to increase their recycling rates. They could not keep pace with new packaging types introduced into the marketplace every few months, increased waste generation, and rising disposal costs, no matter how much we enticed them financially to try harder.

Then, one day in 1997, I was forced to see the situation from a fresh perspective. HP Hood, a national dairy products company headquartered in Massachusetts, switched its milk containers from clear plastic to opaque white plastic.[44] This was done ostensibly to block light to avoid a breakdown of vitamins, and thus maintain the nutritional value of milk and protect its flavor. While this change made sense from a marketing perspective (e.g., product differentiation), Hood's decision suddenly increased the cost of recycling its milk containers, since opaque white plastic has lower value to

43. I led a team of staff at the Massachusetts EEA and Massachusetts Department of Environmental Protection (MassDEP) to develop the state's first guidebook on pay-as-you-throw programs: "Pay As You Throw: An Implementation Guide for Solid Waste Unit-Based Pricing Programs" (January 1997). This work drew on the pioneering work conducted by Lisa Skumatz for the US EPA. The guidebook was updated in 2004 and can be accessed on the MassDEP website, https://www.mass.gov/lists/pay-as-you-throw-paytsave-money-and-reduce-trash-smart.

44. Dairy Foods, "New Study Validates Light Blocking Efforts" (February 24, 2004), https://www.dairyfoods.com/articles/83251-new-study-validates-light-blocking-efforts.

recycling end markets than clear plastic.[45] This added cost was passed onto recyclers, which passed on the cost to municipalities, resulting in a firestorm of emails and phone calls that ricocheted around the recycling community.[46] It was then that I became aware of the degree to which local recycling costs were impacted by corporate packaging decisions. I later realized that this was only one packaging change among many that increased costs for municipal recycling programs.

Even though this strong sign of an ailing recycling system existed in 2000 when PSI was founded, state waste management officials (including myself) were still more concerned about hazardous products than the amount of packaging waste generated. State government officials in the Northeast and Great Lakes regions had conducted pivotal studies on mercury pollution, implicating consumer products as a significant source.[47] In response, the Northeast Waste Management Officials Association (NEWMOA), led by Terri Goldberg, and its state government members developed model state legislation that banned or phased out mercury from use in thermometers, thermostats, auto switches, and other products, and many states enacted these laws.[48] The model also mandated reporting on the intentional use of mercury in products, labeling mercury-added products, and end-of-life collection for products that were subject to the phase-out but could not be eliminated for various reasons, such as a lack of safer alternatives.

Back in 1993, Michael Bender, an environmental advocate from Vermont, sensed this growing movement to safely manage toxic products and formed the North American Hazardous Materials Management Association (NAHMMA) based on pioneering work by Dana Duxbury of the Massachusetts-based Waste Watch Center. Both organizations promoted the special collection of household hazardous waste (HHW), a term coined in 1983 by NAHMMA's first president, Dave Galvin, a former government official from King County, Washington, and later PSI's board president. I had worked with the Waste Watch Center for many years and was on NAHMMA's founding board, later becoming its third president. By that time, many governments were spending significant resources on programs to capture and safely manage products that contained mercury (including fluorescent lamps, thermostats, and single-use

45. Brian C. Jones, "Hood and Garelick Embroiled in Bottle Battle," *Providence Journal-Bulletin* (April 26, 1998; updated January 11, 2011).

46. Janelle Nanos, "Maine passes nation's first law to make big companies pay for the cost of recycling their packaging," *Boston Globe* (July 19, 2021), https://www.bostonglobe.com/2021/07/19/business/maine-move-make-big-companies-pay-all-their-packaging.

47. "Northeast States & Eastern Canadian Provinces Mercury Study, A Framework for Action," NESCAUM, NEWMOA, NEIWPCC, 1998. To obtain a copy, contact the NEWMOA office.

48. "Revised Discussion Document: Mercury Education and Reduction Model Act," NEWMOA (June 1998), accessed October 23, 2022, https://www.newmoa.org/prevention/mercury/final_model_legislation.pdf.

batteries), as well as other HHW (e.g., paint, gas cylinders, pesticides, and rechargeable batteries).

Even so, in contrast to the recycling of packaging in established curbside and drop-off systems, special collection networks had not yet developed for managing HHW, or even bulky wastes (e.g., carpet, mattresses, and furniture). The lack of collection and management infrastructure for these products, along with a lack of consumer education, were key reasons why US municipalities considered toxic and bulky wastes to be more impactful than packaging, for which recycling systems existed, even if they were beginning to falter. Further, waste management companies supported the creation of new collection and management systems for toxic and bulky products that led to new business opportunities for them. Although PSI often faced opposition from producers of toxic and bulky products, we had an important waste management ally in the introduction of new EPR systems for these products, with electronics emerging in 2000 as the fastest-growing consumer product requiring special collection. By this time, a growing number of government officials knew that the producers of toxic products were responsible for measurable environmental impacts and understood how the concept of EPR could relieve them of the financial and management burden of product take-back.

As the years rolled by, US state and local government officials turned their attention to the largest component of the waste stream: packaging. Municipal recycling costs were fluctuating unpredictably, recycling rates were stagnant, and local governments became more aware of their lack of control in managing an ever-more-complex packaging stream. Decisions made by manufacturers and brands directly influenced the costs that municipalities paid for recycling and waste management, and governments had little ability to change the system. While these challenges were not new for local officials, they were mounting. Compared to 2000, however, something significant had occurred: municipal government officials now understood there was another way to approach the problem, one that shifted the financial and management burden away from them and onto those companies that decided which packaging materials to use.

By 2007, PSI had already worked extensively on electronics, paint, radioactive devices, gas cylinders, thermostats, lighting, pharmaceuticals, carpet, and other toxic or problematic wastes, and helped to enact an increasing number of EPR laws in Maine, Vermont, California, Connecticut, Oregon, Washington, and other states. By this time, local governments realized that they could no longer just try harder, spend more money to educate their citizens, or force them to recycle. They now viewed themselves as trapped into paying for waste and recycling no matter which products and packages were being manufactured and discarded. EPR offered a new approach that local officials had not previously known existed or contemplated. EPR was no longer an intangible concept imported from another country. It had now taken firm hold in the United States. The lack of influence felt by local governments in managing packaging waste was now in marked contrast to the powerful position they felt in passing EPR laws for a growing number of toxic products.

A paradigm shift in managing waste was fully in progress, and the groundwork had now been laid for the concept of producer responsibility. By listening to the growing concerns expressed by our state and local government members, who were experts in waste management, I knew that our members were ready to focus on packaging EPR. Unfortunately, producers and waste management companies repeated their opposition to having *any* discussion on the potential for EPR to create a more stable recycling system. Nevertheless, in 2008, PSI began the drumbeat for an EPR solution to packaging waste by holding regular strategic calls among our government members and expanding the dialogue externally by convening webinars and organizing conference panels.

To bolster our efforts to explore solutions to manage packaging waste through a product stewardship framework, PSI submitted three grant proposals in 2008 to US EPA to conduct research and convene a multi-stakeholder packaging recycling dialogue. These proposals sought funding from EPA, similar to the EPA grant assistance PSI had previously received to facilitate multi-stakeholder EPR dialogues on paint, thermostats, lamps, and other products. Unfortunately, each packaging proposal was rejected.

As interest spread among state and local governments for EPR on packaging, brand owners began internal industry discussions about sustainability efforts and related public policy, eventually forming AMERIPEN (American Institute for Packaging and the Environment), a material-neutral association that also represented packaging material suppliers, packaging manufacturers, and recyclers. Although a similar association in Europe, EUROPEN, embraced EPR while representing corporate interests in European EPR packaging programs, AMERIPEN members initially opposed EPR, even though many of the same companies were members of both associations. Brand owners were also engaged on policy issues with the Grocery Manufacturers Association, American Chemistry Council, Foodservice Packaging Institute, American Beverage Association, and the Sustainable Packaging Coalition. These associations played multi-faceted roles by conducting research, sharing information, and exploring non-EPR solutions they could support that addressed the impacts from the postconsumer management of their packaging. But they all held the line firmly against EPR or took a neutral position.

In January 2009, with President Barack Obama at the helm, packaging stewardship appeared ready for a boost as the US EPA initiated a national packaging dialogue that included 10 consumer packaged goods companies, two retailers, seven state governments, five local governments, one producer trade association, three environmental groups, and PSI. The goals were to optimize the recycling system; identify mechanisms to address system shortfalls, including the need for long-term financing; and maximize the source reduction, collection, reuse, and recycling of packaging and printed materials.[49] Through four meetings over 18 months, from September 2010 to April 2011, with intensive workgroup activity and other interactions, stakeholders discussed numerous potential financing strategies in theory, but not about how EPR could be applied to the US packaging system. Midway through the dialogue, in an attempt to unify the group, I initiated discussions to jointly develop a refined problem and goals statement. That effort resulted in a two-page agreement[50] with consensus from most dialogue participants, including industry. Even so, it did not lead to a more in-depth discussion that could inform policy, as expected by government and environmental participants. Ultimately, we came to realize that the goal of the effort was purely to have a discussion, and the final report issued was filled with interesting research and data, but no agreements.[51]

At the time that US EPA initiated its packaging dialogue, PSI's state and local government members had extensive recycling knowledge and a decade of experience with EPR. They had explored voluntary options, pay-as-you-throw programs, financial incentives for equipment and education, enforcement, disposal bans, advanced recycling fees,[52] and other options. They knew EPR worked for toxics and bulky products in the United States, and that EPR worked for packaging in Europe, Canada, Asia, and other regions. They wanted to explore how EPR might work for packaging in America and were ready to take that step. Unfortunately, most of the

49. "Final Report of the Dialogue on Sustainable Financing of Recycling of Packaging at the Municipal Level" (September 19, 2011), 10, submitted to the US Environmental Protection Agency by The Keystone Center (prepared under contract EP-W-09–011, Task Order 072). US EPA archives, https://archive.epa.gov/smm/sfmr/web/pdf/packaging-report.pdf.

50. PSI archives, "US EPA Multi-Stakeholder Packaging Dialogue: Talking Points" (April 20, 2011).

51. "Final Report of the Dialogue on Sustainable Financing of Recycling of Packaging at the Municipal Level" (September 19, 2011), 5, submitted to the US Environmental Protection Agency by The Keystone Center (prepared under contract EP-W-09–011, Task Order 072). US EPA archives, available at https://archive.epa.gov/smm/sfmr/web/pdf/packaging-report.pdf.

52. "Advanced recycling fees" (ARFs) are consumer payments to a retailer on top of the product cost and are intended to cover costs associated with recycling. ARFs are legislated, often visible to consumers on a sales receipt, and require retailers to track fee payments and transfer them to a state government agency, which disburses the funds as grants to municipal and private collection sites. "Eco fees" are also consumer payments to retailers but are distinguished from ARFs in that producers fund the recycling program up front by paying the eco fee into a producer-managed fund. Producers then get reimbursed for the eco fee amount through the sale of the product (e.g., paint and mattresses) to retailers, which in turn get reimbursed when selling those products to consumers. PSI considers eco fees as a form of EPR since the program is *managed* by producers even if not *funded* by them, while ARFs are not considered EPR at all since they are consumer funded and government managed.

industry wasn't anywhere near ready for a discussion that would lead to action. Consumer packaged goods companies claimed unfamiliarity with how EPR programs worked in other countries, even though they had been abiding by EPR laws in Europe and Canada for up to 20 years. They were fearful of how EPR might impact their businesses and had no interest in discussing EPR, if they could avoid it.

One industry—beverage brands—apparently *was* ready to act even if its members were in no mood for stakeholder discussion. In 2010, the Beverage Association of Vermont became the chief backer of the country's first state packaging EPR bill, which would have phased out the existing deposit return system (DRS).[53] Since the bill was not expected by most stakeholders, it dropped like a bombshell, resulting in fierce opposition from environmental groups, producers, and others. One of the key issues then (and still disputed today) is whether or not beverage containers, which have high recycling value, should be collected in municipal recycling programs or through the DRS alternative take-back system. Rather than enacting a discordant EPR bill, Vermont legislators chose instead to address waste prevention and diversion through omnibus recycling legislation.[54] With government attention across the country more focused on toxic and bulky products than packaging, as well as continued opposition to packaging EPR among consumer brands, packaging momentum slowed.

LAYING THE GROUNDWORK FOR US PACKAGING EPR

Over the next decade PSI continued to explore product stewardship solutions to the mounting packaging waste problem, while also working on a range of toxic and bulky products. We increased the frequency of our strategic meetings with our government members and developed a 2010 document, *Recommended Components of a Product Stewardship Program for Packaging in the United States*, which we used as a tool to develop a model state packaging EPR bill. In it, we included the roles of brand owners, local and state governments, and other stakeholders; a definition of the responsible party (producer); the types of packaging materials to be included in the program; from which locations recyclables should be collected; performance goals; and

53. Ken Picard, "Is Vermont's Bottle Bill Ready for Recycling?" *Seven Days* (January 26, 2011), accessed April 3, 2022, https://www.sevendaysvt.com/vermont/is-vermonts-bottle-bill-ready-for-recycling/Content?oid=2142407.

54. Vermont General Assembly website (Act 148), accessed October 4, 2022, https://legislature.vermont.gov/bill/status/2012/H.485; Vermont's Universal Recycling Law (Act 148), 2012, accessed April 3, 2022, https://dec.vermont.gov/waste-management/solid/universal-recycling.

other program aspects. We also included policy concepts to promote design-for-environment practices (e.g., increased recyclability, reduction of materials and toxics), including stewardship fees based on environmental attributes of the packaging material. PSI continually refined this document over time into what we now call PSI's *Elements of Packaging and Paper Products (PPP) EPR Legislation.*

We also featured packaging EPR sessions at our national conferences, developed and facilitated technical webinars highlighting the EPR packaging programs in Canada and Europe, and discussed how those programs could translate to the United States. In 2011, I co-chaired an EPR Harmonization Committee of PAC Next, a Canadian corporate packaging stewardship association, and our PSI team wrote a comprehensive research report on a dozen EPR programs in Europe and Canada. In addition, we held government strategic calls to clarify state and local government interests and to develop strategies to transition to a new recycling system.

PSI also conducted a small research project for Nestlé Waters North America, whose president and CEO, Kim Jeffery, supported EPR for packaging by forming Recycling Reinvented, a nonprofit organization run by Paul Gardner, a former Minnesota state legislator. This advocacy group partnered with Future 500, an organization with experience forging agreements between nonprofits and corporations. The group held meetings and introduced packaging EPR bills in a few states, as did Upstream, founded by Bill Sheehan and now run by Matt Prindiville. These bills drew attention to packaging EPR in the states in which they were introduced but failed to gain traction.

As PSI and other groups' efforts were repeatedly snuffed out by growing industry opposition, municipal recycling costs continued to grow. During this time, I was encouraged by officials at one of the US EPA regional offices to submit a grant proposal to educate producers, recyclers, and governments on product stewardship for packaging. PSI had conducted several earlier stakeholder meetings on telephone books funded by the regional office, and that engagement succeeded in persuading directory publishers to develop a website for citizens to opt out of receiving directories.[55]

I thought this was finally our opportunity to convene an objective discussion on producer responsibility. However, to conduct the project, PSI had to allow a regional recycling organization to be the lead contractor, even though PSI designed the project and wrote the proposal. Soon after the grant was awarded, I received a letter from the lead organization replacing PSI with a

55. National Yellow Pages Consumer Choice and Opt-Out Site, operated by the Local Search Association, https://www.yellowpagesoptout.com.

consulting firm. The reason cited was a blog post I had written[56] in response to an anti-EPR policy statement drafted by a packaging industry lobbyist. The organization that finally emerged from this effort became The Recycling Partnership (TRP), a national nonprofit organization funded by the packaging industry to conduct voluntary recycling initiatives. For many years, TRP provided carts for recycling, as well as education and technical assistance to municipalities, mostly via grants.

Over the next five years, PSI and other EPR advocates continued efforts to educate government officials and legislators about the problems with the recycling and waste management system, and how EPR policy could help to solve them. PSI repeatedly brought into our conferences and meetings the government and industry packaging EPR experts from Canada and Europe that operate and oversee these systems. We developed EPR program summaries, showed recycling system diagrams before and after EPR policy implementation, and provided other EPR educational tools.

Our challenge was to convince legislators that packaging was a costly, but avoidable, problem and that the EPR solution is based on the same principles and concepts as laws that they already enacted for toxic and bulky products such as thermostats, fluorescent lamps, paint, batteries, electronics, pharmaceuticals, and mattresses. In fact, EPR policies for all products and packaging can be organized into the same 16 key elements (see chapter 6 and below). Over time, agency officials and legislators began to understand how EPR could save millions of dollars, relieve a management burden, greatly improve the environment, and create recycling jobs. There was also growing acceptance that recycling rates across the country were stagnant, while waste disposal capacity issues and municipal recycling costs were only getting worse. Still, packaging EPR bills were not moving in state legislatures.

GREEN FENCE AND CHINA NATIONAL SWORD

It took the China National Sword policy to do what advocacy alone was unable to accomplish. For many years, China had gobbled up recyclables from the United States and other countries at an unprecedented rate. China was in such need of recycled materials to fuel its economy that the Chinese government overlooked the high level of contamination in the imported loads they received. Over time, however, China's domestic markets demanded higher quality materials.

---

56. Scott Cassel, "7 Reasons Why AMERIPEN's Stance on EPR Is Flawed" (November 9, 2012), https://productstewardshipinstitute.wordpress.com/2012/11/09/7-reasons-why-ameripens-stance-on-epr-is-flawed.

What started as China's increased inspections of imported recyclables in 2013 (known as the "Green Fence") turned into an effective ban in 2017 on imported recyclables that did not meet stringent quality standards: the "China National Sword." The US recycling industry had failed to heed the warning shot of the Green Fence. But the China National Sword policy sent shockwaves around the world. The US recycling industry had relied so heavily on this single market that, when it dried up, recycling costs in America skyrocketed. Municipalities, whose recycling contracts often include clauses that left them paying for market price increases, became the recipient of sudden and extreme rising costs. In some cases, cities and towns faced hundreds of thousands, and even millions, of dollars to maintain their recycling programs.[57] China's restriction on the import of recyclables thus reverberated among every municipality in the United States. When China's market dried up, it was like a tide going back out to sea exposing the underfunded and splintered US recycling system.

China's market for low-grade materials had long been a boon for single-stream recycling that produced a large volume of often contaminated recyclables, but it was also a disincentive to invest in US domestic processing capacity. When the China National Sword came down, it became clear that US recycling facilities were not technologically capable of processing the materials and were sorely in need of an upgrade. The restrictions on the import of recyclables by China exposed long-standing challenges in the US recycling system.[58] The lack of attention paid to US domestic recycling infrastructure left us behind in the global recycling marketplace. Then came another waste management jolt.

PLASTICS POLLUTION BACKLASH

As the China National Sword began to poke huge holes in the efficacy of the US recycling system, a global uprising against plastic pollution that had been building for years finally took hold. Images of turtles, birds, and other aquatic life lying dead from ingesting plastics alarmed the world. Scientists, academics, and environmental activists shined a huge spotlight on an issue that further launched the packaging problem to the top of the global environmental agenda.

---

57. *New York Times*, "As Costs Skyrocket, More US Cities Stop Recycling" (March 2019), https://www.nytimes.com/2019/03/16/business/local-recycling-costs.html.

58. *Resource Recycling*, "From Green Fence to red alert: A China timeline" (February 13, 2018; updated March 22, 2022), accessed January 26, 2020, https://resource-recycling.com/recycling/2018/02/13/green-fence-red-alert-china-timeline.

The outrage against single-use plastics in the US further fueled discontent over rising municipal recycling costs due to the shutdown of Chinese and other overseas markets. What started as voluntary initiatives to stem the tide against plastics pollution quickly spread to legislative efforts by governments and environmental groups. The passion and power displayed by elementary school students, environmental activists, and other community members forged a ready partnership with many government officials and legislators seeking to control whatever aspect of the garbage problem they could. The global anti-plastic movement had finally reached all the way to the state and local levels, resulting in bans enacted around the country on plastic bags, polystyrene (Styrofoam) packaging, plastic straws, cups, and other single-use plastic items.[59]

In a counter-offensive, the American Recyclable Plastic Bag Alliance (ARPBA) supported state laws that preempted local governments from banning single-use plastics. By 2019, these "ban on bans" laws, developed by the American Legislative Exchange Council (ALEC), had been passed in 19 states, including Missouri, Nebraska, and Arkansas.[60]

In 2016, PSI was set to start a project with Washtenaw County, Michigan, to reduce the use of paper and plastic grocery bags, when we too got caught in the industry backlash. That year, the Michigan Legislature passed a law, modeled on the ALEC bill, that abruptly ended Washtenaw County's effort. ALEC's promotion of state bans preempting local action prevented many local governments from protecting their own environment. In turn, grassroots organizations such as the Surfrider Foundation emerged to counter the actions of ARPBA and ALEC by opposing state preemption laws and adding momentum to pass state and local bag bans through grassroots advocacy. These two opposing forces still reflect the national sentiment around plastics pollution, with opposing sides lined up against one another in a high-stakes game of corporate power versus municipal and citizen activism.

---

59. Jennie Romer, PlasticBagLaws.org, https://www.plasticbaglaws.org, last updated October 2021, accessed on April 12, 2022; Surfrider Foundation's Interactive Map of US Plastic Reduction Policies, https://www.google.com/maps/d/u/0/viewer?mid=15xHcYQygDGYQnJPKraz7cInHT6mQYJND&ll=39.30619901383374%2C-81.21427345003117&z=5, last updated August 31, 2022, accessed September 26, 2022.

60. Surfrider Foundation, "Defend Your Right to Reduce Plastic Pollution," accessed October 3, 2022, https://www.surfrider.org/pages/defend-your-local-right-to-reduce-plastic-pollution; Grassroots Change, "Preemption Watch," accessed on October 3, 2022, https://grassrootschange.net/preemption-watch/#/category/plastics.

THE TRANSITION TO EPR IN THE UNITED STATES

China's Green Fence (2013), its National Sword policy (2018), and the plastics pollution backlash—all taking place at the global level—finally revealed what many of us knew all along: The US recycling system was fractured by misplaced financial incentives and nearly broken. During this time, recycling costs continued to rise, and the familiar sense of financial uncertainty rumbled across the municipal landscape.

Since a waste management paradigm shift was long overdue, the three global events galvanized increased interest among PSI's state and local government members to again update our standard *Elements of Packaging and Paper Products (PPP) EPR Legislation*, which we initially developed in 2010 and repeatedly fine-tuned in subsequent years. By 2017, it became evident that we needed to increase the frequency of government strategy calls to share technical information, decipher the myriad EPR packaging policy options, and decide what specific concepts and legislative text should be included in an updated model packaging bill.

To do so, we convened state and local government recycling and waste management experts who were also EPR experts in their respective states. The combined knowledge of these individuals, each with different viewpoints and expertise, provided the policy substance and practical know-how that allowed PSI staff to facilitate rich policy discussions. PSI served as facilitator, technical advisor, and conduit for the free flow of information among these state and local officials. We started by breaking down the model into 16 key policy elements we derived from our own experience with EPR on multiple products. We then identified options and key decision points for shaping programs in America and incorporated aspects of packaging EPR programs from Canada and Europe.

The process we designed served several functions. First, it educated government officials about basic policy elements they needed to consider for all packaging EPR bills. It also helped them to understand available options for situations unique to their state. The result of this one-year process was a revised conceptual policy document, *Elements of Packaging and Paper Products (PPP) EPR Legislation*, which included a definition of each of the 16 elements, a "base model" that represented a consensus of the government experts, as well as policy "options" that each state could decide to include. The base model included legislative text derived from US state EPR product laws and other sources.

Overall, the model provided the conceptual template for any state or federal agency to develop an EPR bill for packaging and paper products (PPP). Since our government members in those states had benefited from years of engagement with PSI and became intimately familiar with our elements model concept that we used for products, they were able to apply it to develop their own packaging EPR bills. In some states, such as New York, Vermont, Connecticut, Minnesota, and Illinois, PSI facilitated discussions using this conceptual policy tool to assist key state and local officials, as well as other key stakeholders, to develop a detailed, comprehensive bill outline. In other states, such as Oregon, we trained staff on how to use the Elements policy document to assist in developing their own bill. In Maryland, we provided the document to the initial legislative sponsor who facilitated key stakeholders to develop her own bill, and in Massachusetts it was used by a key local government policy expert to develop a bill. The tool was also used in other states to varying degrees to supplement their own bill development process.

To turn the policy decisions from the model into legislation, PSI and our government members either wrote the draft bill ourselves or provided our detailed model to legislative counsel from the state environmental committee, who developed a draft bill that we reviewed numerous times until it read as we intended. The legislative staff writing these bills are usually extremely capable in transforming technical concepts that they are learning for the first time into legislation that is consistent for the state. These government officials are critical to the success of legislative efforts, particularly for robust bills like packaging EPR.

The regular and generous advice that we received from Canadian and European government officials, corporate executives, and nonprofit leaders was instrumental in providing not only technical instruction but also encouragement. These colleagues were a constant reminder that the industry opposition we experienced in the United States came from a place of fear of change and associated costs, which companies did not know how to manage. Engaging repeatedly with Canadian and European EPR experts, who worked for international brand name companies that supported EPR, was a reality check. These people participated in our packaging webinars, conference sessions, and meetings with government officials and legislators. By doing so, they provided a constant pulse of an alternative corporate narrative that boosted hope for US government officials. If these corporate executives could implement EPR laws in other countries and show pride in their management of EPR systems, it was only a matter of time before this same sentiment would take hold here. Importantly, though, these foreign EPR experts always knew there was a line over which they could not cross. They could not tell US governments or their US corporate counterparts what to do, but they *could* tell us what they did, how they did it, and the results they achieved. They knew, and we knew, that policy change in America had to come from within, and even though this change was inevitable, those outside the country could only educate us on their programs and provide technical support. The rest was up to us.

VOLUNTARY INITIATIVES

As PSI government members ramped up their legislative work, brand companies selling products in the United States announced voluntary commitments to design and use more eco-friendly, sustainable packaging. For example, P&G committed to offer 100 percent reusable or recyclable packaging by 2030 and reduce virgin petroleum plastic in their packaging by 50 percent by 2030, as measured against a 2017 baseline.[61] Coca-Cola committed to making 100 percent of its packaging recyclable globally by 2025, with at least 50 percent recycled content by 2030.[62] The company also committed to collecting and recycling the equivalent of 100 percent of the packaging it produces[63] and to converting at least 25 percent of its bottles into reusable bottles and other reusable packaging by 2030.[64]

In addition to these pledges, over 100 companies, associations, and organizations, including Coca-Cola and General Mills, joined the US Plastics Pact, as part of a larger global effort led by the Ellen MacArthur Foundation, and committed to achieve the following four plastics packaging targets: (1) define a list of packaging that is problematic or unnecessary by 2021 and take measures to eliminate them by 2025; (2) 100 percent reusable, recyclable, or compostable plastic packaging by 2025; (3) recycle or compost 50 percent of plastic packaging by 2025; and (4) an average of 30 percent recycled or responsibly sourced bio-based content by 2025.[65] The brand companies that are part of the US Plastics Pact, either individually or as part of an association, publicly recognized that they could not reach their sustainable packaging goals in America without significant improvements to the recycling system.

As they did, tens of millions of dollars of corporate financial contributions were made to groups like The Recycling Partnership (TRP), which provided funding to municipalities and recyclers for recycling equipment, infrastructure, education, and other initiatives. The Ellen MacArthur Foundation (EMF) launched its own voluntary New Plastics Economy Global Commitment in October 2018. This initiative sought to eliminate unnecessary

61. P&G blog, "Accelerating the Scale of Recycling and Sustainable Packaging" (March 16, 2021), accessed March 31, 2022, https://us.pg.com/blogs/global-recycling-day-2021/.

62. Coca-Cola website, accessed March 31, 2022, https://www.coca-colacompany.com/sustainable-business/packaging-sustainability.

63. Coca-Cola website.

64. Coca-Cola website.

65. US Plastics Pact, The US Plastics Pact Roadmap to 2025, accessed March 31, 2022, https://usplasticspact.org/roadmap/.

plastics; develop reusable, recyclable, or compostable packaging; and recirculate plastics to avoid waste.[66] The New Plastics Economy also sought to make all plastic packaging free of hazardous chemicals, and to respect the health, safety, and rights of all people.[67]

Some viewed these voluntary initiatives as a conscious strategy to delay the implementation of EPR, since there were no penalties for not reaching the goals. Even so, the initiatives educated many in the industry about gaps in recycling infrastructure and the financing needed to upgrade the system. It also became apparent to many corporate leaders that the quality and quantity of recycled materials needed by companies to meet their pledges were not considered possible within the current fragmented recycling system, especially given high contamination rates and generally low collection rates in a largely municipally run system. These initiatives also allowed brand owners and others in the packaging supply chain to learn about EPR policies and consider what they might be willing to accept.

FLEXIBLE PACKAGING STEPS FORWARD

By 2019, national momentum on packaging EPR surged to a new level. By this time, the full impact of the China National Sword was being felt by high-level municipal and state officials and legislators. This situation was compounded by a rising movement to ban plastic production, particularly for plastic bags and single-use plastic items, and to switch to non-plastic packaging material. Multiple state EPR for PPP bills, as well as a federal packaging EPR and plastics reduction bill, were introduced or were being developed around the country, with most bills directly or indirectly influenced by PSI's EPR for PPP policy model. With this widening government show of political will, the sentiment of some packaging producers, including members of the Flexible Packaging Association (FPA), began to change. As a result, the FPA president and CEO, Alison Keane, approached me to start a multi-stakeholder facilitated dialogue on flexible packaging. Keane was the former paint industry executive who entered into a mediated agreement with PSI on a paint EPR bill (see chapter 8).

The fast-growing flexible packaging segment represents 19 percent of all packaging, including bags, pouches, and other packages comprised of paper,

66. Ellen MacArthur Foundation, https://ellenmacarthurfoundation.org/global-commitment-2022/overview.

67. By December 2021, EMF had more than 400 signatories that committed to publicly report their progress annually, including major consumer packaged goods companies and retailers such as Unilever, Colgate Palmolive, Walmart, Apple, and MARS. By December 2022, that number rose to 500 signatories.

plastic, film, foil, metallized or coated papers, or any combination of these materials whose shape can be readily changed.[68] Unfortunately, due to a lack of collection and processing infrastructure, and few viable end markets, only 4 percent of flexible packaging is currently being recycled.[69] Flexibles have also been swept up in the global outcry against all petroleum-based plastic packaging, as well as their pervasiveness in the environment. As brand owners embraced flexibles due to their lower cost and other benefits, environmental and social activists sought to ban problematic materials, particularly plastics, including flexible packaging.

Before engaging in a dialogue with FPA, I first needed to take the temperature of our state and local government members, whose job is to balance the interests of businesses, environmental groups, other interested parties, and the general public. Each local and state government leans in a particular political direction, and these positions can change with the political winds. We needed to find out if PSI's government members in states considering an EPR solution for packaging wanted to ban flexibles. Many of our state and local officials were already engaged in banning plastic bags, polystyrene, and other problematic packaging. Even so, they recognized that flexible packaging has environmental life-cycle benefits and were not quite ready to ban it. Nor were they willing to accept a 4 percent recycling rate, along with the costly interruptions that flexible packaging caused at material recovery facilities (MRFs) and the common sight of sachets and wrappers littering streets, rivers, and shorelines. Ultimately, they approved of PSI's engagement with FPA under the condition that our work would seek to develop an EPR system that FPA members could endorse. They also wanted that system to lead to significant reuse or recycling of flexible packaging as measured against performance goals, and left open the option to ban flexibles in the future if those goals were not met.

The PSI-facilitated dialogue with FPA and other stakeholders started in October 2019 after months of upfront planning, research, and in-depth interviews. Within the first few months, the group of packaging manufacturers, state and local government agencies, and two environmental advocacy organizations reached consensus on four foundational aspects. First, the group concurred on the importance of nine beneficial attributes of flexible

68. Product Stewardship Institute, *Flexible Packaging Dialogue Briefing Document*, prepared for the Flexible Packaging Association (November 2019). According to FPA, flexible packaging includes rollstock, bags, pouches, liners, shrink sleeves, shrink wrap and stretch wrap, bulk shipping sacks, flexible lidding, and film wrap around labels. *Not* included are labels (other than shrink sleeve or film wrap around style), photographic film or paper, photo paper, tape, toilet and facial tissues, diapers, feminine products, and thermoforming materials, the shape of which can't be readily changed.

69. The Closed Loop Foundation, "Film Recycling Investment Report," page 1, undated, prepared by RSE USA, accessed April 3, 2022, https://www.closedlooppartners.com/foundation-articles/investment-opportunities-in-film-plastic-recycling/.

packaging that keep products safe in delivery, reduce cost through efficiency, and minimize environmental impacts (primarily through the manufacturing process and its lighter weight than traditional packaging types). Participants also agreed that any new policies developed should maintain or improve these attributes. For example, developing a package that is more recyclable but fails to protect a product, resulting in more waste, would not be an acceptable outcome.

From the numerous individual interviews that PSI staff conducted in advance of the dialogue, I understood that it was important for those in the packaging industry to hear that government officials recognized the benefits of this rapidly emerging packaging type. Many FPA members worked hard to design and manufacture these packages, and they took comments personally. This first agreement was an acknowledgement by governments of the benefits of flexible packaging and established a baseline from which to engage on more difficult issues. It was also a tacit statement from governments that flexible packaging was not on their current list of materials to ban.

From the interviews, I sensed that the industry wanted affirmation about their past efforts to make better packaging before they would be willing to engage with government officials about supporting EPR. If agency officials did not acknowledge those benefits from the start of the dialogue, I knew that it was likely that companies would continue to bring up this point. It could get us stuck. Once acknowledged, however, FPA members were willing to engage in addressing the problems that resulted from their packaging. This is a classic case in negotiation when one party (in this case government) gives something of value to the other party (producers) that is not a cost to them. This agreement became our first step in the dialogue and laid a firm foundation on which to build.

Following PSI protocol, we next discussed the multi-faceted problem caused by flexible packaging. The consensus we reached was an aggregate of the many stakeholder views expressed during the individual interviews conducted prior to the meetings. The problem statement (containing 14 distinct aspects and three sub-aspects) includes "lost resources from a lack of material recovery, ocean pollution and litter, governments and taxpayers having to bear the primary cost of postconsumer management and mismanagement, governments lacking adequate funding for recycling and handling increased waste loads, and producers and consumers not bearing the true life-cycle costs of the goods they buy."[70] If acknowledging the attributes of flexible packaging was

70. Product Stewardship Institute and Flexible Packaging Association, "Shared Elements of EPR Legislation for Packaging and Paper Products (PPP) November 2020," Attachment 2, 3, https://productstewardship.us/wp-content/uploads/2023/01/fpa-psi-2020-dialoguesummary.pdf.

more important for producers, acknowledging the problem statement played a similar role for governments. It became our second joint agreement.

The group developed its third agreement on the "Desired End State," which articulated joint goals to achieve, including a common interest in developing a model system that incentivizes reduction in material use and environmental impacts, maximizes the collection and environmentally beneficial postconsumer management of flexible packaging in the United States, and minimizes costs to government and industry. Finally, we reached a fourth consensus on the attributes of an effective recycling system for managing flexible packaging, which included "reduces environmental impacts and costs, including from litter (e.g., externalities); keeps materials out of the open environment (e.g., oceans, rivers); and provides sustainable funding, including funding for research and development, and infrastructure."[71]

This dialogue was the first time that US packaging manufacturers had engaged with government officials, along with several environmental advocates and recyclers. After four months of working together, and four initial agreements, the group was ready to embark on discussing the elements of a model EPR for PPP program for flexible packaging. By this time, PSI already had developed a refined EPR packaging model with its state and local government members, and bill activity was moving forward in Maine and Oregon. Since a rapport had developed among stakeholders, the PSI/FPA group began commenting together on state bills. We also began to consider each element of the PSI packaging EPR model.

Moving to this next stage is like pulling a chair up to the negotiating table from the back of the room. Engaging in the development of a bill starts by understanding each key element, from "covered materials" (what type of packaging is included), to the "definition of responsible party" (who pays the fees to manage packaging and must adhere to meeting other major program requirements), to how we define "consumer convenience." (These elements are also discussed further below and in chapter 6.) These concepts must first be understood before they can be discussed, debated, and resolved. PSI's Elements document, which is updated regularly based on best practices, is the tool we use to educate stakeholders about basic concepts and, at the same time, discuss their positions and interests relative to the EPR system. Through this process, each stakeholder is provided numerous opportunities to discuss every aspect of bill development. If those at the table are earnestly seeking agreement, the chances are greater that consensus can be achieved.

---

71. Product Stewardship Institute and Flexible Packaging Association, "Shared Elements of EPR Legislation for Packaging and Paper Products (PPP) November 2020," Attachment 2, page 3, https://productstewardship.us/wp-content/uploads/2023/01/fpa-psi-2020-dialoguesummary.pdf.

It is important to keep a discussion moving fluidly and to know when to put aside an issue for later resolution. Such an instance occurred during the group discussion on the transition needed to move from the current situation—in which nearly all flexible packaging is disposed, with only a small amount collected at store drop-off locations for recycling—to an end state with an alternative vision. For some, this meant that producer funding would be allocated toward new technologies at MRFs to sort flexibles from other plastics and packaging, thus expanding markets. For others, the vision was to invest in a range of purification, depolymerization, and conversion technologies known as "chemical recycling," which use various sources of plastics (e.g., plastic bags, water bottles, carpet, plastics in medical waste) to produce marketable plastic products or fuel.

If you are familiar with this issue and the controversy it has engendered, you might be tempted to jump out of your seat to advocate for one side of this debate or the other. But before you do, I will tell you that consensus was *not reached* on this issue. While most FPA members supported these new technologies, government officials unanimously opposed plastics-to-fuel technologies but expressed interest in learning more about technologies that produced marketable plastics. These officials knew they would be required to respond to questions about whether, and how, these technologies should be permitted. This issue was thus put in the "parking lot" for further research, clarification, and decision making; doing so highlighted the importance of the issue while allowing the group to move forward and address other issues. Eventually, the group agreed that the criteria for deciding which, if any, of these technologies would be eligible for funding under an EPR system would be addressed later during the bill development stage in each state.

After the full dialogue group spent several months discussing the EPR bill elements, FPA members asked for a caucus—an opportunity to meet outside the stakeholder meetings—to seek a unified industry position. At the same time, I also caucused with government and environmental participants to help them to clarify their positions. These confidential meetings are an important part of a mediation process. When participants are actively engaged in a stakeholder meeting, they don't always have time to process what they are hearing. It is also important that members of an industry association, like FPA, or a group of government officials, have the chance to check in with their colleagues about key policy positions. If stakeholders are not aligned within their own constituency, their comments will represent fragmented positions, making it more difficult to reach consensus.

After a two-month hiatus, FPA and its members were ready to resume discussions. Since FPA members are mostly packaging *manufacturers* that sell to *brand owners* (most often the "responsible party" in EPR legislation), during their caucus they concluded that only eight of the 16 elements were relevant to them from a regulatory perspective. This meant that, for eight elements, they would continue to refine the agreement within the PSI dialogue, but on the other elements they would leave it to brand owners to negotiate for their own interests. After several additional meetings, dialogue participants reached consensus on eight elements of a packaging EPR system.[72] This agreement became the first in the country between government officials and packaging manufacturers on packaging EPR, and later became a starting place for PSI-facilitated multi-stakeholder packaging EPR dialogues in multiple states.

Significantly, the FPA/PSI dialogue marked the first time in 15 years that an association of packaging manufacturers agreed to *discuss* an EPR system for packaging with government officials. When we began discussions, the path to consensus was not clear and the stakeholder engagement process had to be designed based on information that slowly unfolded. For example, while government officials and environmental advocates were at the table to develop a joint packaging EPR model, not all FPA company representatives had reached that conclusion. FPA's Keane and I both knew that government and environmental participants would not accept anything less than a dialogue focused on a model EPR system. Even so, Keane also had a few members who were not yet convinced about EPR or didn't know enough to have an opinion. They needed to warm up to the process and start engaging with others to become comfortable with that notion.

For the first few meetings, Keane and I worked with that understanding and did not require participants to publicly state their intent to work on EPR. We avoided the words *bill* and *legislation*, while at the same time discussing "EPR programs and systems." This approach worked because Keane trusted PSI's process, and I trusted her ability to speak on behalf of her members as well as her intent to develop elements of an EPR bill. I also knew through the in-depth stakeholder interviews we conducted prior to the meetings that many FPA members were already operating under EPR packaging systems in multiple countries. I understood that their hesitancy in the dialogue had less to do with EPR and more to do with not having experienced a civil policy conversation with government and environmental representatives. I was also assured by Keane that she was aware that one of her members from a prominent company was known for being outspoken against EPR, but that person would not sway the dialogue outcome.

72. Product Stewardship Institute and Flexible Packaging Association, "FPA and PSI Reach Agreement on Legislative Elements of an EPR Bill for Packaging and Paper Products" (December 2020), https://productstewardship.us/wp-content/uploads/2022/10/FPA-PSI-release-Dec-2020.pdf; Product Stewardship Institute and Flexible Packaging Association, "Shared Elements of EPR Legislation for Packaging and Paper Products (PPP) November 2020," https://productstewardship.us/wp-content/uploads/2023/01/fpa-psi-2020-dialoguesummary.pdf.

Multiple conversations between Keane and me, along with additional stakeholder interviews and extensive preparation, resulted in a two-hour preliminary conference call to test the waters. After a productive and respectful call, stakeholders breathed a sigh of relief and agreed to a subsequent two-day, in-person meeting that included an informal reception designed to foster personal relationships. After the meeting, I was told by one of FPA's most vocal participants that the meeting was the first time he was able to have a reasoned dialogue with government officials and environmental advocates. After informally checking in with other FPA members, I sensed a similar feeling that we were on the right track. Government and environmental stakeholders, who had more experience in policy dialogues, were pleased with the earnest engagement by FPA members and appreciated the technical discussions regarding EPR policy. Meeting by meeting, the rapport among stakeholders grew as they became comfortable with each other, the issues, and the process.

No process can guarantee success. It takes many factors, including leadership among participants representing key stakeholder groups. It also takes time, resources, and a willingness to pursue a joint path through collaboration. What occurred with FPA is what can happen when a process is designed with the needs and interests of multiple stakeholders in mind. PSI's job was to construct a forum for honest, constructive dialogue. This feeling of security emerged by design, starting with ground rules for considerate discussion and acknowledgment of the beneficial attributes of flexible packaging. Trust among participants was further built by reaching agreements on the problem, the end state, and the attributes of an effective system for managing flexible packaging. Our time together also needed to be efficient and effective for participants to feel hopeful that the group was continually making progress toward a common policy goal.

Keane's goal was to lead her members into a dialogue process to which she had become accustomed through the previously successful PSI dialogue on paint. She already understood the rhythm of dialogue including how it starts, how it evolves, and its potential for reaching a common goal. Together, our group developed a shared vision that was enabled by a process design in tune with their needs. By the time participants started to focus on the vision and work together, they had lost their apprehensions and were able to begin building joint agreements. They were then ready to do the hard work of going beyond a nice conversation and into key issues that led to decisions.

THE TIDE TURNS TOWARD EPR

As PSI and FPA engaged in dialogue in fall 2019, the constant American and global pressure placed on US packaging brand owners led some companies to consider how they, too, could operate under an EPR system. These companies, along with their trade associations, often developed independent principles of EPR to which they could abide. For example, the Consumer Goods Forum promoted EPR as a main policy approach to achieve a circular economy, establishing their own Principles of EPR in 2020,[73] as did AMERIPEN.[74] Industry positions were further influenced by an EMF report in 2020 that supported EPR for packaging,[75] and another report by the US Plastics Pact, led by TRP and the World Wildlife Fund (WWF), that promoted EPR along with deposit return systems (i.e., bottle bills) for beverage containers.[76] In 2020, after conducting an extensive engagement process with its industry supporters, TRP came out with a policy platform that leaned toward EPR for packaging, with a fee paid by private-sector brands to support residential recycling infrastructure and education, and a disposal surcharge on waste generators to help defray recycling operational costs for communities.[77]

Little by little, the haze of uncertainty in the United States about EPR for packaging began to lift. No longer did we hear "under no conditions will we accept EPR for packaging!" The language changed to "we could consider being part of an EPR system that meets *these* criteria." Although the type of EPR system they sought was often vague, these companies still made the point that they were ready to discuss what that system might look like. Fifteen years of adamant opposition finally softened into a new willingness to explore. This momentum was amplified by EPR interest from environmental groups such as the Natural Resources Council of Maine, Zero Waste Washington, and the National Stewardship Action Council, national groups including NRDC and Sierra Club, and international groups like Greenpeace and WWF. The need for industry to find a way to support EPR while protecting their interests became inescapable.

---

73. The Consumer Goods Forum, "Building a Circular Economy for Packaging" (August 2020), https://www.theconsumergoodsforum.com/wp-content/uploads/Building-a-Circular-Economy-for-Packaging-July-15-2022.pdf.

74. "AMERIPEN Financing Principles and Objectives for Advancing Packaging Recycling in the US" (May 19, 2020), https://www.ameripen.org/news/508260/New-from-AMERIPEN-Advancing-Packaging-Recycling.htm.

75. Ellen MacArthur Foundation, "Extended Producer Responsibility: a necessary part of the solution to packaging waste and pollution" (2021), https://plastics.ellenmacarthurfoundation.org/epr.

76. US Plastics Pact, "The US Plastics Pact Roadmap to 2025" (June 15, 2021), 9, 17, 19, 23, https://usplasticspact.org/roadmap.

77. The Recycling Partnership, "Accelerating Recycling: Policy to Unlock Supply for the Circular Economy" (September 2020), 4, https://recyclingpartnership.org/accelerator-policy/.

By the time some packaging producers and brand owners were ready to talk about EPR, PSI and numerous state and local agency members had already developed legislation or were in the process of doing so. Among that expert group were government officials from Maine and Oregon, each of whom had agency or gubernatorial support to develop an EPR solution. The packaging EPR laws enacted in Maine and Oregon in 2021 were a stark message that the financial and environmental costs of packaging waste required immediate action. The laws enacted in Colorado and California in 2022 confirmed that state governments were serious about solving the waste management crisis.

PSI's model eventually informed state EPR for PPP legislation, either indirectly, as in Maine and California, or somewhat more directly as in Oregon, Washington, Maryland, Massachusetts, and Colorado. In states like New York, Connecticut, Vermont, Minnesota, and Illinois, PSI convened multiple stakeholders—including government officials, national producer associations, waste management companies, and environmental groups—to develop bills based on our model. In Maryland, we were technical advisors to the House and Senate bill sponsors throughout the two-year process. The 2022 Maryland bill, championed by Delegate Brooke Lierman and Senator Malcolm Augustine, became the first packaging EPR bill in the country to garner support from local governments, environmental groups, and packaging brands and manufacturers, including FPA, AMERIPEN, and the Consumer Brands Association. PSI's model also helped form the basis for the EPR component of a comprehensive federal bill introduced in February 2020 by Senator Tom Udall of New Mexico and Congressman Alan Lowenthal of California.[78] That bill was reintroduced in 2021 by Senator Jeff Merkley of Oregon with co-sponsorship again by Congressman Lowenthal.

With nearly two dozen state bills enacted, introduced, or in development, the inevitability of change continues to sink in for all stakeholders—not just producers but also waste management companies, environmental groups, and recycling organizations. Even so, there remain significant differences in interests among these stakeholders. Producers seek greater control over the system into which they pay, while waste management companies want to retain their existing contractual relationships, to be assured that their infrastructure will be used, and to have the ability to find markets for recycled materials. Some environmental groups and zero waste advocates remain distrustful of industry and seek strong state agency management. They also want to include specific targets in statute for waste reduction, reuse, and recycled content, measures to eliminate toxic chemicals and plastics, and a new or expanded DRS.

---

78. Tom Udall, Senator for New Mexico, "Udall, Lowenthal Release Outline of Legislation to Tackle Plastic Waste Pollution Crisis" (July 2019), https://www.tomudall.senate.gov/news/press-releases/udall-lowenthal-release-outline-of-legislation-to-tackle-plastic-waste-pollution-crisis.

A moving train draws attention, and one as powerful as EPR for packaging has generated a major buzz. While this paradigm shift is truly inspirational to experience, it also creates a significant challenge to harmonize the EPR programs that are currently being established with new bills that will continue to be enacted. For companies operating in multiple states, policy consistency is important to reduce compliance cost and complexity. The best way to develop a consistent national legislated program is either through a model state bill or through federal legislation. Both can be accomplished through a well-constructed process that forges agreements.

PSI's model elements document can help to achieve policy consistency by offering common elements and definitions, while also allowing for state variation. In this way it has become a tool for multi-stakeholder decision making. The model provides an opportunity for key stakeholders to jointly gain a basic understanding of key concepts from which they can negotiate for their interests. As described in chapter 6, this method is how PSI develops all its EPR policy models and bills across the roughly 25 products on which we work. Let's now look at the key elements of packaging EPR legislation.

KEY ELEMENTS OF EPR FOR PPP

I often hear the phrase, "the devil is in the details," when people discuss EPR policies. But who wants to confront the devil? It is a term of caution and apprehension, and using it casts a dark shadow over the technical aspects of policy. I prefer to think that *solutions* are in those details. Only when stakeholders understand the specifics of EPR policies will they be able to choose the type of policy they want to implement. Taking it one step further, to coordinate EPR packaging programs across the country, we need to fully understand the common elements of an EPR for PPP bill, and where options exist for state or regional variations.

Over the past two decades, PSI has continually refined its understanding of the basic elements of effective EPR systems for PPP and how they evolve over time. No matter where they are enacted, these systems share the same fundamental principles. Producers are required to take responsibility for financing and managing their postconsumer packaging. Local governments choose whether they want to collect recyclables and provide public education about the program or shift these responsibilities to producers. Recycling companies can continue to provide collection and processing services to municipalities, producer responsibility organizations (PROs), individual households, or a combination of clients. Finally, the state (or federal) government oversight agency ensures effective and compliant program implementation.

EPR for PPP is an approach to waste management *within* which, or *around* which, other complementary policies can be built; EPR is widely recognized as the center of the circular economy. But it is by no means the only policy needed to guide our use of materials to become more sustainable. Reduction, reuse, recycling, and recycled content standards for packaging, material and/or product and packaging bans, container deposit laws, and pay-as-you-throw programs are all complementary policies that can be directly integrated into an EPR law or enacted through separate laws. Canada, for example, has enacted a single-use plastics prohibition on the manufacture, import, and sale in Canada of checkout bags, cutlery, foodservice ware, stir sticks, and straws.[79] The Canadian government has also proposed a requirement that plastic packaging in Canada contain at least 50 percent recycled content by 2030.[80] These two national laws supplement packaging EPR laws in multiple provinces.

Certain elements are standard for all EPR systems, however, regardless of the materials covered. Below is a summary of PSI's *Elements of an Effective EPR for PPP bill*, which is based on EPR programs in the United States and around the world. For a discussion of the basic EPR elements as they pertain to all products and packaging, see chapter 6.

Covered Materials

This element specifies the types of packaging, paper products, and single-use products that are subject to the EPR program, regardless of their recyclability or compostability. In other words, producers pay into these programs for the PPP materials they supply to the market, regardless of whether those materials are eventually recycled, composted, littered, or disposed in the garbage. Specifying the types of packaging, paper products, and packaging-like products are important decisions to include in EPR systems. In fact, many EPR bills direct state oversight agencies, PROs, and/or other stakeholders to create a uniform list of recyclable materials that ensures consistency across municipal programs and provides a uniform statewide basket of materials for recycling markets, while also decreasing consumer confusion.

There are three basic types of packaging: primary, secondary, and tertiary. For example, a single deodorant bought in a store often comes in a plastic casing (primary packaging), but it is delivered to the store in a box (secondary

79. Government of Canada, Single-use Plastics Prohibition Regulations—Overview, accessed March 26, 2023, https://www.canada.ca/en/environment-climate-change/services/managing-reducing-waste/reduce-plastic-waste/single-use-plastic-overview.html.

80. Government of Canada, Technical issues paper: Recycled content for certain plastic manufactured items Regulations, accessed March 26, 2023, https://www.canada.ca/en/environment-climate-change/services/canadian-environmental-protection-act-registry/technical-issues-paper-recycled-content-plastic-manufactured-regulations.html.

packaging) with other deodorant sticks, which might be shipped in a large, shrink-wrapped cube (tertiary packaging) containing many other such packages. Most US EPR for PPP bills regulate packaging handled directly by consumers (i.e., primary packaging), with some also incorporating packaging further up the supply chain (secondary or tertiary packaging), including cardboard from deliveries. "Paper products" includes items such as newspapers, flyers, and marketing brochures, while "packaging-like products" includes sandwich bags, aluminum foil, and single-use products such as plastic cutlery, plates, cups, and straws (also called "foodservice ware"[81]).

Each state packaging EPR law has included a different combination of covered materials: Maine includes packaging only, Colorado includes packaging and some paper products, California includes packaging and packaging-like plastic products, and Oregon includes all categories—packaging, paper products, and packaging-like products. Some bills also include service packaging intended for the consumer market (e.g., carry-out, produce, and bulk food bags), take-out food service ware, and prescription bottles. Exemptions are typically provided for paper products that may become unsafe or unsanitary to recycle by virtue of their use, such as tissues and toilet paper, and for packaging materials that are already covered under other programs, such as containers managed through deposit return systems (DRSs) or paint cans collected and recycled through an existing paint EPR program.

In states where residential and commercial recyclables are currently managed together, the state will often include both materials under covered materials, although the four US packaging EPR laws show the variety of approaches taken thus far. Oregon covers commercial recyclables, Maine includes some commercial recyclables but not distribution packaging, and Colorado covers most commercial packaging materials but not business-to-business transport and distribution packaging, and not exclusively industrial manufacturing materials. In California, commercial packaging is exempt if it can repeatedly demonstrate a high recycling rate (65 percent before 2027 or 70 percent after 2027) and is recycled at a responsible end market.

Industrial, commercial, and institutional (ICI) packaging is often covered in European programs, but only Belgium has a separate PRO (VALIPAC) that manages the ICI sector. In addition, there are separate accreditation systems in Austria for the residential and ICI sectors, and a PRO can seek accreditation in both sectors. Ontario is the only Canadian province that currently requires all ICI program materials to be collected, and Québec's PRO,

81. According to the Foodservice Packaging Institute (FPI), "Foodservice packaging primarily includes single-use products such as cups (beverage and portion), plates, platters, bowls, trays, beverage carriers, bags (single portion and carry-out), containers, lids and domes, wraps, straws, cutlery and utensils for the service and/or packaging of prepared foods and beverages in foodservice establishments," accessed September 28, 2022, https://fpi.org/about/foodservice-packaging-history/.

Éco Enterprises Quebec (ÉEQ), is required to reimburse municipalities that choose to collect ICI materials. Beginning January 1, 2025, ÉEQ will be obligated to begin granting recycling access to additional ICI sources, ensuring full ICI sector access to recycling by 2030.[82]

Covered Entities[83]

These are the places from which covered packaging materials are collected at no additional cost, and often include single and multi-family residences, drop-off sites, and public places. If the existing local government recycling service combines residential service with service to ICI sectors, and the EPR program includes more than residential PPP in covered materials, the level of service in the recycling program should account for these sectors (e.g., if commercial materials are covered, commercial entities would be covered by the program). The principle behind this element is that the new EPR program should cover what is currently collected to maintain the current level of service, but also to strengthen it by collecting from additional places that will increase material recovery and consumer convenience. A key aspect is that the new system should be consistent with how the current recycling system is structured but allow for its expansion to handle new materials and a greater quantity of existing materials. By taking this approach, this element seeks to harmonize service levels statewide.

Collection Convenience

This element specifies the minimum level of convenience that producers are required to provide for collection from covered entities. Clearly, we don't want people to have to travel to a drop-off location under an EPR system if they currently have curbside collection. We want the collection of all program materials from covered entities to be convenient, which can include curbside recycling, clearly marked containers for drop-off facilities, an adequate number of strategically placed and well-marked bins in public places, and similar arrangements at schools, municipal buildings, and other locations covered by the program.

Convenient and standardized collection service will stabilize and enhance the current infrastructure so that a maximum amount of recyclable material is recovered. The starting place for convenient systems is to ensure that current collection opportunities are met or exceeded, and that recycling is at least as

82. Québec's PRO, Éco Enterprises Quebec (ÉEQ) estimates that ICI materials represent approximately 15 percent of total municipal reimbursements province-wide. Mathieu Guillemette, ÉEQ, email communication, January 21, 2020.

83. Canada uses an alternative designation, "eligible sources," rather than "covered entities."

convenient as garbage disposal. But it also seeks to create new infrastructure for materials not currently recovered, such as flexible packaging, reusable packaging, and compostable materials. Most packaging EPR bills will allow alternative collection programs for these or other specified material types if producers meet the same level of convenience as for other covered materials. This option allows flexibility for producers of a specific material that might be recovered in a unique way, which could lower costs by not being lumped into a comprehensive system designed for producers of multiple material types.

Responsible Party

This section identifies who is responsible for funding and managing the EPR program, usually one of three entities: the brand owner that manufactures the packaged product, the licensee of the product, or the first importer or distributor of the packaged product into the jurisdiction with the EPR program. The brand that sells the packaged product into the state is the primary responsible party and can be a retailer that has its own store or private brand. An example of a brand name manufacturer of packaged goods that sells under another brand is Unilever, which manufactures a shampoo under the Dove brand. A brand owner that is not a product manufacturer is Walmart, which sells Walmart-branded shampoo manufactured by another company. And an importer/distributor of wine that was bottled and packaged in France but sold into a US state will also be considered the responsible party under a standard EPR packaging law.

Regarding online sales, packaging that is in contact with the product to protect it (e.g., a plastic Aveda shampoo bottle sold by Aveda into a state) would follow the same three-tiered definition above. However, if that bottle was shipped by Amazon to a consumer, then Amazon would be the responsible party for the Amazon shipping package and Aveda would be responsible for its own bottle. If a product is imported from another country into a state, the responsible party is the first one to import or distribute the packaging into the state.

For paper products, the definition of responsible party is a bit more complicated. It could be the publisher of a newspaper (e.g., the local *Star Tribune*) or a widely distributed catalog (e.g., RH, Land's End); the manufacturer of copy/office paper that sells under its own brand (e.g., Hammermill); or the brand owner of the copy/office paper that sells paper made by another company (e.g., Staples). Again, an importer or distributor of copy/office paper that sells paper into a state for the first time would be the responsible party in the absence of a brand owner.

Most US packaging EPR laws and bills include exemptions for small businesses based on the total weight of all materials they place on the market annually or their total annual gross revenues. For example, Maine's packaging EPR law exempts small businesses with under $2 million in annual gross revenue; it also exempts medium-size businesses with under $5 million in annual gross revenue, but only for the first three years. Other laws and bills exempt businesses producing less than one ton of packaging or $1 million in gross revenue per year. Some programs provide mid-size businesses with the option to either pay a flat fee based on the total weight of covered materials they place on the market annually (with no requirement to produce a detailed annual report) or pay fees on a regular schedule and back it up with a detailed report. These exemptions reduce small business compliance obligations as well as the oversight agency's program administrative burden. Additional exemptions may be deemed politically necessary to pass a bill and could be phased into or out of the program several years after implementation.

Governance

One of the most important aspects of an EPR law is who is responsible for particular system components and who may be directly or indirectly impacted by those components. Another way to view it is, which entity has greatest influence over those components and to what degree. There are five key players in packaging EPR systems, each of which has strong interests in influencing aspects of the system. Producers are generally most concerned about how much money they pay into the recycling system, what that money will fund, and the degree to which they have managerial control over the recycling system and its efficiency. If they are required to fully fund the system, they will usually seek greater operational authority to determine what to fund and how to reduce costs and increase efficiency.

For their part, recycling companies (collectors and processors) seek assurance that the existing collection and processing infrastructure in which they have invested will be used. Collectors of recyclable material want to maintain their existing market share and compete for services in a fair process, while processors want predictable payments to maintain and upgrade their facilities; clear and consistent consumer education; and collection protocols to reduce contamination and increase material value. They would rather continue to get paid by a household, municipality, or business than by a PRO, which has greater negotiating power and could tack on reporting requirements. Although PROs typically use existing infrastructure rather than invest in an entirely new system, recycling companies seek to have this intent included in statutory language.

Governments and nonprofits also have strong interests. Local governments want their costs covered for residential (and sometimes commercial) collection and processing without losing their ability to ensure that quality services are provided. Environmental groups want assurance, through specific performance goals in statute and strong government oversight and enforcement, that waste, toxics, and pollution will be prevented. And state governments seek a system that takes unwanted financial and management burdens off state and local governments, meets performance goals, and allows for efficient and effective agency oversight.

Many of these stakeholder interests overlap of course, while others compete against one another, resulting in different outcomes in each state depending on the local recycling system, political interests, and other variables. The governance element in PSI's packaging EPR model includes three main features that define key roles for producers and other stakeholders regarding program operations: the producer responsibility organization (PRO), the advisory council, and government oversight and enforcement.

Producer Responsibility Organization (PRO)

A primary function of a PRO is to meet producers' legal obligations to finance and manage the system to varying degrees depending on the type of EPR system established. In the United States, it is common practice for EPR laws to allow producers to comply either as individual companies or collectively under a PRO, which can reduce costs by coordinating services such as data collection, stewardship plan development, and annual report submission. Beyond that, there are many considerations, such as which entities besides producers (e.g., collectors, recyclers) are allowed to serve on the PRO board of directors, whether a PRO must be a nonprofit organization or can be a for-profit entity, whether one PRO is required or if multiple PROs are allowed and, if so, whether to limit the number of PROs (e.g., by requiring each to represent a minimum market share).

Although most US EPR laws allow producers the option to form one or more PROs, only a handful of pharmaceutical and electronics programs have competing PROs. Government agencies overseeing the two PROs that currently operate drug take-back programs—Med-Project and Inmar—require coordinated plans that are perceived to be one "seamless" program by residents (e.g., one website, consistent branding). States have also taken other steps to ensure that producers provide efficient and coordinated services to the public. For example, in Illinois' electronics program, the state oversees a coordinating body called a clearinghouse, which is funded by electronics manufacturers and organizes electronics collections across the state in multiple service areas that include producers and at least one recycler. The clearinghouse seeks to equitably assign the responsibility (and cost) of collections among six service areas based on trucking distance to collection sites, amounts of electronics collected, and other factors.

The four US packaging EPR laws include slightly different aspects related to the PRO. Maine's law requires the state agency to issue a competitive bid after which it will choose a single entity called a stewardship organization (SO) to operate the system under a 10-year contract; the state does not specify if the SO needs to be nonprofit or should only include producers. The Oregon packaging EPR law takes the more common approach in allowing multiple PROs but requires them to be nonprofits. Both Colorado and California packaging EPR laws require producers to establish one nonprofit PRO in the beginning but allows for multiple PROs after several years of implementation. They also require the PRO boards to include non-voting members from material trade associations (e.g., those representing suppliers of plastics, paper, glass, metal). Other US states are still figuring out which arrangement will work best for them. However, there seems to be a preference for nonprofits—and indeed many packaging bills require PROs to be registered as 501(c)(3) nonprofit organizations—due to their inherent public purpose and perceived transparency.

In Canada, there is no harmonized way in which provincial EPR laws prescribe the nature of a PRO, although there are some similarities. For example, most provinces allow competing PROs but only allow producers on the governing board of PROs since they pay for the recycling system, and to avoid a conflict of interest with those receiving funding.[84] Also, some provinces require nonprofit PROs while others do not. Similarly in Europe, there is a wide range of systems regarding the composition of a PRO, including whether they are comprised only of producers, if they also include waste management companies, and are for-profit or nonprofit. For example, the Czech EPR packaging law has the strictest limits by requiring that only producers can join a PRO, which must be a nonprofit. By contrast, German law is free of any restrictions on the PRO, allowing 11 for-profit PROs run either by waste management companies, retailers, or other private entities.[85]

Advisory Council

A multi-stakeholder advisory council (or committee or board) offers a valuable opportunity to provide non-binding recommendations to PROs and the state oversight agency, with council members most often appointed by the

84. Catherine Abel, Vice President, Strategic Initiatives and Policy, Resource Recovery Alliance, email communication, July 28, 2022.

85. Joachim Quoden, Managing Director, Extended Producer Responsibility Alliance, email communication, September 5, 2022.

head of the state environmental agency to avoid overtly political appointments. The council should include a balance of representatives from local governments, collectors, recyclers, environmental groups, environmental justice organizations, academic institutions, and those in the packaging supply chain. The council usually does not include consumer brand owners on the PRO board or those whose interests are otherwise represented by the PRO. Advisory councils meet on an ongoing basis (e.g., quarterly) to review the stewardship plan and annual reports (see below), and as needed throughout the program.

Although the concept of advisory councils is still new in America, they can be a valuable addition to an EPR program if managed effectively. This is particularly important for packaging EPR given the diverse stakeholders in the complex supply chain, including material sourcing, packaging manufacture, brand owners/producers, and importers/distributors. Others whose input is critical are the collectors and processors that market recyclable materials. For this reason, PSI recommends that a state agency, which has the responsibility to staff the council, consider using internal or external trained facilitators, particularly those with technical expertise, to manage the strong and passionate competing interests vying for control on the council.

Government Oversight and Enforcement

A basic aspect of governance involves whether performance goals are specified in the law, established through rulemaking, or proposed by producers in a stewardship plan submitted to the state oversight agency for review and approval. Those who distrust producers, as well as state agencies that prefer stronger oversight (e.g., Maine and Oregon), may want the state to set the goals through statute or regulation. Other state laws and bills (e.g., Colorado, Maryland, Connecticut, and Vermont) would allow producers to propose specific targets to the state for review and approval within the stewardship plan, delaying debate on actual rates and dates until after bill enactment.

Regardless of how much flexibility a state law gives to producers, however, the state agency always maintains oversight authority in an EPR law. It is the state's job to ensure that producers comply with the law and, if they don't, to take enforcement action. State officials are legally responsible to review and approve (or reject and re-review) a PRO's stewardship plan within a set timetable. They need to verify that goals are met and, if not, work with the PRO and other stakeholders to develop additional strategies to ensure they *are* met. They may also need to keep a list of all obligated producers and handle conflicts over reimbursement transactions. An increasing number of state packaging EPR laws and bills require the state agency to establish a list of recyclable materials that will be collected in all communities to seek consistency in local education and outreach. This government oversight role might also include developing regulations, planning, conducting research, approving stewardship plans and annual reports submitted by producers, ensuring compliance through enforcement, evaluating the program, ensuring data accuracy, and related tasks.

With all environmental laws, the buck stops with the government oversight agency. If they don't do their job, there will be free riders and the system will not provide a level playing field for fair competition. Without an ability to penalize companies and PROs for noncompliance, the state oversight agency would be giving a competitive advantage to those who do not play by the rules over those who strive to follow the rules and achieve intended results. It is impressive how much pride some companies feel about the EPR programs they operate and the goals they achieve. By contrast, I have seen some industries drag out program implementation and make it as difficult as possible. This compliance and enforcement provision is meant for those latter industry sectors and companies.

There are at least four basic enforcement options available to government oversight agencies under an EPR program. The first three pertain to occasions when a PRO or company fails to meet a performance goal, convenience standard, or other such requirement. In these cases, agencies typically follow an enforcement progression from least to most severe: (1) require additional measures in an approved stewardship plan, such as more collection sites or enhanced education; (2) require a stewardship plan amendment or a new plan to be submitted for review and approval; (3) issue financial penalties; and (4) enforce a ban on the sale of covered materials. More recently, those concerned about a state environmental agency's enforcement ability have sought provisions for enforcement by the state attorney general or inspector general. The authority and discretion for an oversight agency to act if a PRO or company is out of compliance must be clearly stated in statute. If financial penalties are sought, the statute must specify what those penalties will be.

A common provision in all EPR laws prohibits a producer from selling, using, or distributing covered materials in the state unless the producer is part of an approved stewardship plan. By extension, this prohibition would necessarily apply to products contained in the packaging covered by the EPR laws. Over the past two decades, however, I do not recall any agency taking this type of enforcement action related to EPR programs in America. Having worked with state and local government officials for most of my career, I know that they use a methodical enforcement process that gives multiple opportunities to the regulated community to comply, including sending notices of noncompliance, convening meetings, and allowing incremental improvement. But if infractions are not corrected, financial penalties have been levied against producers and stewardship organizations, with funds going back into the program and not to the often-raided state general fund. Some bills also allow individuals (e.g., consumers) or producers that are part of an approved plan to take legal action against non-compliant producers or the PRO, known as "private right of action." These steps are rare but are intended to enforce a level playing field and eliminate free rider companies. Other bills and laws (e.g., Colorado) include language that clearly states that nothing in the relevant section of the law "creates a private right of action."

Funding Inputs

This element refers to the way that a PRO funds the cost of providing the EPR program and incentivizing upstream design changes by charging fees to its producer members and managing payments. To establish these fees, the PRO needs to determine how much it will cost to collect, reuse, recycle, and compost all covered materials that are collected from covered entities. It must also estimate the costs of consumer education and outreach, state government oversight and administration, PRO program administration, anticipated consulting services, and other costs as specified in statute, subsequent regulation, or in the stewardship plan.

In the four packaging EPR laws, as well as all US packaging EPR bills, the PRO is required to fund a "needs assessment" that describes the system and costs needed to achieve the intent and goals of the legislation. This report is sometimes conducted by the state (or its independent contractor) and other times by the PRO with government oversight. Since Oregon conducted extensive system analysis prior to proposing legislation, its law requires that PROs fund a needs assessment only for specific elements of the system that were not previously studied by the state (e.g., litter and marine debris). The needs assessment is a key step that will inform the development of the stewardship plan. Although needs assessments have not been specified as such in EPR *product* laws (e.g., paint), PROs operating those programs still conduct a similar assessment prior to program implementation. However, the needs assessment in packaging EPR laws is unique in that it is mandated and follows guidance in statute for what needs to be covered. It is also typically revised every 5 to 10 years.

Once the total system cost is estimated, a PRO can develop the formula to calculate how much each producer is required to pay into the system. Producer fees are calculated based on the types and amounts of packaging materials they use and the cost to manage those materials in the system. These fees are also intended to provide additional financial incentives for producers to use packaging that is reusable, recyclable, or compostable; contains recycled content; eliminates toxics; and otherwise maximizes environmental outcomes and lowers net life-cycle impacts. Since a key intent is to provide a strong incentive for producers to use sustainable packaging, these payments are most often referred to as "eco-modulated fees." For example, brand owners selling a package made from a highly recyclable material that contains a high amount of recycled content will usually be charged less than another company using a nonrecyclable material that also adds system cost through litter. Although they are called fees, they are *not* fees paid by consumers, nor do they end up in government coffers. Producers internalize these costs into their general cost of doing business by paying into a fund most often managed by a PRO on behalf of regulated producers that pay a fee on all covered materials, regardless of whether they are recycled, composted, disposed, or littered.

Eco-modulated fees are outlined broadly in statute in the Oregon, Colorado, and California packaging EPR laws and in most packaging EPR bills. The three laws require the PRO to submit a proposed fee schedule in the stewardship plan for approval by the government oversight agency. The laws require that fees be based on the cost to manage the materials, as well as provide incentives for material attributes such as recyclability, postconsumer content, reuse, reduction, reduced toxicity and litter, and clear labeling. In the Maine packaging EPR law, the fees are set through rulemaking and based on the median per ton cost for municipalities to manage each material.

In Canada, three EPR for PPP programs use a fee-setting methodology developed by the former Canadian Stewardship Services Alliance (subsequently the Resource Recovery Alliance and now Circular Materials) that ensures that difficult-to-recycle materials contribute sufficient funds to advance their adoption in the recycling system or are deterred from use in favor of more sustainable materials.[86] The methodology also incentivizes producers to use materials with higher end-market value that is, therefore, less costly to recycle. For example, the methodology would typically attribute a share of commodity revenues to corrugated cardboard because this material earns value in end markets, whereas no share of commodity revenues is attributed to plastic laminates for which there are few end markets due to the recycling challenges they represent.

Materials that are easy to sort at MRFs, such as steel cans, also often incur lower fees in an EPR system, while those that are harder to sort incur higher fees. In France, for example, dark plastics that are not visible to optical sorting machines trigger a 10 percent "malus" fee, which adds 10 percent to

86. This is known as the Four-Step Fee Methodology. Canadian Stewardship Services Alliance (now Circular Materials) also conducts a Material Cost Differentiation (MCD) study to more accurately assess the cost impacts of materials on the system. The MCD study generates a Material Cost Index (MCI), which is a key input to the Four Step Fee Methodology. For further reading on the MCD study, see: CSSA, "Material Cost Differentiation (MCD) Project," © 2020, https://www.circularmaterials.ca/mcd/.

the base fee that producers must pay into the program.[87] France's PRO for PPP, CITEO, has also implemented a 50 percent fee bonus (i.e., discount) for producers that integrate a minimum of 50 percent recycled HDPE into LDPE films.[88] In Québec, the PRO (Éco Entreprises Québec, or ÉEQ) offers a 20 percent credit to producers for the inclusion of recycled content in certain materials at specified thresholds.[89] ÉEQ also recently established an Ecodesign and Circular Economy team that is developing a model to incorporate environmental criteria into its fee structure.[90] Eco-modulated fees are expected to spark more upstream design changes in the coming years and became mandatory in all EU member states as of January 5, 2023. The European Commission requires PROs to connect fee modulation strictly to packaging recyclability that is differentiated into six categories.[91]

Funding Allocation

PROs use the fees they collect from producers to invest in maintaining and/or expanding the current recycling system in order to accept new materials and a greater quantity of recyclables. Funding under US packaging EPR laws generally covers capital improvements to curbside collection infrastructure, drop-off and public space recycling access, transportation, sorting, processing, and public education. However, each of the four packaging EPR laws, as well as bills introduced, vary somewhat.

For example, under Maine's law, producers will fund the median net cost of municipal collection, transportation, sorting, and processing of a reduced scope of products. In Oregon, however, producers will pay for a wider product scope, contamination reduction and removal, long-distance transportation, some system expansion, and processing, but not collection. Oregon's law also requires producers to establish new depots to collect materials not currently collected by local governments and to ensure that all materials flow to "responsible end markets," which it defines as "a materials market in which the recycling or recovery of materials or the disposal of contaminants is conducted in a way that benefits the environment and minimizes risks to public

87. CITEO, "The 2020 rate for recycling household packaging" (July 2019), https://bo.citeo.com/sites/default/files/2019-10/20191008_Citeo_2020%20Rate_The%20rate%20list.pdf.

88. Axel Darut, CITEO, "Fee modulation in France," March 2019 presentation.

89. Éco Entreprises Québec (ÉEQ), "FAQ: Am I entitled to the credit for post-consumer recycled content?" © 2020, https://www.eeq.ca/en/faq/prepare-report/am-i-entitled-to-the-credit-for-post-consumer-recycled-content.

90. Éco Enterprises Québec (ÉEQ), "2018 Annual Report," released 2019, https://www.ÉEQ.ca/rapportannuel2018/en/index.php#intro.

91. Joachim Quoden, Extended Producer Responsibility Alliance (EXPRA), personal email communication, January 21, 2020. See European Commission's Waste Framework Directive, accessed October 25, 2022, https://environment.ec.europa.eu/topics/waste-and-recycling/waste-framework-directive_en.

health and worker health and safety."[92] The state estimates that producer payments will cover about 28 percent of all system costs in a typical year.[93] In addition to covering many of the basic costs, Colorado's law also covers the cost of disposing of nonrecyclable materials. And while California's law includes funding for many of the basic costs, including enhancements to existing infrastructure, it might not include ongoing operational costs for recycling programs in place before the law took effect.

Another key decision regarding funding allocation is whether, and how much, stakeholders will invest in infrastructure for flexible packaging or compostable packaging made from organic materials such as mushrooms or marine organisms. Unfortunately, standards are currently lacking for the criteria under which funding could be allocated to collect and process these materials under an EPR program. Recycling relies on material strength for repeated trips through the circular economy, whereas compostable materials break down and will likely require a separate collection, transportation, and processing network to make marketable compost that does not contaminate the recycling stream.[94]

One of the most basic characteristics of EPR programs is whether, and to what degree, a PRO will reimburse local governments for their costs to run the system or hire their own contractors to perform those services. Municipalities with robust recycling programs tend to want to maintain at least initial control over their programs to guarantee continued service to residents. They prefer to get reimbursed by producers for their system costs, whether they use municipal staff or contract for services. Since local governments in Maine, Oregon, and California played a significant role in establishing their states' decades-old recycling systems, the packaging EPR laws they enacted are based on municipal reimbursement.

In most states, at least some municipalities do not provide any recycling service for their residents so they cannot get reimbursed by producers. This arrangement is known as "subscription service" because individual households contract directly with a company to collect their recyclables. Under some EPR reimbursement systems, these households will continue paying for that service and won't be reimbursed by the EPR program, but in other such systems, these subscription households could be provided service by their contracted hauler, which would then be reimbursed by the PRO. In Colorado,

92. Oregon State Legislature, ORS 459A.863, https://www.oregonlegislature.gov/bills_laws/ors/ors459A.html.

93. David Allaway, Oregon Department of Environmental Quality, email communication, May 27, 2022.

94. As of July 2022, only Italy has a special sub-PRO (called BIOREPAK) within the CONAI PRO that manages compostable and biodegradable plastics in a separate collection system, https://biorepack.org/.

most households have subscription service. Under that state's full producer responsibility system, producers will cover all recycling system costs and assume full responsibility for managing the system, including hiring companies to collect and process recyclables from households. Since producers were provided greater control over the Colorado system as compared to programs in Maine, Oregon, and California, they were willing to provide more system funding.

Other states, including New York,[95] introduced a "hybrid model" that would reimburse municipalities with recycling programs a "reasonable rate" but also allows households with subscription service to choose to either maintain their current service and continue to pay for it themselves or have the PRO pay for and provide the service. The reasonable rate that a PRO pays to municipalities is determined on a state-by-state basis, usually either set by regulation or proposed by a PRO in the stewardship plan and approved by the state oversight agency after advisory council input. This hybrid model emanated from PSI's packaging EPR system model and multiple facilitated sessions with board members of the New York Product Stewardship Council. It has been useful for stakeholders working on state legislation to understand these three basic conceptual frameworks—municipal reimbursement, full producer responsibility, and a hybrid approach—to create their own state bills.

Some Canadian systems started long ago with producers partially reimbursing municipalities for their cost to operate their recycling programs. The provinces of Ontario and Québec provide a possible path for some US states in that they started their programs with partial municipal cost reimbursement and have since moved to full producer responsibility, with full payment and greater system control by producers over time. In Ontario, the rate at which producers reimbursed municipalities has been 50 percent since 2002, but it is set to increase to 100 percent starting in 2023 as producers assume full financial and operational control of a newly designed system.[96] In Québec, producers initially funded 50 percent of the municipal recycling program cost until 2009, 75 percent by 2010, and 100 percent by 2013, all under a municipal reimbursement system.[97] In 2019, the provincial government announced that it would amend its law to full producer responsibility, giving more system

95. New York packaging EPR legislation was first introduced in 2020 by State Senator Todd Kaminsky.

96. "The story of Ontario's Blue Box," Stewardship Ontario (undated), https://stewardshipontario.ca/wp-content/uploads/2013/02/Blue-Box-History-eBook-FINAL-022513.pdf; Ontario Resource Productivity and Recovery Authority, Blue Box Regulation, under the 2016 Resource Recovery and Circular Economy Act, accessed October 26, 2022, https://rpra.ca/programs/blue-box/regulation.

97. Éco Entreprise Québec (ÉEQ), 2014 Annual Report, 12, https://www.eeq.ca/wp-content/uploads/EEQ_RA2014_NUM_vfa.pdf.

management control to producers starting in 2025.[98] Having followed these provincial trends for years, the British Columbia Ministry of the Environment jumped straight to a system at the outset of its program, launched in 2014. European packaging EPR systems, many of which have been in operation over 35 years, typically follow a full producer responsibility model.

PSI and its government members initially anticipated that the United States could learn from the European and Canadian provincial programs and move directly to a full producer responsibility system, immediately relieving local governments of both a financial and management burden. However, many state and local governments were reluctant to relinquish control of their decades-old recycling systems to packaging producers. Waste management companies also sought to stop or delay the implementation of EPR programs because of their interest in maintaining their current municipal and individual household service contracts. To most of these companies, any change was perceived as a threat to their business model. And, more recently, some environmental groups and other organizations have expressed a strong distrust of producer control, seeking to maintain local and state government control over the recycling system and at times placing a greater degree of responsibility on already overburdened governments. Maine's and Oregon's 2021 packaging EPR laws reflected these dynamics when they began to shape their policies in 2018 to 2019, resulting in strong state and local government control over system operations and a lower degree of producer authority.

As each stakeholder group gets more comfortable with how EPR systems operate and becomes more familiar with the differences among program models, attitudes are likely to evolve. Producers in Canada, Europe, and other areas of the globe still seek greater harmonization of their programs, which will likely influence their expectations for American programs. They also want to establish more uniform operating procedures that incentivize lower material contamination and more efficient and effective services. Further, they may seek a standardized list of materials to be collected and managed across multiple jurisdictions to realize economies of scale. In Europe and Canada, service providers help multiple PROs calculate and set fees, and often offer a suite of additional services. For example, Canadian PPP PROs in four provinces contract with Circular Materials. This organization provides administration, compliance, and material management services to roughly 3,000 packaging producers. Similarly, about 30 PROs in Europe, as well as several others in non-European countries, subscribe to the Extended Producer Responsibility Alliance (EXPRA), which facilitates cooperation between

98. Éco Entreprise Québec (ÉEQ), accessed October 26, 2022, https://www.eeq.ca/en/enactment-of-the-regulation-for-the-modernization-of-curbside-recycling-companies-and-partners-invited-as-of-now-to-engage-with-eco-entreprises-quebec/.

members, including best practices. In America, we may yet transition in a similar direction as programs established earlier in Canada, Europe, and around the world. And, at the same time, as every country likes to do, we will establish our own unique brand of packaging EPR programs.

Design for Environment

Although eco-modulated fees will incentivize producers to use packaging that has lower life-cycle environmental impacts, other statutory provisions can further minimize the environmental and health impacts of covered materials. Some US packaging EPR laws or bills reference the Toxics in Packaging Act[99] (passed in 19 states), which prohibits the intentional use of cadmium, lead, mercury, hexavalent chromium, and per- and polyfluoroalkyl substances (PFAS). Other provisions require the state agency to develop a standard list of toxic substances or chemicals of concern; if packaging contains one or more of these listed substances it will not be considered recyclable, compostable, or reusable, and may be banned.

Performance Goals

There are many aspects of an EPR law that can lead to successful programs, but we will only know whether our goals are being met when we evaluate the results. Designing an EPR law is akin to building a house, with the law being like the architectural plan that outlines the foundation, the size and layout of rooms, and the shapes and proportions of the structure. But until we see the completed job and have lived in it for a while, we won't know if the house meets our expectations. Do the rooms flow as depicted in the plans? Is it well constructed and insulated? Are the materials strong and sustainable? Performance goals are the way we will know if an EPR bill provides the benefits we expect.

While it is considered best practice to establish performance goals in the law itself, a secondary option is to develop them by regulation following bill enactment. A third option, which gives more leeway to producers, is for the PRO to propose goals in the initial stewardship plan. The aim of policy architects is to establish goals that are ambitious but achievable. To do so, they need to have baseline data, which is best established before introducing an EPR bill, but is sometimes developed as part of a needs assessment or in a rulemaking process. If too much time is spent debating what goal is achievable, program implementation may be delayed.

---

99. Toxics in Packaging Clearinghouse, https://toxicsinpackaging.org/the-clearinghouse/state-members.

Standard goals included in US packaging EPR laws include the amount of covered material to be collected (i.e., recovered) from covered entities, as well as the amount that is recycled, which refers to the amount sent from recycling facilities to recycling markets to be used for new products and packaging. Other standard goals include material reuse, recycled content, and toxics reduction, which can be measured across all, or a subset, of covered materials. Additional goals can be set for source reduction of packaging and single-use products. For reuse, recovery, and recycling targets, performance is calculated relative to the amount of material that producers place on the market. The difference between the amounts recovered and recycled indicates the rate of contamination (i.e., compromised, degraded, or non-recyclable material), representing an important metric to determine the efficiency of program operations. We clearly don't want to fill recycling trucks with junk to be landfilled or incinerated.

A statute or state regulatory process often will include achievable minimum/baseline material-specific targets for reuse, recovery, and recycling as a starting point for the program, with the PRO to propose updated material-specific targets over time via the stewardship plan. If performance targets need to be revised, the PRO will explain why they could not be reached or why they were exceeded. If the state and the PRO do not agree on performance targets, the state has the authority to modify the submitted stewardship plan by incorporating new performance targets and/or requiring that additional measures be taken by the PRO (e.g., investing in new or upgraded technology, additional collection sites, or enhanced education and outreach). An annual independent audit will verify progress made toward meeting performance standards and inform the development of updated targets.

Performance targets for packaging are relied on by policymakers to a much greater degree than targets for consumer products (e.g., electronics) because the amount of packaging sold into the market more closely represents the amount available for collection. One purpose of packaging is to protect a product during shipping and until it is purchased and used by a consumer. Products, by contrast, are used by consumers for much longer periods of time, which could be months (e.g., batteries), years (e.g., carpet), or decades (e.g., thermostats). The factors that determine how many of those products are available for collection requires a bevy of assumptions. For example, home thermostats are often changed out due to building renovations, which are influenced by the strength of the economy, as well as new technologies that provide greater efficiency and convenience. Those calculations can be modeled but the results can also be fraught with uncertainty.

Products that have a reuse market, such as paint and mattresses, add yet another variable that further complicates the amount of those products available for collection. These assumptions are often challenged by PROs concerned about falling short of goals and incurring financial penalties, which can delay program implementation and environmental protection. For this reason, convenience standards have become more prevalent in the measurement of EPR programs for consumer products, while packaging EPR laws still contain both convenience standards and performance targets.

Outreach and Education

These requirements include awareness and informational campaigns that ensure that consumers, retailers, and other key stakeholders are informed about the EPR program and how to take action supported by reasons why packaging reduction, reuse, and recycling are important. This element requires producers to inform and educate consumers across the state about proper end-of-life management for covered materials, as well as the location and availability of curbside and drop-off collection locations. Recycling instructions should be consistent statewide, easy to understand, and easily accessible. This includes consistent labeling of packaging as recyclable or compostable in accordance with the actual ability in the state for it to be recycled or composted.

The stewardship plan is the place where producers propose how they will evaluate the reach and effectiveness of multimedia consumer education campaigns over time. Educational materials will typically include a website with recycling instructions specific to the state and be accessible to diverse populations. Since so much involves on-the-ground outreach and education, PROs coordinate these activities with local solid waste management entities. It behooves these entities to periodically survey residents (e.g., through intercept interviews) to determine if their actions can in fact be attributed to outreach efforts and, if so, which efforts were more effective. Producers will reveal in an annual report whether performance targets are being met and, if not, the oversight agency has enforcement options, as explained in the governance section above, ranging from requiring additional measures, an amended or new plan, or financial penalties. The agency's main goal is to ensure program effectiveness and compliance.

Equity and Environmental Justice

While there is significant US and global experience with many of the other EPR bill elements, there is still limited but growing knowledge about how to achieve equity and environmental justice in the context of EPR laws. These provisions might include a requirement for a state to conduct an equity study to improve access to recycling for underserved communities (e.g., multifamily residences) and another study on how to improve social equity within the recycling system (e.g., by ensuring equitable employment and economic development opportunities). It could also include a requirement that all contracts entered into by the PRO include language guaranteeing a livable wage and quality benefits to workers. Other considerations include educational materials that are adequately translated in all major languages spoken by local populations, and that the collection infrastructure be accessible to disabled residents (e.g., ADA compliant).

An effective EPR system should reduce environmental, health, and social impacts across the entire life cycle of products and packaging, upstream and downstream. Therefore, these laws should ideally also include a requirement that the covered materials used in packaging be certified as responsibly sourced (e.g., paper certified by the Forest Stewardship Council) and that the PRO (or municipality under some cost reimbursement systems) certify that materials are processed and transferred to responsible facilities and end markets that meet recognized environmental and public safety standards, such as the Ten Principles of the UN Global Compact.[100] The inclusion of environmental justice organizations on the advisory council will add ongoing expertise and insights into how to best serve residents so that EPR programs can produce results that are equitable and just for all.

Stewardship Plans

Under a US packaging EPR law, producers are required to submit a stewardship plan to the state oversight agency, either individually or collectively (under a PRO), describing how they will meet their obligations under the law. While an EPR statute can be viewed as the architect's structural blueprint for a program, the stewardship plan is akin to a general contractor's implementation plan that provides greater detail to guide program operation. A plan must be approved by the governing authority, which will do so only if the plan meets the requirements and intent of the law. Since this must occur before the EPR program can be launched, it is a critical aspect of the system. A typical stewardship plan includes the following:

- how covered materials will be collected from covered entities and where they will be processed;
- responsible parties covered under the plan and the stewardship organization structure;
- funding mechanisms, including how material fees will be structured and collected;

100. United Nations Global Compact, accessed March 10, 2022, https://www.unglobalcompact.org/what-is-gc/mission/principles.

- how program funds will be dispersed to pay for all covered costs, including to municipalities;
- methods for reaching performance targets, including plans for market development;
- how materials will be collected to meet consumer convenience and geographic coverage;
- consumer education and outreach;
- sound management practices for worker health and safety;
- design-for-environment provisions;
- how producers will work with existing recycling programs and infrastructure; and
- how producers will consult with state and local governments, the advisory council, and other important stakeholders.

It is typical for plans to be discussed at length between the PRO and the oversight agency prior to plan approval. Those on the advisory council provide input on the plan directly to the PRO and/or to the state. Since the council is usually only advisory, those heavily impacted by the law, particularly recycling companies, often seek representation on the PRO board of directors as non-voting members. The PRO usually submits a draft plan first for review to the advisory council, after which the state oversight agency must decide if it will approve the plan, require changes, or reject it. It is best practice to require the PRO to respond in writing to recommendations it does not accept from the advisory council. This should take place prior to the oversight agency's decision on the plan.

A four- or five-year initial plan is standard, with subsequent plan updates submitted for approval at four- or five-year intervals, or as requested by the oversight agency. The agency can require a plan amendment prior to the five-year interval if targets are not being met or circumstances change in a significant way. For EPR systems in which the PRO manages day-to-day operations (e.g., Colorado), the stewardship plan will also include a customer service plan, including answering citizen or customer questions and resolving issues.

Annual Report

To know whether an EPR program is successful, it needs to be evaluated against the goals and expectations set out in the law and stewardship plan. For this reason, producers are required to report annually on a range of metrics and methods included in the law and stewardship plan. In this way, the report allows stakeholders and the public to determine how well the program performed during the past year relative to set performance goals. The report should include the quantity of materials processed domestically and in overseas markets, the location of processing facilities, and a certification of compliance with internationally recognized health, environmental, and worker safety standards. The report should disclose the amount of contamination in the recycling stream; the amount and destination of disposed materials; collection service vendors, collection locations, population coverage, and accessibility; program expenses; educational efforts and results; and customer service results.

Annual reports should also include an independent financial audit. Since data tracking and reporting are also program costs and a time commitment, the law should seek to require producers to only provide data that are truly needed to evaluate the program. A good annual report will be one that balances the interests of the full stakeholder group, including producers, governments, environmental justice and equity groups, and others. Packaging EPR laws are new to the United States, and we should view them as subject to change based on new data, insights, and understanding. Therefore, the law should allow the oversight agency to require additional information it deems appropriate to include in the annual report for the purpose of program evaluation.

A good EPR law will be one that is flexible enough to add metrics and methods for program operation and evaluation over time so that it provides the data that stakeholders need to evaluate the program and to modify operations as needed to meet stated goals. Ideally, the PRO(s) will meet with the advisory council and the oversight agency to jointly determine a basic annual report format that can be used by PROs in all states to harmonize methods, reduce costs, and enhance program comparison. PSI facilitated the development of such an annual report template with PaintCare, a PRO that operates paint EPR programs, and a range of states with existing paint EPR laws. A similar template could be used by PROs operating other EPR programs, including packaging.

ADDITIONAL COMPONENTS

- *Implementation timeline*: Not only do EPR laws specify stakeholder roles, but they lay out the time frame in which each step is to take place. Packaging EPR laws include a host of interconnected activities that rely on multiple partners for successful implementation. Dates are specified for the completion of a needs assessment, producer registration, submission of a stewardship plan and the agency's response timeline, regulatory dates, frequency of annual report and stewardship plan submission, and other important program aspects.

- *Anti-trust and competition*: EPR laws allow, and sometimes require, producers to collaborate and determine costs and fees, although they are also competitors. To protect a producer or stewardship organization from liability for any claim of a violation of antitrust, restraint of trade, or unfair trade practice (e.g., "price fixing"), the law will provide an exemption for a violation of antitrust laws to the extent the producer or stewardship organization is exercising authority to carry out the provisions of the law. In return, the state will supervise and oversee the PRO. This provision is standard in all EPR laws and does not in any way promote unlawful or monopolistic behavior, as some EPR opponents have claimed.
- *Preemption and related laws*: The law should be as compatible as possible with existing state programs, regulations, and laws, including a deposit return system, pay-as-you-throw, toxics in packaging, and other related systems. It should also strive for statewide consistency. However, EPR for packaging laws should not automatically preempt local legislative authorities from imposing additional standards or restrictions on products and packaging unless it is part of a strong negotiated state agreement. My experience has been that municipalities will accept local preemption, and the consistency and certainty that it provides producers, in exchange for a strong statewide law. However, if the state bill does not contain certain provisions that locals want to impose (e.g., plastic bag bans), many local authorities will not want to be preempted. Legislation should also address any regulatory hurdles that existing laws may impose that would inhibit collection, transportation, and recycling of packaging. It should also not intentionally or inadvertently incentivize disposal over recycling, or recycling over upstream waste prevention or reuse.
- *Authority to promulgate regulations*: Many EPR laws include a provision that recognizes an oversight agency's ability to promulgate regulations necessary to implement the program. In reality, some states show a proclivity to develop regulations while others stay clear of regulations due to the added time and cost, perceived complexity, or aversion to government control.
- *Financial data and proprietary information protection*: The PRO and the state must protect confidential proprietary information provided by customers, service providers, responsible parties, and any other commercial entities participating in the program.

KEY TOPICS

Policy Flexibility

A fundamental aspect of developing policy is whether a provision is specified in statute, regulation, or the stewardship plan. Each of these three options represents a degree of policy flexibility and a time period in which decisions are made and the system is implemented. Typically, the most basic aspects of a bill are specified in statute (e.g., covered materials, covered entities, and producer definition). Policy set in statute provides ultimate authority to the government oversight agency, and if legislative text is clear, stakeholders can implement the program more quickly.

But if the legislature and oversight agency are not certain about the policy details, authority can be delegated to the agency to shape the policy further through a regulatory (i.e., rulemaking) process, often with guidelines laid out in statute. Since Maine and Oregon were the first states to pass EPR packaging laws, these pioneers wanted more time to delve into certain specifics. For example, Oregon is developing regulations to determine standards for consumer collection convenience, which materials will be collected for recycling, the reduction of contamination of recyclables, and certain aspects of funding. In Maine, regulations are being developed to determine the formula by which municipalities will be reimbursed by producers, as well as performance targets for reuse, recycling, and collection. Since the environmental agencies in these two states will determine implementation procedures through the rulemaking process, the agencies have a greater degree of control of the outcome.

Alternatively, if an agency does not want to develop regulations to elucidate policies, they may prefer that producers propose policy and program details through the PRO stewardship plan submitted to the agency for approval. For example, if a state wants more time to develop an appropriate standard for waste reduction, they could put the onus on the PRO to propose in the stewardship plan a specific rate at which packaging will be reduced by a specific date (e.g., "a 25 percent reduction of covered materials as measured from the baseline rate, by 2027"). The PRO would also have to propose how the baseline rate would be calculated. This approach gives producers a greater opportunity to provide the rationale for their proposal but also takes a burden off the agency to be the lead, as it would be in a rulemaking process. In this case, while producers have more flexibility to propose policy, the administrative agency still maintains ultimate authority during the approval process before the program can begin, often after input from a multi-stakeholder advisory council.

Cost of Consumer Goods

One of the biggest arguments levied against EPR packaging laws is that they raise consumer prices. This theory has arisen from those who strongly oppose a change to the recycling system. Their argument is simple: EPR will increase costs on brand owners, which will raise the price of goods, and this will disproportionately impact low-income households. It may sound convincing to some, but it is a false narrative.

Under the current recycling system, consumers pay three times: *first* for the packaging of the products they buy (included in the cost of goods), and *a second time* for the collection, recycling, or disposal of that packaging through their municipal taxes or private subscription costs. Consumers, as well as all citizens, also pay *a third time* for the cost of pollution associated with the production and postconsumer management of packaging waste.

Further, there are few controls over these costs, which have only increased and are already being passed on to consumers, taxpayers, and ratepayers. Under EPR, these costs can decrease because producers will have financial incentives to use more sustainable materials that have less environmental impact and greater postconsumer value. In addition, the associated costs of disposal, litter, and other pollution will decrease. EPR laws will also pump tens of millions of dollars into modernizing and expanding state recycling systems, which will translate into a more efficient system, providing consistent product labeling and consumer education, and stabilizing end markets. These investments will create significant value for the entire recycling system, including consumers. It is no different than a company investment in a facility retrofit to reach great efficiency.

In addition, there is no evidence that consumer prices have risen noticeably due to packaging EPR laws in Canada and Europe, where these systems have been operating for up to 35 years. There has been one Canadian study, from York University in Ontario, that used a modeling approach—not actual data—to predict that consumer prices will rise due to EPR.[101] This study assumes that producers will pass 100 percent of their costs onto consumers in the province where EPR is located, with no spreading of costs across regional or national customer bases. By contrast, a 2021 study conducted for the Oregon Department of Environmental Quality by Resource Recycling Systems (RRS), a US consulting firm, compared actual product prices from

101. Dr. Calvin Lakhan, York University, "Modeling impact on consumer packaged goods pricing resulting from the adoption of Extended Producer Responsibility in New York State" (undated), accessed October 12, 2022, https://wastewiki.info.yorku.ca/study-examining-the-economic-impacts-of-epr-legislation-for-packaging-waste-in-new-york-state/.

store shelves in several Canadian provinces.[102] The RRS study found that in most cases consumer prices for the same goods remained constant and were not dependent on whether a province had an EPR law. A more recent study by Columbia University in 2022 found that, even if EPR compliance costs were to lead to the doubling of packaging costs for producers, the upper bound of the cost increase to consumers would be roughly 0.69 percent of grocery spending, or a maximum of $4 per household per month.[103]

Those operating packaging EPR programs in Europe over the past 35 years also did not find verifiable increases in consumer goods prices resulting from EPR policies. Joachim Quoden, Managing Director of the Extended Producer Responsibility Alliance (EXPRA), told me the following: "In Europe, we have not seen any noticeable increases in costs for consumers. At the same time, municipalities have avoided having to increase local waste taxes, even while services for the collection of packaging waste increased."[104] EXPRA is an organization of 31 nonprofit packaging and packaging waste recovery and recycling systems from 29 countries, including 19 European Union member states and Québec in Canada. EXPRA members are PROs that represent regional and global consumer brands that must comply with EPR laws.

Finally, it is always a producer's choice whether to pass costs on to consumers, just like any other cost they incur in product manufacture, whether fuel prices, labor, rent, or material costs. Consumer product companies weigh numerous factors in deciding how much, if any, cost to pass on to consumers, particularly the degree to which a cost increase will reduce consumer demand for a particular packaged product.

Chemical Recycling

The terms *advanced recycling*, *chemical recycling*, and *molecular recycling* are often used interchangeably to refer to a wide range of technologies that process recovered plastic packaging and products containing plastics (e.g., flexible packaging, carpet, textiles), into multiple materials. Although there are no universally agreed-upon definitions, "chemical recycling" is the most

102. RRS, "Impact of EPR Fees for PPP on Price of Consumer Packaged Goods" (revised February 19, 2021), accessed October 12, 2022, https://www.oregon.gov/deq/recycling/Documents/rscRRSconsumer.pdf.

103. Satyajit Bose, "Economic impacts to consumers from extended producer responsibility (EPR) regulation in the consumer packaged goods sector," Columbia University (June 27, 2022), accessed October 12, 2022, https://academiccommons.columbia.edu/doi/10.7916/n2af-vv87; funded by The Recycling Partnership.

104. Joachim Quoden, EXPRA, email communication, November 15, 2022.

commonly used term[105] and refers to one of three technology types: purification, depolymerization, and conversion. Purification facilities use chemical solvents to produce plastic resins that are free from additives and dyes and produce only plastics. Depolymerization breaks the molecular bonds of plastics to recover building blocks that can be reconstructed into "like-new" resins. These facilities can create both plastics and materials that can be converted to fuel. Conversion technologies (i.e., pyrolysis and gasification) convert plastics into refined hydrocarbons and petrochemicals, which are used to produce mostly fuel today but could also produce plastics.[106]

These technologies have become a lightning rod in the United States. The lack of clarity and understanding about these three technology categories, the strong and divergent interests of stakeholders, and the sheer magnitude of the ramifications of permitting decisions have resulted in what can resemble a cartoonish brawl that has whipped up a cloud of dust while fists fly and heads bob. On one hand, some brands and plastics manufacturing companies have already invested millions of dollars into the development of these technologies, claiming that they expand end-of-life options for plastics and exceed the limitations of traditional mechanical recycling by enabling infinite processing without loss of quality.[107] Many companies seek even greater investments, including public funding at the federal, state, and local levels, to accelerate the pace of these developments.[108] Others go even further and seek state legislation (already enacted in 20 states)[109] to classify facilities using any of the three technologies as manufacturing plants, which would require a less onerous permitting process than a standard recycling facility.

Many environmentalists, however, decry these technologies as "greenwashing," viewing it as a way for the plastics industry to claim environmental

105. Industry practitioners have largely coalesced behind the following definition of "chemical recycling" for plastics: "includes, but is not limited to, manufacturing processes such as pyrolysis, gasification, depolymerization, solvolysis, catalysis, reforming, purification, hydrogenation, dissolution, dehydrochlorination, and other similar existing or newly developed technologies or processes that convert waste plastic materials into a feedstock to be used in the production of new polymers, monomers, intermediates, or other materials." Eastman, *Circular Economy Glossary*, updated August 15, 2021.

106. "Transitioning to a Circular System for Plastics: Assessing Molecular Recycling Technologies in the United States and Canada" (undated), Center for Circular Economy at Closed Loop Partners, https://www.closedlooppartners.com/wp-content/uploads/2022/09/Molecular-Recycling-Report_FINAL.pdf.

107. Alexander H. Tullo, "Companies are placing big bets on plastics recycling. Are the odds in their favor?" *Chemical & Engineering News* (October 11, 2020), https://cen.acs.org/environment/sustainability/Companies-placing-big-bets-plastics/98/i39.

108. Plastics Industry Association, *RECOVER ACT: Realizing the Economic Opportunities and Value of Expanding Recycling* (2019), accessed January 4, 2022, https://www.plasticsindustry.org/sites/default/files/2019%20Recover%20Act%20Flyer.pdf.

109. Megan Smalley, "Two states pass advanced recycling legislation," *Recycling Today* (July 5, 2022), https://www.recyclingtoday.com/article/missouri-new-hampshire-pass-advanced-recycling-legislation/.

benefit while continuing to expand and undermine efforts to eliminate single-use plastics.[110] Some even seek to prohibit the permitting of all three technologies because, to them, none make economic or environmental sense and, therefore, are not worth the gamble. Many also believe that investments into chemical recycling infrastructure will prolong the use of petrochemicals to manufacture new plastics, and that investments in plastics recycling continue our dependence on plastics; rather, they want to focus on upstream waste prevention and product/packaging redesign.[111]

Permitting a facility using one of the three technologies is a heavy responsibility for government regulators. State and local government agencies have, for the most part, taken a middle path, prohibiting technologies that process plastics into fuel from being considered recycling, while being willing to consider allowing the designation for materials that can be made into marketable plastics. These officials want to encourage technologies that convert plastic packaging and products into marketable plastics, as long as the net environmental benefit is greater than current alternatives (e.g., mechanical recycling in a typical materials recovery facility). They want to know that the investments, energy, and resources needed to scale up chemical recycling technologies will result in a more sustainable economy with reduced environmental impacts. Toward this end, PSI has begun to develop criteria that government officials can use to determine whether, and how, to permit "chemical recycling" technologies that produce marketable plastics.

In Europe, where EPR has been active for over 35 years, PROs running EPR systems are still at the stage of investing in research and development of various technologies.[112] Determining the right path for the United States will require continued interaction with our international colleagues and the development of new permitting approaches. Ongoing dialogue among stakeholders—consumer brands, plastics production companies, recyclers, environmental advocates, government officials, and others—will help to determine whether, and if so how, to regulate these emerging technologies. Our goal should be to lower overall environmental, social, and economic impacts, including greenhouse gas emissions, and allow flexibility as new

---

110. GAIA, *All Talk and No Recycling: An Investigation of the US "Chemical Recycling" Industry* (2020), https://www.no-burn.org/wp-content/uploads/All-Talk-and-No-Recycling_July-28.pdf.

111. Ivy Schlegel, *Deception by the Numbers,* Greenpeace (September 9, 2020), https://www.greenpeace.org/usa/wp-content/uploads/2020/09/GP_Deception-by-the-Numbers-3.pdf.

112. For example, CITEO (the packaging producer responsibility organization in France) formed a consortium in 2019 to "promote eco-design, recycling and recovery projects for plastic and paper." Involvement includes petrochemicals company Total, U.K. plastic recycling company Recycling Technologies, and global brands Nestlé and Mars. "The partners will examine technical and economic feasibility of recycling complex plastic waste such as small, flexible and multilayered food-grade packaging." "Cross-industry consortium to study plastic chemical recycling in France," *Plastics News* (December 10, 2019), https://www.plasticsnews.com/news/cross-industry-consortium-study-plastic-chemical-recycling-france.

information is acquired. We certainly do not want to burden the next generation by locking into a bad decision that is made hastily today.

DEPOSIT RETURN SYSTEMS (BOTTLE BILLS)

In America, there are 10 state Deposit Return Systems (DRSs), with redemption rates ranging from 42 to 75 percent;[113] of the DRS states, only Maine, Oregon, and California have also enacted packaging EPR laws. Although the first state EPR for packaging laws were enacted in 2021, DRSs by contrast were enacted much earlier in the United States, starting over half a century ago in Oregon (in 1971) and well before the US EPR movement started. DRS policies have proven extremely effective at reducing litter and increasing beverage container recycling, providing a reliable, high-quality stream of recyclables for end markets. Over the past 50 years, however, the beverage industry has introduced new types of containers into the market, and the impact of inflation on redemption values has significantly decreased the return rate of containers.

In the 1990s, when I managed the Massachusetts DRS, the redemption rate for containers was about 72 percent. By 2023, that rate was about 42 percent, the lowest in the nation, and in that year the buying power of that original nickel deposit was only two cents. Massachusetts is not unique: Many state deposit return programs have become ineffective, and an immense number of beverage containers continue to be disposed. Or worse, littered. These deficiencies have led advocates in several states (e.g., Container Recycling Institute, US Public Interest Research Group [US PIRG], state PIRGs, Sierra Club) to seek to "modernize" DRS systems by expanding what is covered (e.g., adding water, spirits, and/or wine), raising the redemption value (e.g., from five to at least 10 cents and adjusting periodically for inflation). The beverage industry has responded by proposing that producers take greater management control of the system, much like they do in packaging EPR systems.

Initiatives to promote, enact, and expand DRSs are still contentious among the beverage industry, environmental advocates, waste management companies, and local governments. A key issue is that a DRS would transfer valuable beverage container materials from the current recycling system to an alternative collection system, reconfiguring the current financial arrangement as to which entity gains or loses revenue and control. Another key challenge to integrating packaging EPR with a DRS is that multiple constituencies vie

113. CIRT, National Bottle Redemption Legislation (July 29, 2022), https://www.cirt.tech/blog-posts/national-bottle-redemption-legislation.

for political attention and compete for their bill to pass. In fact, some legislators may not want to consider the two bills in the same session, perceiving the two approaches as competing rather than complementary.

There is no reason why a DRS and EPR for packaging cannot exist side by side, and there are at least three options for integrating the two policies: (1) combine a DRS with a packaging EPR bill and introduce them as one bill; (2) introduce two separate but complementary bills in the same legislative session—a packaging EPR bill and either an amendment to an existing DRS or a new DRS bill; and (3) introduce either a DRS bill or a packaging EPR bill in one session, then introduce the other bill in the subsequent session. Integrating the two policies through any of these approaches requires planning, stakeholder engagement, good communication, and perseverance.

Some advocates in non-DRS states have sought to exclude *all* beverage containers from packaging EPR bills because they believe that including them reduces the political will to enact a new DRS. However, if *all* beverage containers were excluded from packaging EPR bills, many of those containers would still be put into recycling bins. They would then become "free riders": recycling of those materials would be paid by producers of other packaging materials that are covered under the EPR law, while producers of the beverage containers pay nothing. To address this concern, some EPR bills specify that nothing in the bill shall preclude the future passage of a container deposit/return system. Another strategy might be to place stringent recovery, reuse, and/or recycling targets on beverage containers that, if not met, would trigger the requirement to implement a new or expanded DRS.

Due to the heightened global popularity of DRSs and packaging EPR, stakeholders in the United States have an opportunity to knit together these two highly complementary and powerful policies. In fact, in 2023, Washington became the first state to file a bill that would integrate a DRS with a packaging EPR bill. An option to state-by-state bills is a national bottle bill that can harmonize the system of beverage container return across the country. Such a provision was included in the Break Free from Plastic Pollution Act, a federal bill introduced in 2020 that included EPR at its core, along with other provisions aimed at reducing plastics production, banning single-use plastic packaging, and other items. There is certainly a growing demand for consistent, high-quality recycled materials, and a great deal is riding on how quickly we can meet this demand.

COMPOSTABLE PACKAGING

The global push-back against plastic packaging has resulted in some companies pivoting away from petroleum-based plastics and embracing packaging that is made from plants and bio waste, and that also may be compostable and/or degradable. These packaging types look and perform like plastic packaging, but many require specialized conditions to break down. Some feature labels that mislead consumers into thinking that they are structurally similar to fruits and vegetables, giving the impression that they are environmentally friendly. In reality, the conditions often needed for some of this packaging to break down may only exist in commercial or industrial facilities, such as anaerobic digesters.[114] Even some certified compostable items do not break down in some types of compost facilities, leaving a finished compost that can be contaminated with bits of partially degraded material.[115]

More recently, packaging has also been made from mushrooms, shrimp shells, and other bio-based materials. The technologies for developing these materials are so new that waste management regulations have not fully caught up, often leaving companies in a state of regulatory limbo. Although there are national standards for bio-based plastics—including ISO (International Organization for Standardization), CEN (European Committee for Standardization), and ASTM (American Society for Testing and Materials)—as well as labeling laws in Minnesota, California, Maryland, and Washington State[116]—there are still many challenges to ensuring that bioplastics are truly "compostable."[117]

Since compostable packaging is designed to decompose, it must be kept separate from materials meant to be recycled, which are designed for strength. Given that most people cannot discern these materials, the two types of packaging often mix, resulting in a contaminated load that has reduced value and often must be disposed. For these and other reasons, compostable packaging, particularly degradable plastics, have largely been scorned by local governments, composters, and recyclers due to the extra cost they add to composting and recycling systems.[118]

EPR systems offer an on-ramp to compostable packaging, much like they do for materials not currently recycled, such as flexible packaging.

---

114. US Environmental Protection Agency, accessed October 18, 2022, https://www.epa.gov/trash-free-waters/frequently-asked-questions-about-plastic-recycling-and-composting.

115. "A Message from Composters Serving Oregon," accessed June 6, 2022, https://www.oregon.gov/deq/mm/Documents/MessagefromComposter-En.pdf.

116. Labeling laws pertaining to use of the terms *biodegradable*, *compostable*, and related claims about plastic packaging and/or plastic products have been enacted in Minnesota (https://www.revisor.mn.gov/statutes/cite/325E.046), California (https://oag.ca.gov/sites/all/files/agweb/pdfs/environment/ag_website_environmental_claims.pdf), Maryland (https://mgaleg.maryland.gov/mgawebsite/legislation/details/hb1349?ys=2017rs), and Washington (https://lawfilesext.leg.wa.gov/biennium/2019-20/Pdf/Bills/Session%20Laws/House/1569-S.SL.pdf).

117. European Bioplastics, "What are the required circumstances for a compostable product to compost?" accessed October 20, 2022, https://www.european-bioplastics.org/faq-items/what-are-the-required-circumstances-for-a-compostable-product-to-compost/.

118. "A Message from Composters Serving Oregon," accessed June 6, 2022, https://www.oregon.gov/deq/mm/Documents/MessagefromComposter-En.pdf.

Unfortunately, the collection and processing infrastructure for these new materials is not widespread.[119] Even so, under EPR, producers of compostable packaging must still pay into the system. And they, too, want to benefit through investments in new collection and processing infrastructure for their packaging. One of the key challenges as this issue relates to packaging EPR laws is to develop language in statute, regulation, and stewardship plans to guide the outcomes we want to achieve: bio-based packaging materials that are collected and processed in a system that results in high-quality compost, and is low-impact from a life-cycle perspective.

ELIMINATING TOXIC SUBSTANCES IN PACKAGING

Another key interest—the elimination of toxic substances in packaging—can either be addressed in a packaging EPR bill or in a stand-alone complementary bill. As mentioned in an earlier section, multiple states developed model legislation in 1989 to prohibit the intentional use of four heavy metals—lead, mercury, cadmium, and hexavalent chromium—in packaging and packaging components sold or distributed throughout the states. Legislation based on this model has been enacted in 19 states, and the latest model revision occurred in 2021 to add the class of perfluoroalkyl and polyfluoroalkyl substances (PFAS) and ortho-phthalates as regulated chemicals. The update also included new processes and criteria for identifying and regulating additional chemicals of high concern in packaging.[120] The Toxics in Packaging Clearinghouse manages the data generated as a result of the laws, for use by state officials, and packaging EPR laws and bills often require packaging to be in compliance with the legislation.

Integrating toxics reduction into EPR packaging and product bills is still evolving. The 19 states that have enacted the toxics in packaging model bill already have the ability to add toxic substances to the list of regulated materials. Other states should consider introducing that model bill. Since reducing toxics is a complex and substantive component to include in packaging EPR bills, most producers and government agencies seek parallel but separate efforts—allowing the packaging EPR bills to proceed while continuing to map out actions to reduce toxics. One thing most agree on is that we don't want a packaging EPR law to enable the reuse or recycling of toxic materials. Instead, these materials should be removed from the waste stream and

---

119. Biodegradable Products Institute, accessed October 18, 2022, https://bpiworld.org/Infrastructure.

120. Toxics in Packaging Clearinghouse, accessed October 20, 2022, https://toxicsinpackaging.org/model-legislation/.

disposed, and provisions put in place to eliminate the reintroduction of toxics into new products and packaging.

MATERIALS MANAGEMENT

One of the more subtle technical aspects of solid waste management emerged as a challenge to the concept of a circular economy, which assumes that materials used are reused, recycled, or composted in a circular loop, reducing the need to extract raw materials. Since EPR laws include performance goals to measure reuse, recycling, and composting, these laws promote a circular economy. That circular concept can sometimes be at odds with another concept known as sustainable materials management, or SMM which, according to US EPA is "a systematic approach to using and reusing materials more productively over their entire life cycles."[121]

Taking an SMM approach could mean that, under certain circumstances, reusing or recycling materials can create *more* environmental impacts than disposing of those materials. For example, using an analytical tool such as life-cycle assessment, one might find that collecting a type of packaging (e.g., polystyrene), transporting it to a recycling facility, and processing it might create greater environmental impacts than disposing of that packaging in a landfill or waste-to-energy plant. To some dedicated recyclers, this type of analysis appears to lead in the wrong direction, especially if pollution impacts from packaging disposal or litter are not considered. To many, simultaneously taking into account two similar but potentially dissonant viewpoints can be difficult, even confusing. While it is important to acknowledge that not all materials should be recycled if an independent and credible life-cycle assessment shows disposal is a better option, it can also lull us into a static perspective of conditions. Rather, we should consider changes that are needed to the system to alter analytical results and bring the circular economy in line with an SMM approach.

121. US Environmental Protection Agency, accessed November 1, 2022, https://www.epa.gov/smm.

Conclusion

Recycling old stuff into new stuff is magical, and worms composting food scraps into soil is miraculous. Turning back the clock to reuse is fascinating, and using less material is intriguing. All are gratifying, hopeful even. These actions can reduce impacts from material use to achieve a balance with the capacity of earth and its inhabitants to replenish those resources—to reach a "materials use equilibrium." Once we do, we will have created a sustainable social contract between industry and the public. Today, we are out of balance amid an escalating erosion of our social fabric and a pervasive distrust of industry and government.

Throughout history, improvements in waste management have been incremental. We used to throw waste in a field outside of town. We then stuck it in an unlined pit or burned it in an incinerator without pollution controls. We reused and recycled materials only if their worth was greater than the value of a person's time to collect and transport them to a buyer.

The environmental movement arose as an outcry against pollution and human health impacts: waste chemicals seeped into drinking water and were emitted into the air; abandoned tire piles burned uncontrollably; unwanted pesticides polluted our neighborhoods and farmlands; and used motor oil flowed into rivers from storm drains. We responded by regulating landfills and waste-to-energy plants. We started voluntary recycling programs, then municipalities assumed that responsibility. We added fees to the purchase or sale of new tires, pesticides, and motor oil, and required state agencies to collect those funds and dole out grants for local government cleanup programs.

These examples are proof that US companies, governments, and organizations can act together to protect the environment and public health when a threshold of safety has been breached. We are at a similar point now. Plastics are in our oceans and bodies. Local government recycling costs are sky-high and unpredictable, with waste disposal capacity widely contested. Companies are being hounded by the public to take responsibility for their products. We are, once again, beyond the threshold of safety. And environmental issues are part of daily current events because they can no longer be denied.

Consumer product waste has passed the point where caution has given way to fear. It was prudent to reduce, reuse, and recycle at the start of the environmental movement. Now it is imperative. It is basic human survival. Those who sense trouble cannot wait to act before those who are unaware, or just don't care, become involved.

Expecting those responsible for product impacts to restrain their own business practices has failed. For decades, people in positions of power have denied the undeniable—from pesticides to tobacco to opioids to climate change—those who benefitted from unregulated activity have resisted imposed restraint. Denying responsibility has resulted in lost and damaged lives and a polluted environment that has cost billions of dollars to remediate. The increased severity of forest fires, hurricanes, and floods has let us know that climate change is real, and our habitat is out of balance.

While researching this book, I have been struck by the power of one idea—extended producer responsibility—that was conceived by one person, and how it spread around the world. Thomas Lindhqvist's EPR policy concept was not copyrighted, patented, or held secret. It was hatched in Sweden, implemented first in Germany, and then transported to the rest of Europe, Canada, Asia, the United States, South America, Australia, and around the globe. While there are many commonalities, EPR has not been implemented in the same way in each country because policies must account for the uniqueness of each place.

EPR has spread because it is based on the simple concept of responsibility. But many are reluctant to take responsibility if others do not. Our societal fabric hinges on the concept of individual and collective responsibility. Collectively, it is possible to overcome the inertia of opposition and to account for the business interests at stake in developing solutions. The concept of EPR has been increasingly accepted in the United States over the past two decades. States that have and continue to enact new EPR laws are leading a paradigm shift in the degree of responsibility assigned to companies, governments, and consumers of products and packaging. Once enacted, though, these states face a lengthy implementation process that will allow stakeholders to acclimate to a new normal. At some point in the future, EPR systems as implemented today will seem like the past era of early recycling programs. And it will be time once again to move forward with new approaches, new policies, and new technologies.

EPR systems are an attempt to govern conflicting human desires—to use nature for our own needs while protecting nature from ourselves. Over time, EPR systems will evolve into more refined policies. Societies are built on laws that attempt to balance individual freedoms with social responsibility.

As with all environmental laws, regulations, and policies, however, we need to determine what is important enough to restrict individual or company freedoms for the sake of the common good. There is an art in finding the right balance when regulating activity for the protection of human health and the environment.

In writing the conclusion for this book, I realized that there really are no conclusions. EPR is part of a continuum of ideas, like clay that has been remolded again and again, each time adding a new element, a new twist, a new look. Bob Dylan explained this concept well when he talked about his songwriting with David Remnick, editor of *The New Yorker*: "All these songs are connected [to traditional music]. I just opened up a different door in a different kind of way. . . . I thought I was just extending the line."[1] I, too, feel like I am just extending the line. As we balance the human need to control nature for our own benefit and to protect nature to sustain life on Earth, we can learn from those who came before us, both the successes and the setbacks. And may we always be in awe of the natural beauty around us—of which we are a part.

1. *The New Yorker*, October 24, 2022.

Acknowledgments

Writing a book for four years while running an organization required a vast support network. I owe tremendous gratitude to my longtime colleague, Amy Cabaniss, for the genesis of this book. A solid waste management practitioner, educator, and scholar, Amy recognized the timely need for a book such as this and offered me the opportunity to write about this emerging policy concept and share my personal journey. Amy remained an ardent supporter in many ways throughout the evolution of the book, not the least of which was lending her editing talent. Her patience and contributions are greatly appreciated.

Two educators provided me with the skills I needed to start an organization like the Product Stewardship Institute (PSI). University of Pennsylvania Professor Bob Giegengack unlocked for me the science and wonder of the geologic world, introduced me to environmental issues, and encouraged me to challenge prevailing norms. I will miss his unflagging support. MIT Professor Larry Susskind showed me facilitation and mediation techniques that I could apply in new ways to engage multiple stakeholders in policy disputes. Science, policy, and stakeholder engagement—together—allowed me to create my own path in the vast world of product sustainability.

PSI's staff and board of directors have nurtured the growth of PSI over the past 22 years. They worked closely with me to sustain a nonprofit advocacy organization that forged a new environmental path in the United States. Erin Linsky, PSI's first Boston-based employee, and Sierra Fletcher, former policy director, helped me to lay the foundation for PSI's work on communications and early policy initiatives, including pharmaceuticals and packaging.

Amanda Nicholson, PSI's chief operating officer for more than a decade, has played vital roles in countless ways—providing fiscal responsibility, business savvy, policy and program knowledge, human resources, and more. Amanda also provided key book insights, edited most chapters, and took on greater responsibility to allow me to finish the book. Suna Bayrakal, policy director for nearly a decade, provided important information and key insights on the many product sectors in which she is an expert, including batteries, mercury-containing products, household hazardous waste, mattresses, and carpet. Thanks also to Sydney Harris, former policy director, for reviewing the packaging chapter. Rachel Lincoln Sarnoff, PSI's communications director, offered keen insights and edited several chapters; Julia Wagner, marketing and communications coordinator, created and upgraded multiple graphics; and Jess Atkinson, former communications director, provided both editorial assistance and graphic support. Finally, Brendan Adamczyk, former associate, provided fact checking and citations, always with a smile.

I have been extremely fortunate to have had members of PSI's board of directors who have strongly supported me and the organization for more than two decades. These state and local government officials are the policy entrepreneurs and engines of the US EPR movement. They work with me and PSI staff to develop policy models, drive EPR within their agencies and states, and enact legislation. Several PSI board members provided extremely useful and timely feedback on the book chapters, particularly Tom Metzner, PSI's president from the Connecticut Department of Energy and Environmental Affairs, and Scott Klag, PSI board member for all of our 22 years. Other government members who provided important chapter reviews include Elena Bertocci, Maine Department of Environmental Protection, and David Allaway, Oregon Department of Environmental Quality. Former PSI presidents Rich Berman, Dave Galvin, and Jen Holliday provided steadfast leadership over multi-year terms, each time extending our reach and strengthening our impact.

Much of my EPR knowledge has been learned through interactions with colleagues from Canada and Europe. Since packaging has been a PSI priority for 15 years, I have had the great pleasure of working with Joachim Quoden, Managing Director of the Extended Producer Responsibility Alliance, an association of 31 producer responsibility organizations (PROs) operating packaging EPR programs in Europe and other countries. For the book, Joachim provided me with insights into the history and development of packaging EPR across Europe, as well as multiple updates on the European Commission's packaging directive. He also provided a steady corporate voice of reason, letting US EPR supporters know that US packaging EPR laws were just around the corner, as did John Coyne, formerly of Unilever and the Canadian Stewardship Services Alliance (CSSA). I would also like to thank Peter Borkey of the Organization for Economic Cooperation and Development for including me in global EPR events that greatly informed my understanding of EPR and enabled me to share it with those in the United States.

Pascal Leroy, Director General of the WEEE Forum, an international association representing 46 electronics PROs across the globe, provided several useful updates on electronics EPR programs in Europe. Catherine Abel, former Vice President at CSSA, offered key information and insights about Canadian programs across provinces, and Duncan Bury, former Environment

Canada official, provided a useful framework for the progression of Canadian EPR programs. My knowledge of Canadian programs was rounded out by Mathieu Guillemette and Philippe Cantin of Éco Entreprises Québec, a packaging PRO; and Mary Cummins of Ontario's Resource Productivity and Recovery Authority.

I would also like to thank those who provided important information, data, and perspectives in their review of the case studies on paint, batteries, and packaging: Heidi McAuliffe, American Coatings Association; Marjaneh Zarrehparvar and Brett Rodgers, PaintCare; staff at Call2Recycle; Alison Keane, Flexible Packaging Association; Dan Felton, AMERIPEN; and Scott Mouw, The Recycling Partnership.

Of course, even with all the generous support I received, any errors or omissions in this book are my own, and I welcome comments.

Finally, I am extremely grateful to my wife, Susan, who supported me in innumerable ways, including talking through concepts, contributing significant comments on the manuscript, and offering much encouragement. Above all, Susan made tremendous sacrifices that afforded me the time to write this book. A big thanks also to our daughter, Sarah, who provided great comments on the manuscript and many times said, "You got this, pops!" And to my son-in-law Yaron, for his unflagging ability to lighten the moment and pull me into a much-needed game of Mexican Train Dominoes.

Index

About the Author

Scott Cassel is CEO and founder of the Product Stewardship Institute, a national policy advocate and consulting nonprofit, and serves on the founding board of the Marine Debris Foundation. Scott has served as Director of Waste Policy for the Massachusetts Executive Office of Energy and Environmental Affairs, among other positions, across a 40-year waste management career. He holds a bachelor of science degree in geology and environmental science from the University of Pennsylvania and a master's in city planning with a focus on environmental policy and dispute resolution from the Massachusetts Institute of Technology.